Power Analysis Attacks
Revealing the Secrets of Smart Cards

T0142817

Power Analysis Attacks
Revealing the Secrets of Smart Cards

by

Stefan Mangard
Elisabeth Oswald
Thomas Popp
Graz University of Technology
Graz, Austria

 Springer

Stefan Mangard
Institute for Applied Information Processing
 & Communications
Graz University of Technology
Inffeldgasse 16a
8010 GRAZ
AUSTRIA

Elisabeth Oswald
Institute for Applied Information Processing
& Communications
Graz University of Technology
Inffeldgasse 16a
8010 GRAZ
AUSTRIA

Thomas Popp
Institute for Applied Information Processing
 & Communications
Graz University of Technology
Inffeldgasse 16a
8010 GRAZ
AUSTRIA

Power Analysis Attacks: Revealing the Secrets of Smart Cards
by Stefan Mangard, Elisabeth Oswald and Thomas Popp

ISBN-13: 978-1-4419-4039-1   e-ISBN-13: 978-0-387-38162-6
e-ISBN-10: 0-387-38162-7

Printed on acid-free paper.

9 8 7 6 5 4 3 2 1

springer.com

Contents

Foreword

"We don't understand electricity. We use it."
— Maya Angelou

When I started researching tamper resistance, I couldn't afford the chip reverse engineering equipment needed for physical attacks. This presented a significant problem: my company's clients wanted results, but I couldn't use the attack techniques that were well known at the time.

I started with a simple question: what information is available to attackers but is not assumed in the cryptographic protocols? I had previously discovered that subtle timing variations could compromise keys, so I decided to find out if the power consumption could reveal anything useful.

Hours after lugging home a $500 analog oscilloscope (on sale at Fry's Electronics), my research partner Joshua Jaffe and I saw our first power traces. We started with a smart card performing RSA, and found that we could identify major features of the algorithm. Before long, we learned to discern differences in the squaring and multiplication steps and found our first key using SPA. We also analyzed DES implementations, and found that they could also be broken, though only at night—the daylight in our office drowned out the screen of our cheap oscilloscope.

In the months that followed, we built digital data capturing systems, which permitted much more sophisticated analysis. We also developed visualization software and implemented many analysis techniques, including DPA. We tested numerous products, and every smart card and other tamper resistant device we looked at was breakable. The power of DPA was both amazing and terrifying.

In parallel, we also put a tremendous amount of effort into developing countermeasures. Solving the problem is tricky, since the statistics behind DPA can identify tiny correlations buried in noise. Similarly, eliminating information leakage is nearly impossible since the movement of electrons invariably consumes electricity and generates electromagnetic fields. Today, my company licenses the patents that came out of this work. If you're building products with DPA countermeasures, please be aware that these patents exist—I don't want people or companies to be surprised.

It's interesting that SPA and DPA went undiscovered for so long. The main reason is simple: it wasn't anybody's job to look for the issue. Cryptographers focused on achieving mathematical strength, while engineers were responsible for the hardware and software. Attack research was also neatly compartmentalized, with work on algorithms separated from physical defenses. Few cryptographers were working on real products, and few engineers understood the math. As you read this book, I hope you'll think about other security problems that may be going unnoticed.

In the years since DPA became public, research on tamper resistance has thrived. Whole conferences are now focused on the topic. Researchers such as the authors of this book and others have taken DPA attacks in many new directions. Products have also improved greatly. Although most products reviewed by my company today still have DPA vulnerabilities, the best products now have very good defenses. Vendors have also learned that tamper resistance is an incredibly hard problem and are far more realistic about their security claims.

Looking ahead, tamper resistance will remain a tremendously interesting and important problem. In 2007, over two billion tamper-resistant chips will be manufactured for securing communications, payments, printer consumables, pay TV systems, government IDs, and countless other applications. Attacks against tamper resistant semiconductors have resulted in billions of dollars of fraud and piracy. As a rule, cryptographic protocols are only as secure as the underlying secrets, and undoubtedly there is a great deal that is not yet known about how to store and manage keys safely.

You will find this book fascinating, interesting, and alarming. As you study, roll up your sleeves. Ask skeptical questions. Attend research conferences. Obey the laws. And, above all else, have fun.

Paul Kocher
President and Chief Scientist of Cryptography Research Inc.
San Francisco, September 2006

Preface

Power analysis attacks are cryptanalytic attacks that allow the extraction of secret information from cryptographic devices. In contrast to other cryptanalytic attacks, they exploit power consumption characteristics of devices rather than mathematical properties of cryptographic algorithms. Power analysis attacks are non-invasive attacks that can be performed with off-the-shelf equipment. Therefore, they pose a serious threat to the security of cryptographic devices like smart cards.

Smart cards are the most popular kind of cryptographic devices. According to Eurosmart, an international association of smart card companies, the market for smart cards with microprocessors has more than doubled during the last few years. In 2003, less than one billion cards have been shipped, while in 2006 about two billion cards are expected to be sold. Most of these cards are used in security-sensitive applications like telecommunications, financial services, governmental services, corporate security, and pay TV. Smart cards are vital components of these applications, and hence, their security is crucial.

Driven by the need to develop countermeasures, researchers have started investigating power analysis attacks in detail. Power analysis attacks have turned out to be a fascinating field of research. Understanding them requires know-how from different disciplines such as cryptology, statistics, measurement technology, and microelectronics. They have attracted the attention of researchers from all these fields. As a result, a large number of research articles has been published over the previous years. In fact, it has become quite challenging to keep track of all these publications and to understand how the presented ideas are related to each other. Furthermore, there is no introductory literature available to familiarize learners with all the different types of power analysis attacks and countermeasures. This book aims to fill this gap.

In this book, we provide a comprehensive introduction to power analysis attacks. Starting with a discussion of cryptographic devices, their design, and their power consumption, we bring together statistics and electrical engineering in order to explain and analyze different types of power analysis attacks and countermeasures. This book is written for researchers as well as for practitioners having a background in cryptology, security, or microelectronics.

Structure of This Book

This book consists of eleven chapters and two appendices. The first two chapters are introductory and aimed at readers who are either new to power analysis attacks or who have little engineering background. The next two chapters provide important information about the power consumption of cryptographic devices, how it is measured, and how it is characterized statistically. These two chapters are written for readers who are interested in detailed background information. They are not required for understanding the basics of the six subsequent chapters that deal with power analysis attacks and countermeasures. However, they are necessary for the advanced topics. Conclusions are provided in the last chapter of this book. The following list provides a more detailed overview of the topics that are discussed in different chapters of this book.

Chapter 1 explains how cryptography and cryptographic devices are linked. It surveys different types of attacks on cryptographic devices, provides a simple example of a power analysis attack, and classifies countermeasures against power analysis attacks.

Chapter 2 discusses the design and implementation of cryptographic devices. It explains how a typical semi-custom design flow works. Furthermore, it discusses logic cells in general and complementary metal-oxide semiconductor (CMOS) cells in particular.

Chapter 3 focuses on the power consumption of cryptographic devices that are built using CMOS cells. It explains how the power consumption can be simulated and which power models are appropriate for power analysis attacks. Furthermore, it discusses measurement setups for power analysis attacks.

Chapter 4 shows how power traces can be characterized by statistics. First, it discusses the characteristics of individual points within a power trace. In this context, the chapter also discusses how to quantify the leakage of individual points. Next, it discusses the relation of points within a power trace. Last, it briefly shows how confidence intervals and hypothesis tests work and how they can be used in power analysis attacks.

Chapter 5 introduces simple power analysis (SPA) attacks. It shows that visual inspections of power traces often provide useful information. Furthermore, it introduces template attacks and collision attacks. Several examples illustrate the presented ideas.

Chapter 6 introduces differential power analysis (DPA) attacks. It discusses the basics of these attacks and presents several examples of DPA attacks on software and hardware implementations of AES. All these attacks are based on the correlation coefficient. The chapter also shows how to simulate DPA attacks and how to determine the number of needed power traces. In addition, it surveys alternatives to the correlation coefficient and discusses template-based DPA attacks.

Chapter 7 deals with hiding countermeasures. The chapter surveys countermeasures like shuffling, the random insertion of dummy operations, the flattening the power consumption, and noise engines. Furthermore, the chapter provides an in-depth discussion dual-rail precharge logic styles.

Chapter 8 analyzes the effectiveness of hiding countermeasures. Formulas are provided to calculate the effect of the hiding countermeasures on the number of power traces that is needed for successful DPA attacks. Subsequently, DPA attacks on misaligned power traces are discussed. Last, the effectiveness of two dual-rail precharge logic styles is analyzed.

Chapter 9 deals with masking countermeasures. It explains the different kinds of masking and it discusses the most important issues that need to be considered when implementing this countermeasure in software and hardware. The chapter also discusses methods to mask logic cells.

Chapter 10 analyzes the effectiveness of masking countermeasures. It shows how second-order DPA attacks can break masked implementations in software and hardware. Several different methods to perform second-order DPA attacks are compared. Furthermore, the chapter shows that template-based DPA attacks are a very powerful method to break masked implementations.

Chapter 11 provides conclusions for this book.

Appendix A contains the original research article of Kocher *et al.* [KJJ99] about power analysis attacks.

Appendix B describes the Advanced Encryption Standard (AES). In addition, it sketches one AES implementation in software and one in hardware. These two implementations are used for the examples of power analysis attacks that are presented in this book.

While the introductory Chapter 1 and the background Chapters 2 to 4 have a short summary at the end, the Chapters 5 to 10, which are about attacks and countermeasures, end with notes and further reading. These notes and further reading are an important part of this book. They link other research to the ideas that we explain within the chapters. Note that although the notes and further reading are comprehensive, they are not exhaustive. In order to make reading this book more convenient, we also provide a glossary and an overview of the notation that we use throughout the book. In addition, we provide an author index and a subject index at the end of this book. There is also a website for this book:

http://www.dpabook.org

The website provides power traces and analysis scripts for Matlab and Octave. The aim of the website is to allow readers to experiment with power analysis attacks and countermeasures.

Acknowledgments

Writing this book would not have been possible without the financial support of several institutions. We would like to thank the Institute for Applied Information Processing and Communications (IAIK) of the Graz University of Technology for its support. Furthermore, we would like to thank the Secure Information Technology Center - Austria (A-SIT), the Austrian Science Fund (FWF), and the European Commission (EC) for funding various projects related to power analysis attacks.

The software and hardware implementations of AES that we use to demonstrate power analysis attacks throughout the book are mainly achievements of two of our co-workers. Christoph Herbst deserves the credits for implementing AES in various ways on the microcontroller. Norbert Pramstaller deserves the credits for implementing AES in dedicated hardware.

Many of our colleagues have devoted a significant amount of their time in order to proofread this book. We would especially like to thank the following people for their help: Martin Feldhofer, Christoph Herbst, Mario Lamberger, Karl Christian Posch, Norbert Pramstaller, Vincent Rijmen, Martin Schläffer, and Stefan Tillich.

We would like to thank Vincent Rijmen and Joan Daemen for providing us with the pictures that describe AES. Thanks also go to Christoph Herbst for making the cover art of this book.

Furthermore, we would like to thank the three inventors of power analysis attacks, Paul Kocher, Joshua Jaffe, and Benjamin Jun, for allowing us to include the original "DPA paper". In addition, we would especially like to thank Paul Kocher for writing the foreword of this book and Joshua Jaffe for his valuable comments.

Last, we would like to thank all our colleagues and friends who have supported us scientifically and morally while writing this book.

STEFAN MANGARD, ELISABETH OSWALD, AND THOMAS POPP

GRAZ, SEPTEMBER 2006

Notation

The meaning of each variable and function is described in the chapter that uses it for the first time. The most important global conventions are that we use bold capital letters for matrices and bold small letters for vectors. Unless transposed, vectors are always column vectors. The rows, columns, and elements of matrices are denoted as follows. Column j of a matrix $\mathbf{T}$ is denoted by $\mathbf{t}_j$, row i is denoted by $\mathbf{t}'_i$, and the element i of column j is denoted by $t_{i,j}$. We now provide a list of the most important variables and functions used in this book.

ck	Index of the correct key in $\mathbf{k}$
ct	Index of the position in the power traces that leaks information about the attacked intermediate result in a DPA attack
C	Covariance matrix (size $N_{IP} \times N_{IP}$) and the estimator of it
$Cov(X,Y)$	Covariance of X and Y
D	Number of power traces used for a DPA attack
$E(X)$	Expected value of the random variable X
$\mathbf{d}$	Vector of input or output values that are used for a DPA attack (size $D \times 1$)
$\mathbf{H}$	Matrix of hypothetical power consumption values in a DPA attack (size $D \times K$)
h	Template
$\mathcal{H}$	Matrix of Templates

k	Vector of key hypotheses in a DPA attack (size $1 \times K$)
k	Key
k_{ck}	Key used by the attacked cryptographic device
K	Total number of key hypotheses
μ	Mean of a normal distribution
m	Mean vector and the estimator of it (size $N_{IP} \times 1$)
m	Mask
m_d	Mask for a data value d
M	Number of different values a mask can have
n	Calculated number of needed power traces for a DPA attack
N_{IP}	Number of interesting points
$p(X = i)$	Probability that $X = i$
$\Phi(x)$	Probability density function of the standard normal distribution
ρ	Correlation coefficient
r	Estimator for the correlation coefficient
R	Result of a DPA attack (size $K \times T$); usually a matrix of estimated correlation coefficients
σ	Standard deviation of a normal distribution
s	Simulated power trace
S	Matrix of simulated power consumption values
s	Estimator for the standard deviation
$S(x)$	AES S-box function
t	Power trace (size $T \times 1$)

$\tilde{t}$	Preprocessed power trace
T	Matrix of power consumption values (size $D \times T$); each row corresponds to one power trace
$\tilde{\mathbf{T}}$	Matrix of preprocessed power consumption values
T	Number of points in a power trace
V	Matrix of hypothetical values in a DPA attack (size $D \times K$)
$Var(X)$	Variance of the random variable X
$\bar{x}$	Estimator for the mean

Glossary

3sDL	3-State Dynamic Logic
AES	Advanced Encryption Standard
ALU	Arithmetic Logic Unit
ASIC	Application Specific Integrated Circuit
CML	Current-Mode Logic
CMOS	Complementary Metal-Oxide Semiconductor
CRT	Chinese Remainder Theorem
DES	Data Encryption Standard
DIP	Dual In-Line Package
DPA	Differential Power Analysis
DPDN	Differential Pull-Down Network
DPTR	Data Pointer
DPUN	Differential Pull-Up Network
DR	Dual Rail
DRP	Dual-Rail Precharge
DSA	Digital Signature Algorithm
DSDR	Dual-Spacer Dual-Rail
DyCML	Dynamic Current-Mode Logic
ECC	Elliptic Curve Cryptography
ECDSA	Elliptic Curve Digital Signature Algorithm
EDA	Electronic Design Automation
EM	Electromagnetic
EMC	Electromagnetic Compatibility

EMV	Europay/Mastercard/Visa
FFT	Fast Fourier Transformation
FIPS	Federal Information Processing Standard
FPGA	Field-Programmable Gate Array
GALS	Globally-Asynchronous Locally-Synchronous
GF	Galois Field
GPIB	General-Purpose Interface Bus
HD	Hamming Distance
HDL	Hardware Description Language
HF	High Frequency
HMAC	Hash-Based Message Authentication Code
HSM	Hardware Security Module
HW	Hamming Weight
IC	Integrated Circuit
IDEA	International Data Encryption Algorithm
IEC	International Electrotechnical Commission
I/O	Input-Output
LCD	Liquid Crystal Display
LSB	Least Significant Bit
LSQ	Least Square
MCML	MOS Current-Mode Logic
MDPL	Masked Dual-Rail Precharge Logic
ML	Maximum Likelihood
MOS	Metal-Oxide Semiconductor
MSB	Most Significant Bit
NED	Normalized Energy Deviation
NIST	National Institute of Standards and Technology
NMOS	N-Channel Metal-Oxide Semiconductor
NSA	National Security Agency
PC	Personal Computer; Program Counter
PCB	Printed Circuit Board
PLCC	Plastic Leaded Chip Carrier
PMOS	P-Channel Metal-Oxide Semiconductor
PS/2	Personal System 2

RAM	Random Access Memory
RC4	Rivest Cipher 4
RC6	Rivest Cipher 6
RFID	Radio-Frequency Identification
ROM	Read-Only Memory
RS	Recommended Standard
RSA	Rivest-Shamir-Adleman
RTL	Register-Transfer Level
SABL	Sense Amplifier Based Logic
SFR	Special-Function Register
SHA	Secure Hash Algorithm
SNR	Signal-to-Noise Ratio
SPA	Simple Power Analysis
SPICE	Simulation Program with Integrated Circuit Emphasis
SR	Single Rail
TEM	Transverse Electromagnetic
TOE	Time-Of-Evaluation
TDPL	Three-Phase Dual-Rail Precharge Logic
USB	Universal Serial Bus
VHDL	VHSIC Hardware Description Language
VHSIC	Very High Speed Integrated Circuit
VML	Voltage-Mode Logic
WDDL	Wave Dynamic Differential Logic
ZV	Zero Value

Chapter 1

INTRODUCTION

Smart cards are frequently used as cryptographic devices to provide strong authentication of users and to store secret information securely. Smart cards are among the most critical components of modern security systems.

When Kocher *et al.* [KJJ99] showed in 1998 that power analysis attacks can efficiently reveal the secrets of smart cards, the belief in the security of cryptographic devices was shattered. This book describes why power analysis attacks work, how they can be conducted, and how they can be counteracted.

In this chapter, we briefly explain the use of cryptographic devices and cryptography in modern security systems. We survey different kinds of attacks on cryptographic devices, and we present a concrete example of a power analysis attack. Furthermore, we provide an overview of countermeasures against power analysis attacks.

1.1 Cryptography and Cryptographic Devices

Modern security systems use cryptographic algorithms to provide confidentiality, integrity, and authenticity of data. Cryptographic algorithms are mathematical functions that typically take two input parameters: a message (which is also called plaintext) and a cryptographic key. The cryptographic algorithm maps these parameters to an output, which is called ciphertext. This process is called encryption. In modern cryptography, the cryptographic algorithm itself is assumed to be known. This means that all details about it are publicly available and only the cryptographic key is kept secret. This important principle goes back to Auguste Kerckhoffs who was a Dutch cryptographer of the 19th century.

We distinguish between symmetric and asymmetric cryptography. In symmetric cryptography, the entities that communicate share a common secret key. A well-known example for a symmetric encryption algorithm is the *Advanced*

Encryption Standard (AES) [Nat01]. It is a block cipher, which means that it encrypts blocks of texts of a fixed size. In the case of AES, the block size is defined to be 128 bits. The key size can be 128, 192, or 256 bits. The versions of AES are called accordingly: AES-128, AES-192, and AES-256. In this book we use the abbreviation AES to refer to AES-128. An overview of the working principle of AES is given in Appendix B. Due to its widespread use, all examples of power analysis attacks in this book target implementations of AES.

In asymmetric cryptography, every user has a key pair. The key pair consists of a public parameter, which is called the public key, and a secret parameter, which is called the private key. There are many asymmetric cryptographic algorithms in use today. The most popular algorithm is the *Rivest-Shamir-Adleman* (RSA) algorithm [RSA78]. RSA keys have at least 1 024 bits in today's applications. This prevents practical attacks on RSA.

Breaking a cryptographic algorithm typically means finding the secret key based on some public information, which can be for instance pairs of plaintexts and ciphertexts. A cryptographic algorithm is considered to be secure in practice if there is no attack known that can break it within a reasonable amount of time and with a reasonable amount of computing power. A cryptographic algorithm is considered to be computationally secure if breaking it requires computing power that is not available in practice. Many algorithms are designed such that the effort of breaking them grows exponentially with the number of bits of the key. Consequently, the length of the key is an important factor in the security of a cryptographic algorithm. The notion of computational security is stronger than the notion of practical security. Algorithms that are popular today are considered to be computationally secure.

One property of modern symmetric and asymmetric cryptographic algorithms is that they can be efficiently performed by a computer. This implies that the key that is used for such an algorithm needs to be stored on the computer as well. However, typical personal computers (PCs) are rather insecure platforms. Viruses and worms spread via the Internet and frequently infect a large number of PCs within a short time. Hence, unless special attention is paid to its configuration, a PC is not an adequate platform to store valuable assets. Cryptographic keys are such valuable assets. For instance, knowing somebody's secret key might allow someone to pose as that person. Closed platforms such as smart cards are more suitable for storing cryptographic keys. Typical smart cards are not connected to the Internet and do not allow software to be installed. They can be considered as a protected environment that stores keys and that performs cryptographic operations. A smart card is an example of a cryptographic device. Other examples of cryptographic devices are USB (Universal Serial Bus) tokens or contactless devices such as RFID (Radio-Frequency Identification) tags.

> *Cryptographic devices* are electronic devices that implement crypto-graphic algorithms and that store cryptographic keys.

Cryptographic devices are capable of performing cryptographic operations using the keys that are stored on them. They are able to communicate the results of the cryptographic operations to the outside.

The fact that cryptographic devices are used to perform cryptographic algorithms leads to a new issue for the practical security of the algorithms. In practice, not only the security of the cryptographic algorithm is of interest. The security of the whole system, *i.e.* the cryptographic device that implements the cryptographic algorithm, needs to be considered.

Breaking a cryptographic device means extracting the key of the device. A person who tries to extract the key of a cryptographic device in an unauthorized way is called an attacker. Any attempt to extract the key in an unauthorized way is called an attack.

In order to evaluate the security of a cryptographic device, it is necessary to make assumptions about the knowledge that an attacker has about it. The strongest assumption that can be made extends Kerckhoffs' principle. It is that the attacker is assumed to know the details of the cryptographic device.

> The security of a cryptographic device should not rely on the secrecy of its implementation.

1.2 Attacks on Cryptographic Devices

In recent years, several kinds of attacks on cryptographic devices have become public. The goal of all these attacks is to reveal secret keys of cryptographic devices. However, the techniques that are used to reach this goal are manifold.

Attacks on cryptographic devices differ significantly in terms of cost, time, equipment, and expertise needed. Consequently, there are also several ways of categorizing these attacks. In the context of this book, we follow the most commonly used approach of categorizing attacks on cryptographic devices. This categorization is essentially based on two criteria. The first criterion is whether an attack is passive or active.

- **Passive Attacks:** In a passive attack, the cryptographic device is operated largely or even entirely within its specification. The secret key is revealed by observing physical properties of the device (e.g. execution time, power consumption).

- **Active Attacks:** In an active attack, the cryptographic device, its inputs, and/or its environment are manipulated (tampered with) in order to make the device behave abnormally. The secret key is revealed by exploiting this abnormal behavior of the device.

Although the term "tamper" rather points towards active attacks, the term "tamper resistance" is often used in order to refer to active and passive attacks on cryptographic devices.

The second criterion for categorizing an attack on a cryptographic device is the interface that is exploited by the attack. Cryptographic devices have several physical and logical interfaces. Some of these interfaces can be accessed easily, while others can only be accessed by special equipment. Based on the interface that is used for an attack, it is possible to distinguish between invasive, semi-invasive, and non-invasive attacks. All of these attacks can be either passive or active.

- **Invasive Attacks:** An invasive attack is the strongest type of attack that can be mounted on a cryptographic device. In such an attack, there are essentially no limits to what is done with the cryptographic device in order to reveal its secret key.

  An invasive attack typically starts with the depackaging of the device. Subsequently, different components of the device are accessed directly using a probing station. This part of an invasive attack is passive if the probing station is only used to observe data signals (e.g. signals on a processor bus). It is active if signals in the device are changed to alter the functionality of the device. For this purpose, devices like laser cutters, probing stations, or focused ion beams can be used.

  Invasive attacks are extremely powerful. However, they typically require quite expensive equipment. Consequently, only very few publications on this topic exist. Among the most important ones are [KK99], [And01], and [Sko05].

- **Semi-Invasive Attacks:** In semi-invasive attacks, the cryptographic device is also depackaged. However, in contrast to invasive attacks, no direct electrical contact to a chip surface is made—the passivation layer stays intact.

  The goal of passive semi-invasive attacks is typically to read out the content of memory cells without using or probing the normal read-out circuits. A successful attack of this kind was published in [SSAQ02].

  The goal of active semi-invasive attacks is to induce faults in the device. This can be done by X-rays, electromagnetic fields, or light. For example, an optical fault induction attack was published in [SA03].

  Semi-invasive attacks typically do not require as expensive equipment as invasive ones. However, the total effort that is needed to conduct a semi-invasive attack is still relatively high. In particular, the process of locating the right position for an attack on the surface of a modern chip requires quite some time and expertise. The most comprehensive publication on semi-invasive attacks is the PhD thesis of Skorobogatov [Sko05].

- **Non-Invasive Attacks:** In a non-invasive attack, the cryptographic device is essentially attacked as it is, *i.e.* only directly accessible interfaces are exploited. The device is not permanently altered and therefore no evidence of an attack is left behind. Most non-invasive attacks can be conducted with relatively inexpensive equipment, and hence, these attacks pose a serious practical threat to the security of cryptographic devices.

  In particular passive non-invasive attacks have received a lot of attention during the last years. These attacks are often also referred to as *side-channel attacks*. The three most important types of side-channel attacks are timing attacks [Koc96], power analysis attacks [KJJ99], and electromagnetic (EM) attacks [GMO01, QS01]. The basic idea of these attacks is to determine the secret key of a cryptographic device by measuring its execution time, its power consumption, or its electromagnetic field.

  Besides side-channel attacks there are also active non-invasive attacks. The goal of these attacks is to insert a fault in a device without depackaging it. Faults can for example be induced by clock glitches, power glitches or by changing the temperature of the environment. A survey of this kind of attacks can be found in [BECN+04].

This book focuses exclusively on power analysis attacks. Readers who are interested in more information on other ways to attack cryptographic devices are referred to [ABCS06]. There is also an online "Side Channel Cryptanalysis Lounge" [Cha06], which provides a list of scientific articles on all kinds of attacks on cryptographic devices. The large majority of these articles covers power analysis attacks and countermeasures. These attacks have received by far the most attention of all attacks on cryptographic devices. In fact, the number of publications on this topic has become so big that it is difficult to keep track of all the research going on in this field. Providing a comprehensive overview of power analysis attacks and countermeasures is one of the goals of this book.

Power analysis attacks have received such a large amount of attention because they are very powerful and because they can be conducted relatively easily. Consequently, they pose a serious threat to the security of cryptographic devices in practice. For the design and development of modern cryptographic devices, it is crucial to be familiar with power analysis attacks and countermeasures. Unprotected devices can be broken with minimal effort.

Table 1.1. Basic properties of the attacked microcontroller.

Processor type:	8051-compatible microcontroller
Bus width:	8 bits
Internal memory:	256 bytes
Supply voltage:	5 V
Communication interface:	RS-232
Clock frequency:	11 MHz

1.3   Power Analysis Attacks

The basic idea of this kind of attack is to reveal the key of a cryptographic device by analyzing its power consumption. Essentially, two dependencies of the power consumption are exploited: the data-dependency and the operation-dependency.

> *Power analysis attacks* exploit the fact that the instantaneous power consumption of a cryptographic device depends on the data it processes and on the operation it performs.

In this section, we present a simple example that illustrates the principles of power analysis attacks. The example is based on an 8051-compatible microcontroller that executes a software implementation of AES. Table 1.1 shows the basic properties of this microcontroller. We use this device as standard platform throughout the book. It is small and simple, and therefore ideally suited to introduce power analysis attacks and countermeasures.

In the current example, the microcontroller is programmed to perform an AES encryption. The microcontroller receives a plaintext from a PC over its RS-232 interface, encrypts it, and sends the result back to the PC. During the time the encryption is performed, the power consumption of the microcontroller is measured. For this purpose, a 1 Ω resistor has been inserted in the ground wire of the power supply of the microcontroller. The voltage drop along this resistor is measured and recorded using a digital oscilloscope.

Figure 1.1 shows such a recorded voltage drop. This recorded voltage drop is proportional to the power consumption of the microcontroller. Hence, we refer to the voltage drop as power consumption and to the corresponding trace as power trace. The shape of a power trace strongly depends on the operations that are executed by the device and on the data it processes. A close look at Figure 1.1 reveals that there is indeed a close relationship between the executed AES implementation and the power consumption.

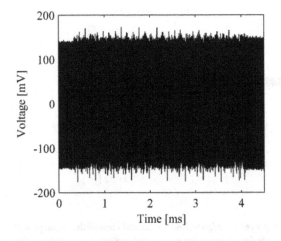

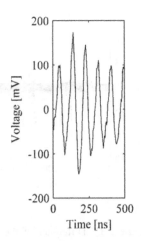

Figure 1.1. The voltage drop (power consumption) of the microcontroller while it performs an AES encryption.

Figure 1.2. Zoomed view of the recorded power trace.

The power trace shown in Figure 1.1 is quite uniform during the first 0.35 ms. During this time, the microcontroller waits for the plaintext. After having received the plaintext, nine full AES encryption rounds are performed. The execution of each round leads to a pattern of approximately 0.4 ms in the power trace. In particular, the negative peaks of the nine repetitions of this pattern can be observed clearly in Figure 1.1. After the nine full AES rounds, the final AES round is executed. In the final round, the MixColumns operation is omitted (see Appendix B), and hence, round ten leads to a shortened version of the round pattern in the power trace. At approximately 4.1 ms, the power trace becomes uniform again. At this moment of time, the encryption is finished and the microcontroller waits for a new plaintext.

An attacker can of course also zoom in closer on a power trace. When zooming in on a trace, the power consumption of individual clock cycles becomes visible. Figure 1.2, for example, shows the power trace in a small interval of approximately six clock cycles. Each peak that can be seen in this trace corresponds to the power consumption of the microcontroller in one clock cycle. Power analysis attacks work because the peaks look different for different operations and different data. Attacks that exploit this property based on just one power trace are referred to as simple power analysis (SPA) attacks [KJJ99].

In our example, we now use more power traces to take a closer look at the data dependency of the power consumption of the microcontroller. In fact, we first analyze how the power consumption of an AES encryption depends on the most significant bit (MSB) of the first byte of plaintext. We refer to this bit as d. In order to determine the effect of d on the power consumption, we have measured the power consumption of the microcontroller while it has encrypted

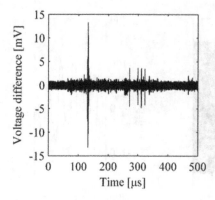

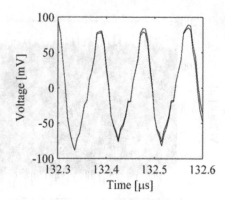

Figure 1.3. Difference of the mean power trace for $d = 1$ and the mean power trace for $d = 0$.

Figure 1.4. Zoomed view of the mean power trace for $d = 1$ and the mean power trace for $d = 0$.

1 000 random plaintexts. Hence, we have obtained about 500 power traces of encryptions where $d = 0$ and about 500 power traces of encryptions where $d = 1$.

A simple method to determine the effect of d on the power consumption is to calculate a difference of means. This is done by calculating the mean of those power traces where d equals 1 and the mean of those power traces where d equals 0. In this way, a mean power trace for $d = 1$ and a mean power trace for $d = 0$ is obtained. Subsequently, these two mean traces are subtracted. Figure 1.3 shows the difference between the mean traces for $d = 1$ and $d = 0$ during the first round of AES. This difference is close to zero at almost all moments of time. However, there are also a few significant peaks in the difference trace. These peaks reveal the moments of time where the power consumption of the microcontroller depends on d. At these moments of time, instructions are executed that either process d directly or that process some data that depends on d.

The highest peak that can be observed in Figure 1.3 is located at approximately $132 \, \mu s$. When checking the assembly code of the program that is executed by the microcontroller, it turns out that at this moment of time the first byte of plaintext is loaded. The value d is the MSB of this byte, and hence, the power consumption of this loading operation strongly depends on d.

Figure 1.4 shows a zoomed view of the mean trace for $d = 1$ and of the mean trace for $d = 0$ at the position of the loading operation. At the beginning, the two traces are almost identical. However, then there are three clock cycles where the height of the power consumption peaks is different for $d = 1$ and $d = 0$. This difference represents the dependency of the power consumption on *d*. The difference of the height of the second peak is the difference that causes the significant peak shown in Figure 1.3.

The smaller peaks in the difference trace of Figure 1.3 can be analyzed in a similar manner. At each of these positions, instructions are executed that depend directly on *d* or on data that is related to *d*. Obviously, the power consumption of the microcontroller contains information about the data it processes. We now exploit this property in a power analysis attack to determine the secret key that is used by the microcontroller.

At the beginning of an AES encryption, the microcontroller loads a plaintext. Subsequently, the plaintext is exclusive-ored with the secret key and ten encryption rounds are performed. The first operation that is performed within each round is the SubBytes operation. SubBytes is a byte-level operation. This means, SubBytes applies a function $S()$ to each byte of the state. Consequently, each output byte of this operation in the first round can be calculated based on one byte of plaintext and one byte of the secret key. We refer to the first byte of the plaintext as *p* and to the first key byte as *k*. The corresponding output byte of SubBytes in the first round can be written as $S(p \oplus k)$. In our attack, we exploit the fact that the power consumption of the microcontroller at some moment of time depends on the MSB of this output byte: $v = MSB(S(p \oplus k))$.

Performing such an attack is very similar to the analysis we have previously performed for the value *d*. First, 1 000 random plaintexts are encrypted and the power consumption is recorded. Subsequently, the power traces are split into two groups: the traces where $v = 1$ and the traces where $v = 0$. Unlike in the previous analysis of the plaintext bit *d*, the intermediate value *v* now depends on the secret key byte *k*.

Initially, an attacker does not know *k*. However, an attacker can guess this value, and actually, there are only 256 possible values for *k*. Therefore, an attacker can easily run through all possible values of *k* and can use these values to calculate *v*. In practice this means that the attacker first guesses that $k = 0$. Based on this guess, *v* is calculated for each of the 1 000 encryption runs. Afterwards, the attacker calculates the mean of those power traces where *v* was supposedly 1 and the mean of those power traces where *v* was supposedly 0. The difference between these two mean power traces is then plotted. The same procedure is repeated for all other 255 possible key values, and in this way, the attacker obtains 256 difference plots. There is one plot for each key guess.

We have determined 256 such difference plots based on guessing key byte number one of our microcontroller. Figure 1.5 shows the difference plots for the key guesses 117, 118, 119, and 120. It can be observed that there are very high peaks in the difference plot for the key guess $k = 119$. In fact, the plot looks somehow similar to the difference plot we previously obtained when we analyzed the effect of *d* on the power consumption (see Figure 1.3). A closer

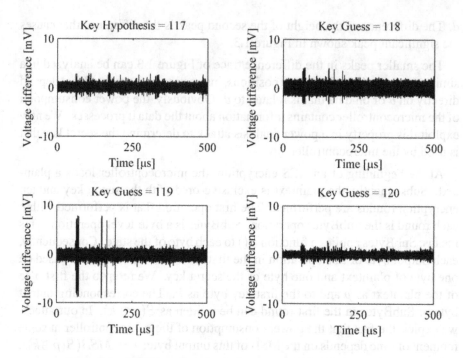

Figure 1.5. Difference plots for the key guesses 117, 118, 119, and 120.

analysis reveals that these plots not only look similar, but also that they actually display similar effects.

The value of key byte number one in the microcontroller is 119. Hence, we have calculated v correctly when we guessed that $k = 119$. The difference plot for this key guess therefore actually shows how the power consumption of the microcontroller depends on v. There are significant peaks in this plot because the power consumption strongly depends on v. Remember, the power consumption of the microcontroller depends on all data values that are processed—just like the value d, the value v is one of them.

An important question is: What happens if v is calculated based on a wrong key guess? In these cases, the difference plots are not calculated based on data that is actually processed in the device. The calculated values v that are used to split the power traces into two groups ($v = 1$ and $v = 0$) are not processed by the microcontroller. Therefore, the power consumption does not depend on these values, and consequently no large peaks occur in these difference plots.

There only occur some smaller peaks, as for example in the plot for key guess 118. These small peaks are a consequence of the fact that the values v that are calculated based on key 118 are not completely statistically independent from the values v that are actually processed in the device. However, the dependency is significantly smaller than in case of the correct key guess. Therefore, an attacker can easily identify the correct key by simply looking for the difference plot with the highest peaks.

Attacks using this strategy are called differential power analysis (DPA) attacks. DPA attacks were introduced by Kocher *et al.* in their pioneering article [KJJ99]. These attacks exploit the fact that the power consumption of cryptographic devices depends on intermediate values that are processed during the execution of a cryptographic algorithm. In our example, we have exploited the fact that the power consumption of the microcontroller depends on the MSB of the first output byte of the SubBytes operation in round one. Attacks can also be mounted based on other intermediate values. In particular, it is possible to use intermediate values that depend on other key bytes. In this way, the whole AES key can be revealed easily. In fact, the same power traces can be used for all key bytes—no new traces need to be recorded.

1.4   Countermeasures Against Power Analysis Attacks

DPA attacks work because the power consumption of cryptographic devices depends on the intermediate values of the executed cryptographic algorithms. Hence, it is clear how these attacks can be prevented. The goal of countermeasures is remove this dependency.

> The goal of countermeasures against DPA attacks is to make the power consumption of the cryptographic device independent of the intermediate values of the executed cryptographic algorithm.

The countermeasures against DPA attacks that have been published so far can essentially be categorized into two groups: hiding and masking. Figure 1.6 illustrates how these two types of countermeasures break the link between the intermediate values of a cryptographic algorithm and the power consumption of the device that executes this algorithm.

The basic idea of hiding is to remove the data dependency of the power consumption. This means that either the execution of the algorithm is randomized or the power consumption characteristics of the device are changed in such a way that an attacker cannot easily find a data dependency. The power consumption can be changed in two ways to achieve this goal: the device can be built in such a way that every operation requires approximately the same amount of energy, or it can be built in such a way that the power consumption is more or less random. In both cases, the data dependency of the power consumption is reduced significantly. However, in practice the data dependency cannot be removed completely—there always remains a certain amount of data dependency.

It is important to point out that implementations protected by hiding countermeasures process the same intermediate results as unprotected implementa-

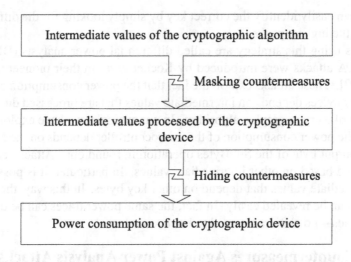

Figure 1.6. The basic concept of hiding and masking countermeasures.

tions. Resistance against power analysis attacks is solely achieved by altering the power consumption characteristics of the cryptographic device.

This is different in the case of masking. The basic idea of masking is to randomize the intermediate values that are processed by the cryptographic device. The motivation behind this approach is that the power that is needed to process randomized intermediate values is independent of the actual intermediate values. A big advantage of masking is that the power consumption characteristics of the device do not need to be changed. The power consumption can still be data dependent. Attacks are prevented because the device processes randomized intermediate values only.

The concepts of hiding and masking can also be used to prevent SPA attacks. Both types of countermeasures are discussed in this book based on many examples and case studies. Besides these two main types of countermeasures against power analysis attacks, there is also a general principle that helps to prevent SPA and DPA attacks: the key that is used by a cryptographic device should be updated as frequently as possible. The fewer power measurements an attacker can obtain for a fixed key, the more difficult it becomes to reveal the key.

Keys that are only used for a few cryptographic operations are usually called session keys. Using session keys can make power analysis attacks significantly more difficult. This idea is for example used in the EMV standard [EMV04], and it has been discussed in [Koc05]. Session keys should be used whenever it is possible.

> Session keys are cryptographic keys that are used only for a small number of cryptographic operations. This can make power analysis attacks more difficult. Hence, they should be used whenever it is possible.

However, there are many scenarios where it is not practical or not possible to update the secret keys frequently enough to prevent power analysis attacks. This is why hiding and masking countermeasures are generally the first line of defense for cryptographic devices in practice.

1.5 Summary

Modern security systems depend heavily on the use of cryptography. Cryptographic algorithms are rather complicated functions that require a computer for their efficient computation. A cryptographic device is a device that implements a cryptographic algorithm and that stores the corresponding key. Whereas the algorithm and its implementation are public, the key is kept secret. Hence, it is important that no information about the key leaks during the computation of the cryptographic algorithm. Unfortunately for the security of cryptographic devices, such leakage typically occurs. Power analysis attacks exploit the fact that the instantaneous power consumption of a cryptographic device depends on the data it processes and on the operation it performs. Based on this dependency it is possible to extract the secret key of a cryptographic device. Power analysis attacks can be prevented by adequate countermeasures. Such countermeasures break the link between the intermediate values and the operations of a cryptographic algorithm and the power consumption of the device that executes the algorithm. This book discusses power analysis attacks as well as countermeasures against them.

Chapter 2

CRYPTOGRAPHIC DEVICES

For the discussion of power analysis attacks and countermeasures, it is helpful to have some basic knowledge about cryptographic devices. In particular, it is helpful to have a basic understanding of how they are built. This chapter provides this information in compact form. It is intended for readers without a background in hardware design.

First, we sketch the typical components of cryptographic devices. Second, we talk about the design flow that is used to build them. This means, we discuss the sequence of steps that are necessary to get from the specification of a device to the actual device. Last, we focus on logic cells. In particular, we give a brief introduction to complementary CMOS, which is the most popular technique to implement logic cells for digital circuits. This exposition follows the notation that is used in [RCN03], which is a popular book about the design of digital circuits.

2.1 Components

Cryptographic devices usually consist of several components. Each of these components implements a specific functionality, such as the encryption of data or the storage of cryptographic keys. The components of cryptographic devices can essentially be divided into two groups. The components in the first group perform cryptographic operations, e.g. a digital circuit that performs encryptions. The components in the second group handle data of the cryptographic operations, e.g. non-volatile memory that provides an encryption key. Below we list the most important components of typical cryptographic devices.

- **Dedicated Cryptographic Hardware:** This component includes all hardware that is solely dedicated to performing cryptographic operations, e.g. a dedicated cryptographic circuit that implements AES.

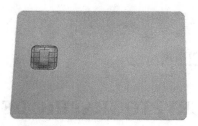

Figure 2.1. AES encryption chip in a
PLCC package.

Figure 2.2. Cryptographic smart card.

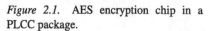

Figure 2.3. Cryptographic device in a USB token.

- **General-Purpose Hardware:** This component includes all general-purpose hardware that is used to perform cryptographic operations, e.g. a microcontroller that is programmed to perform AES encryptions.

- **Cryptographic Software:** This component consists of any type of software that implements cryptographic operations, e.g. software that implements AES.

- **Memory:** This component stores data of cryptographic operations, e.g. AES encryption keys.

- **Interface:** This component handles the data transfer to and from a cryptographic device. Cryptographic applications impose special demands on the interface. It is for example crucial that the interface prevents sensitive data, like a cryptographic key, from unauthorized access from the outside.

The components of cryptographic devices can be implemented either on separate chips or on a single chip. If they are implemented on separate chips, the chips need to be mounted on a printed circuit board (PCB). Suitable packages for chips that are mounted on a PCB are for example the dual in-line package (DIP) or the plastic-leaded chip carrier (PLCC) as shown in Figure 2.1. Examples of cryptographic devices that are built based on multiple chips are cryptographic acceleration cards or hardware security modules (HSMs).

Besides cryptographic devices with multiple chips, there are also many devices that just consist of a single chip. Smart cards are such devices. Figure 2.2 shows a smart card. The smart card contains a chip that is mounted below the visible contacts. Another single-chip cryptographic device is a USB token such as shown in Figure 2.3. In this book, we mainly focus on cryptographic devices that are implemented on a single chip. However, unless explicitly stated otherwise, our results and conclusions also hold for cryptographic devices with multiple chips.

2.2   Design and Implementation

In practice, there are essentially two ways to implement a digital circuit on a chip. The first way is to implement the digital circuit as an application specific integrated circuit (ASIC). In this case, it is necessary to create a layout of the chip. Based on this layout, the chip can then be manufactured by a semiconductor foundry. The manufacturing is done based on a so-called process technology. Cryptographic devices, like smart cards or USB tokens, are often built based on ASICs.

The second way to implement a digital circuit on a chip is to use a field-programmable gate array (FPGA). FPGAs essentially consist of programmable logic cells and programmable wires between these cells. By loading a configuration file into the FPGA, the engineer defines the behavior of the cells and the connections between the cells.

Although ASICs and FPGAs are different, the steps that have to be taken in order to implement a circuit as ASIC or on an FPGA are similar. We provide an overview of the typical design steps in the following.

2.2.1   Design Steps

The design process of a digital circuit is traditionally divided into four steps. The process starts with the specification of the circuit and it ends with the physical design. During each step of the design process, the circuit representation becomes more and more concrete until finally a layout of an ASIC or a configuration file for an FPGA is obtained. The design process is an iterative process.

- **Specification:** In this step, a high-level description of the functionality of the digital circuit is developed. This can be done either by writing a specification document or by using a high-level programming language. The latter has the advantage that the specification is executable. Hence, it can be used to generate test cases. These test cases can be used in the subsequent design steps to verify that the developed designs still meet the specification.

- **Behavioral Design:** In this step, a description of the digital circuit at the so-called register-transfer level (RTL) is created. Such a description precisely defines the sequence of operations and the bit width of all signals. During behavioral design, many important design decisions have to be made. For example, it is necessary to decide which operations are performed in hardware and which operations are performed in software. Furthermore, the clocking strategy (synchronous or asynchronous) has to be fixed during this step.

- **Structural Design:** In this step, the behavioral description is converted into a netlist. Such a netlist is essentially a list of all the cells (or transistors) of the circuit and of all the connections between them. For the conversion, it is necessary to decide whether the digital circuit will be implemented as ASIC or on an FPGA. For an ASIC, a netlist of cells or transistors, which are available for the chosen process technology, needs to be generated. For an FPGA, a netlist, which consists of the cells that are provided by the FPGA, needs to be generated. We refer to the cells that implement the functionality of the digital circuit on an ASIC or an FPGA as logic cells.

  During structural design, fine-grained design decisions have to be made in order to meet specific constraints for the digital circuit. These constraints typically concern area requirements, speed, and power consumption. Meeting these constraints may require modifying or adapting the behavioral description of the circuit accordingly. Hence, the design of a circuit is an iterative process.

- **Physical Design:** This last step in the design process is slightly different for ASICs and FPGAs. In case of an ASIC, the cells (or the transistors) of the netlist are placed. This means, the physical location of the elements on the chip is fixed. This process is called placement. Furthermore, the connections between the elements are created. The process of creating wires between the elements on the chip is called routing. In case of an ASIC, the result of the physical design step is called layout. A layout is a geometric description of all semiconductor elements and wires of the circuit. It can be used by the foundry to actually manufacture the chip.

  In case of an FPGA, the physical design step is very similar, yet it is typically less complex. First, a link between the cells in the netlist and the cells on the FPGA is created, *i.e.* it is decided which cell of the netlist corresponds to which physical cell on the FPGA. Subsequently, it is decided which physical connections on the FPGA are used for the connections listed in the netlist. In case of an FPGA, the result of the physical design step is a configuration file that can be used to configure the FPGA.

The classification of the design steps as we have sketched it is not very rigid in practice. In a specific design flow, the borders between the different steps may become blurred. A design step may also be further split into smaller steps.

The actual steps of the design process significantly depend on the chosen target platform. This is particularly true for the structural and the physical design step.

Now we look at the design process for ASICs and FPGAs in more detail. In case of an ASIC, the four steps that we have just discussed can be done in two ways: It is possible either to use a full-custom design approach or to use a semi-custom design approach. In full-custom design, the logic cells of the digital circuit are handcrafted. This allows for a highly optimized circuit design in terms of speed, power consumption, and area requirements. However, the major drawbacks of full-custom design are the very high design costs and a long time to market. As a result, full-custom design is only economical for digital circuits that are produced in very high volume. Nowadays, only highly performance-critical or area-critical parts of a digital circuit are designed in this way, e.g. memory modules. Most ASICs are built using the semi-custom design approach. This design approach is also used to implement digital circuits on FPGAs.

Semi-custom design is characterized by the fact that digital circuits are built from a limited set of predefined basic cells (combinational cells, sequential cells, I/O cells, *etc.*). In case of an FPGA, these are the cells that are available on the FPGA. In case of an ASIC, these are the so-called standard cells. Standard cells are defined in a library that describes these cells in various ways. The description typically includes the logic function, the timing behavior, the power consumption, and the layout of the cells. A standard-cell library needs to be created only once for a given process technology. This is typically done by the foundry using this process technology.

2.2.2 Semi-Custom Design

In a semi-custom design flow, the specification of the digital circuit is mapped to the logic cells of the standard-cell library or to the logic cells of the FPGA. Many steps can be done automatically due to specific restrictions that are applied to the used logic cells. There exists a large number of so-called electronic design automation (EDA) tools that can be used for this purpose.

> In a semi-custom design flow, the conversion of a design from the behavioral to the structural level and further on to the physical level is mostly done automatically. This conversion process is called synthesis.

We now describe a typical semi-custom design flow for ASICs. The steps of this design flow are also illustrated in Figure 2.4. The left side of this figure indicates how these steps map to the four basic steps that we have discussed before. Semi-custom design flows for FPGAs are similar but less complex.

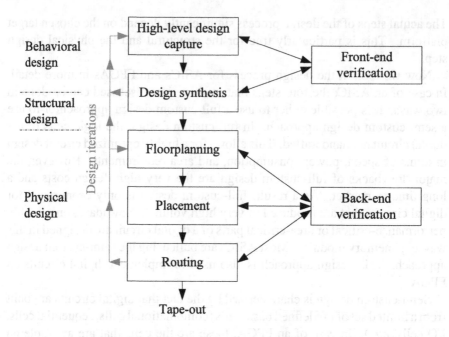

Figure 2.4.   Semi-custom design flow using standard cells.

In particular, many verification steps do not need to be performed in case of FPGAs.

- **High-Level Design Capture:** In this step, a description of the digital circuit is entered into the design system. This can be done at different levels of abstraction. For high levels of abstraction, behavioral description languages like SystemC, SystemVerilog, or SystemVHDL are used. For low levels of abstraction, hardware description languages (HDLs) like Verilog or VHDL are used. Descriptions based on HDLs are usually done at the RTL level. Hence, the combinational and the sequential behavior of the circuit are already defined in a cycle-accurate fashion.

- **Design Synthesis:** Depending on the abstraction level used for design capture, different types of design synthesis are necessary. *Behavioral synthesis* supports the process of taking various high-level architectural decisions like the separation of a design into a hardware and a software part. Often, such a process can be automated only partially. The result of behavioral synthesis can for example be an RTL description of the digital circuit expressed using an HDL. Then, *logic synthesis* is used to continue in the design process. In logic synthesis, the RTL description of the digital circuit is mapped to the actual cells that are available in the standard-cell library. Logic synthesis often also involves different optimizations of the digital circuit. It is nowadays done quite efficiently by EDA tools in an automated way. The result of logic synthesis is a *netlist* that consists of standard cells and sometimes also of so-called macro cells. These are predefined, complex modules like multipliers, memories, or even complete microcontrollers.

- **Floorplanning:** In this step, the overall layout of the chip (height, width, aspect ratio, *etc.*) is defined based on the number and size of the cells in the circuit. Regions for the core logic (logic cells that implement the actual functionality of the digital circuit), the I/O cells, and the power supply cells are defined. Furthermore, the layout of the power supply grid of the chip is designed.

- **Placement:** In this step, the precise placement of the cells of the digital circuit is done. The placement is influenced by the connections between the cells. As a result, the placement tool tries to place cells that are connected closely to each other. During placement, an important subtask is the generation of the clock tree. The clock tree distributes the clock signal to all (sequential) cells in the circuit. In a synchronous circuit, it is important that the clock signal arrives at all cells at the same time.

- **Routing:** In this step, the connections between the cells are established. This is done according to the connections that are specified in the netlist and the actual placement of the cells. Modern routing tools can change the cell placement to some extent in order to solve routing violations like infeasible routes.

- **Front-End Verification:** In this step, the functionality of the behavioral and the structural design is verified. This is mainly done by functional and logical simulations. In addition, a so-called static timing analysis of the digital circuit is done for performance checking. Typically, simulations and static timing analysis are based on the estimated physical properties of cells and wires in the digital circuit. For increased accuracy, it is possible in later design steps to extract more accurate physical information (e.g. signal timing information) from the layout in order to include this information in the simulations and the timing analysis. This process is called *back-annotation*.

- **Back-End Verification:** In this step, many different checks are performed to verify the functionality of the layout of a digital circuit. It relies heavily on the fact that circuit parasitics can be accurately estimated at this level. *Circuit parasitics* are unwanted electronic elements that occur in all fabricated circuits, e.g. capacitances between wires. The main tasks of back-end verification are design-rule checking (checking geometric design rules of transistor structures and wires in the layout), circuit extraction (extraction of the transistor netlist from the layout including exact transistor sizes and circuit parasitics), and layout-versus-schematic checking (checking that the placed and routed netlist is functionally still the same as the synthesized netlist).

- **Tape-Out:** When all design constraints are met, a file containing the circuit layout is generated and sent to the semiconductor foundry. The file contains all information that is necessary for the chip manufacturing process.

In modern semi-custom design flows, the steps are often not performed strictly sequentially as Figure 2.4 might suggest. Today, an integrated approach is necessary that usually includes more than one iteration through the design process. The reason is that it is a very challenging task to accurately estimate the physical behavior of digital circuits that are implemented using modern process technologies.

2.3   Logic Cells

Digital circuits (ASICs or FPGAs) are always built based on logic cells. This is independent of the fact whether a full-custom or a semi-custom design flow is used. Logic cells are the smallest logic building blocks of a circuit. They take one or more logic values as input and map them to logic values at their outputs according to their logic functions. On a chip, logic cells are physical objects, and hence, it is necessary to find a suitable physical representation for the logic values at their inputs and outputs.

The supply voltage of logic cells is typically referred to as V_{DD}, while the ground voltage is typically denoted by GND. In most digital circuits, these voltage levels are also used to represent the logic values that are present in a circuit. Such circuits are called voltage-mode logic (VML) circuits. In this book, we focus on VML circuits due to their widespread use. We use V_{DD} as a representation for the logic value 1 and GND to represent 0.

There exists another type of circuits where the logic values are represented by voltages between V_{DD} and GND. The actual output value of a cell in such circuits is defined by currents that are passing through the cell. Circuits that are implemented in this way are called current-mode logic (CML) circuits.

2.3.1   Types of Logic Cells

In digital circuits, two types of logic cells are used. The first type of logic cells implements basic logic (Boolean) functions like inversion, NAND, XNOR, and multiplexing. These cells are called *combinational cells* because their output values are a logical combination of their input values. Figure 2.5 shows the symbol of a 2-input combinational cell realizing the NAND function. A 2-input NAND cell performs the following Boolean operation: $q = \overline{a \cdot b}$. The output q of the NAND cell is 1 if at least one input a or b is 0. Otherwise, the output is 0.

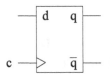

Figure 2.5. Combinational logic: 2-input NAND cell.

Figure 2.6. Sequential logic: edge-triggered D-flip-flop.

The second type of logic cells are *sequential cells* like latches, flip-flops, and registers. Their output values not only depend on their current input values, but also on preceding input values or on their initial state. This means, sequential cells memorize previous input values. Sequential cells have an input for a clock signal. This signal is used to trigger operations of sequential cells.

Figure 2.6 shows the symbol of an edge-triggered D-flip-flop. An edge-triggered D-flip-flop stores the value of its input d at its output q whenever an active edge occurs at the clock input c. The active clock edge can be either the positive edge or the negative edge of the clock signal. A D-flip-flop typically also provides the inverted value $\bar{q}$. A digital circuit is called a synchronous circuit if all sequential cells of the circuit are driven by a common clock signal. Otherwise, we talk about an asynchronous circuit.

The cells of digital circuits are implemented using transistors. *Transistors* are basically electronic switches that are built by a special arrangement of so-called p-type and n-type semiconductor structures. Typical types of transistors are metal-oxide semiconductor (MOS) transistors and bipolar transistors.

In practice, not every process technology allows building all different kinds of semiconductor devices. In fact, the most commonly used process technology only supports p-channel MOS (PMOS) and n-channel MOS (NMOS) transistors. This process technology is called *CMOS* (complementary metal-oxide semiconductor). The CMOS process technology is used for almost all digital circuits in practice. Figure 2.7 shows the symbols for NMOS and PMOS transistors. An NMOS transistor is conducting between its drain and source terminal if the voltage between its gate and source terminal is positive. In contrast to this, a PMOS transistor is conducting if the voltage between its gate and source terminal is negative. In VML circuits, the source terminals of NMOS transistors are in most cases connected to *GND*, while the source terminals of PMOS transistors are typically connected to V_{DD}. Consequently, in VML circuits NMOS transistors are usually conducting when their gate terminals are set to V_{DD} while PMOS transistors are usually conducting when their gate terminals are set to *GND*.

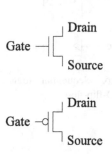

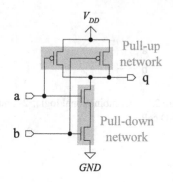

Figure 2.7. Symbol of an NMOS transis-   *Figure 2.8.* Transistor circuit of a comple-
tor (above) and a PMOS transistor (below).   mentary CMOS NAND cell.

2.3.2   Complementary CMOS

The techniques that are used to implement logic cells are called *logic styles*. Complementary CMOS is the most popular logic style to build logic cells based on PMOS and NMOS transistors. The basic idea of this logic style is to arrange PMOS and NMOS transistors in a complementary structure. The PMOS transistors form a so-called pull-up network (located between V_{DD} and the output of the cells) and the NMOS transistors are used for a so-called pull-down network (located between GND and the output of the cells). The pull-up network and the pull-down network are built in such a way that the networks are never conducting at the same time.

In complementary CMOS, all combinational and sequential logic cells are implemented using corresponding pull-up and pull-down networks. Figure 2.8, for example, shows a schematic diagram of a NAND cell implemented in complementary CMOS. The function of this NAND cell can be explained as follows. As long as at least one input a or b is set to GND (or logic 0), at least one NMOS transistor of the pull-down network is insulating and at least one PMOS transistor of the pull-up network is conducting. Thus, the output q of the NAND cell is set to V_{DD} (or logic 1). Only if both inputs of the NAND cell are set to V_{DD}, both PMOS transistors of the pull-up network are insulating and both NMOS transistors of the pull-down network are conducting. In this case, the output is set to GND.

Due to the complementary behavior of the pull-up and the pull-down network, there is almost no current flow between the V_{DD} line and the GND line during static operation. Static operation of a cell means that fixed values are applied to the inputs. As a result, complementary CMOS cells have almost no static power consumption. Another important property of complementary CMOS cells is that their outputs are driven at all moments of time. This means, the outputs are always connected either to V_{DD} or to GND via a low-resistance path. CMOS cells whose outputs are not always connected to V_{DD} or GND are called dynamic CMOS cells. In this case, the output nodes are floating for short periods of time.

Complementary CMOS cells form the basic building blocks of most digital circuits. The cells are interconnected to form modules with more functionality and higher complexity (e.g. adders, counters, state machines). This process continues up the hierarchy levels until complete systems like microcontrollers and encryption modules are constructed. Complementary CMOS is by far the most common logic style when using a CMOS process technology. Therefore, the term "complementary" is often omitted when referring to complementary CMOS cells and circuits.

2.4 Summary

Cryptographic devices consist of several components that perform cryptographic operations or that handle the data used in these operations. The components of cryptographic devices can be implemented either in a single chip or in multiple chips. In this book we focus on single-chip cryptographic devices.

The design process for cryptographic devices (ASICs or FPGAs) can essentially be divided into four steps. During these steps, the abstraction level of the digital circuit is continuously reduced until a circuit layout (ASIC) or a configuration file (FPGA) is generated. A very common implementation approach is semi-custom design. In this approach, the functionality of a digital circuit is mapped to standard cells or to the cells of an FPGA in an automated way.

The most commonly used process technology is CMOS. CMOS can be used to implement PMOS and NMOS transistors. Complementary CMOS is a logic style to implement logic cells in a way such that there is almost no static power consumption and such that the outputs of the cells are driven all the time.

Chapter 3

POWER CONSUMPTION

Digital circuits consume power whenever they perform computations. They draw current from a power supply and then dissipate the received energy as heat. The power consumption of digital circuits is a very important topic. The power consumption determines whether a chip needs to be cooled or not, it determines which kind of power supply is necessary and, in case of cryptographic devices, it determines whether a device can be attacked or not. Obviously, this is the most important property of the power consumption in the context of this book.

In this chapter, we discuss the power consumption of cryptographic devices in detail and we show how it can be measured in practice. Our discussion starts with an analysis of the power consumption of CMOS circuits in general. Subsequently, different simulation techniques and power models for digital circuits are introduced. Finally, we discuss setups to measure the power consumption of cryptographic devices. In particular, we elaborate on the quality criteria of such measurement setups.

3.1 Power Consumption of CMOS Circuits

The total power consumption of a CMOS circuit is the sum of the power consumptions of the logic cells making up the circuit. Hence, the total power consumption essentially depends on the number of logic cells in a circuit, the connections between them, and the fact how the cells are built. These properties are a result of design decisions that are taken at the system level (overall system architecture, used algorithms, hardware/software splitting, *etc.*), the architecture level (specific implementation of hardware and software components), the cell level (design of the logic cells), and the transistor level (semiconductor technology used to implement the MOS transistors of the logic cells).

When operating a CMOS circuit, the circuit is provided with the constant supply voltage V_{DD} and with input signals, as shown in Figure 3.1. The logic

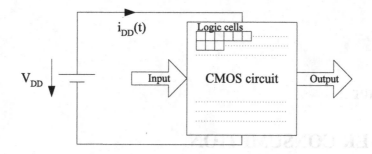

Figure 3.1. Power consumption of CMOS circuits.

cells in the circuit process the input signals and draw current from the power supply. We denote the total instantaneous current by $i_{DD}(t)$ and the instantaneous power consumption by $p_{cir}(t)$. Hence, the average power consumption P_{cir} of the circuit over time T can be calculated according to (3.1).

$$P_{cir} = \frac{1}{T} \int_0^T p_{cir}(t)dt = \frac{V_{DD}}{T} \int_0^T i_{DD}(t)dt \qquad (3.1)$$

As pointed out in Section 2.3.2, logic cells are usually implemented using complementary CMOS. We now use the simplest CMOS cell, the CMOS inverter, to describe when and why CMOS cells dissipate power. The discussion of the inverter is representative for all other cells, because all CMOS cells are built based on complementary pull-up and pull-down networks. In case of an inverter, these networks consist of the two transistors $P1$ and $N1$, see Figure 3.2. In case of more complex gates, more PMOS and NMOS transistors are necessary for these networks.

The power consumption of an inverter can essentially be divided into two parts. The first part is the static power consumption P_{stat}. This is the power that is consumed if there is no switching activity in a cell. The second part of the power consumption is the dynamic power consumption P_{dyn}. In addition to the static power, a cell consumes dynamic power if an internal signal or an output signal of a cell switches. The total power consumption of a cell is the sum of P_{stat} and P_{dyn}.

3.1.1   Static Power Consumption

CMOS cells are built in such a way that their pull-up network and their pull-down network are never conducting at the same time for constant input signals. For example, in case of the CMOS inverter shown in Figure 3.2, $P1$ is conducting and $N1$ is insulating if the input a is set to *GND*. Vice versa, $P1$ is insulating and $N1$ is conducting if the input a is set to V_{DD}. In both cases, there is no direct connection between the V_{DD} line and the *GND* line.

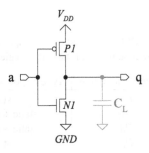

Figure 3.2. Lumped-C model of a CMOS inverter.

Therefore, only a small leakage current is flowing through the MOS transistor that is turned off. We denote this leakage current by I_{leak}. Hence, the static power consumption P_{stat} can be calculated according to (3.2). The leakage currents of MOS transistors are typically in the range of 10^{-12} Ampere $=$ 1 pA, see for example [RCN03]. However, the leakage currents are increasing significantly for modern process technologies that have structure sizes of less than 100 nanometers.

$$P_{stat} = I_{leak} \cdot V_{DD} \tag{3.2}$$

> The static power consumption of CMOS circuits is typically very low. It is increasing significantly for modern process technologies with very small structure sizes.

3.1.2 Dynamic Power Consumption

Dynamic power consumption occurs if an internal signal or an output signal of a logic cell switches. In the following, we neglect the power that is consumed when an internal signal switches. This can be done because the power that is consumed when an output signal switches is typically much higher. At a fixed moment of time, an output signal of a CMOS cell can essentially perform one out of four transitions, see Table 3.1. In the two cases $0 \rightarrow 0$ and $1 \rightarrow 1$, the cell consumes only static power, while in the other two cases also dynamic power is consumed. The power consumption values P_{00}, P_{11}, P_{01}, and P_{10} depend on the type of the cell and on the used process technology. However, in general it holds that $P_{00} \approx P_{11} \ll P_{01}, P_{10}$. In all cases, the dynamic power consumption is data dependent.

> The dynamic power consumption of CMOS circuits is typically the dominant factor in the total power consumption. The dynamic power consumption depends on the data that is processed by the CMOS circuit.

Table 3.1. Output transitions of a CMOS cell and the corresponding power consumption. The power that is required to switch internal nodes of the cell is neglected.

Transition	Power consumption	Type of power consumption
$0 \to 0$	P_{00}	static
$0 \to 1$	P_{01}	static + dynamic
$1 \to 0$	P_{10}	static + dynamic
$1 \to 1$	P_{11}	static

There are essentially two reasons for the dynamic power consumption P_{dyn} of a CMOS cell. The first one is that the load capacitance of the cell needs to be charged. The second one is that there occurs a short circuit for a short period of time if an output signal of a cell is switched.

Charging Current

When performing output transitions, CMOS cells draw a charging current from the power supply to charge the output capacitance C_L. The output capacitance contains the so-called intrinsic capacitances and the extrinsic capacitances of a CMOS cell. Intrinsic capacitances are internal capacitances that are connected to the output of the CMOS cell. Extrinsic capacitances are the capacitances of the wires that are connected to the subsequent CMOS cells and the input capacitances of these cells. The size of C_L depends heavily on the physical properties of the used process technology, the length of the wires to the subsequent cells, and the number of subsequent cells. Typical values for C_L are between 10^{-15} Farad $= 1$ fF and 10^{-12} Farad $= 1$ pF. We use the so-called lumped-C model to describe the charging power consumption of CMOS cells, *i.e.* the intrinsic and the extrinsic capacitances of the cells are lumped together into a single capacitance C_L that is connected to *GND*.

In case of the CMOS inverter (Figure 3.2), C_L is charged via the PMOS transistor $P1$. This charging happens if there is a $1 \to 0$ transition at the input a. As a result of this input transition, the value at the output q changes from 0 to 1. Thus, C_L gets charged and this causes a current flow from the power supply. When the input a is changed from 0 to 1, the value at the output q changes from 1 to 0. In this case, C_L is discharged via the NMOS transistor $N1$. No current is drawn from the power supply.

The average charging power P_{chrg} that is consumed by a cell during the time T can be calculated as shown in (3.3). In this equation, $p_{chrg}(t)$ denotes the instantaneous charging power consumed by the cell, f is the clock frequency, and α is the so-called activity factor of the cell. The activity factor corresponds to the average number of $0 \to 1$ transitions that occur at the output of a cell in each clock cycle. For example, if a cell switches its output value from 0 to 1 once every clock cycle, α is 1.

In (3.3), it can be observed that P_{chrg} is proportional to $\alpha \cdot f$, the load capacitance C_L, and the square of the supply voltage V_{DD}. Thus, the most effective way to reduce the dynamic power consumption of CMOS circuits is to reduce the supply voltage. Less effective options are to reduce the clock frequency, the load capacitance, or the switching activity of the cells.

$$P_{chrg} = \frac{1}{T} \int_0^T p_{chrg}(t)dt = \alpha \cdot f \cdot C_L \cdot V_{DD}^2 \qquad (3.3)$$

Short-Circuit Current

The second part of the dynamic power consumption of a CMOS cell is caused by the temporary short circuit that occurs in a CMOS cell during the switching of the output. In case of a switching event of a CMOS inverter (Figure 3.2), there is a short period of time where both MOS transistors ($P1$ and $N1$) are conducting simultaneously. This happens for $0 \rightarrow 1$ and $1 \rightarrow 0$ transitions.

The average power consumption P_{sc} that is caused by the short-circuit currents in a cell during the time T can be calculated as shown in (3.4). In this equation, $p_{sc}(t)$ is the instantaneous short-circuit power consumed by a cell. The value of the current peak that is caused by the short circuit during the switching event is denoted by I_{peak}. The time for which the short circuit exists is denoted by t_{sc}. Note that two switching events (with a temporary short circuit) occur in a complete switching cycle of an inverter. A complete switching cycle means a cycle in which the output of the inverter performs a $0 \rightarrow 1$ transition and a $1 \rightarrow 0$ transition. Furthermore, for the calculation of P_{sc}, the waveform of the short-circuit current for one switching event is approximated by a triangle with base length t_{sc} and height I_{peak}. This is a commonly used approximation that describes the short-circuit power consumption quite well.

$$P_{sc} = \frac{1}{T} \int_0^T p_{sc}(t)dt = \alpha \cdot f \cdot V_{DD} \cdot I_{peak} \cdot t_{sc} \qquad (3.4)$$

Simulation of the Dynamic Power Consumption

The result of an analog simulation of the CMOS inverter is shown in Figure 3.2. The following parameters have been used: $V_{DD} = 3.3\,\text{V}$, $C_L = 1\,\text{fF}$, $P1_{gate\ width} = 3.5\,\mu\text{m}$, $N1_{gate\ width} = 1\,\mu\text{m}$, $P1/N1_{gate\ length} = \text{minimal}$ structure size of process technology $= 0.35\,\mu\text{m}$.

The upper plot of Figure 3.3 shows the signal at the input a and the resulting signal at the output q. The lower plot shows the supply current $i_{DD}(t)$ that is drawn by the inverter. During the first switching event (starting at $3\,\text{ns}$) the output signal of the inverter performs a $1 \rightarrow 0$ transition. The load capacitance C_L is discharged internally and only short-circuit power is consumed.

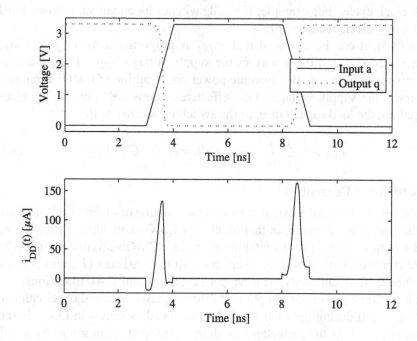

Figure 3.3. Upper plot: Input signal applied to a CMOS inverter (solid line) and the resulting output signal (dotted line). Lower plot: Current drawn by the CMOS inverter.

The second current peak (starting at 8 ns) occurs for a $0 \rightarrow 1$ transition of the output signal of the inverter. This peak is higher than the first one because not only short-circuit power is consumed but also C_L must be charged. The base length of the second peak is in the range of the base length of the first one. Note that analog circuit simulators use very comprehensive models of MOS transistors that already include the parasitic capacitances of the transistors, see Section 3.2.1. This is why there are some negative peaks in $i_{DD}(t)$.

3.1.3   Glitches

In a CMOS circuit, typically many combinational logic cells are connected to each other, *i.e.* the output of one combinational cell is used as input for a second one. The output of this cell is then used by a third one, *etc.* This is called a multistage combinational circuit. An important observation for such circuits is that the inputs of combinational cells in this circuit do not always arrive at the same time.

There are two reasons for this. First, the combinational logic cells and the wires between these cells have non-zero and in general different propagation delays. It takes a certain time until a cell switches its output upon a change of the input, and it also takes a certain time for a signal to propagate from one

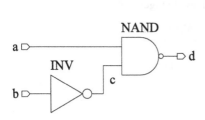

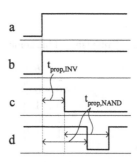

Figure 3.4. Small combinational circuit where glitches can occur.

Figure 3.5. Signal waveforms of the combinational circuit shown in Figure 3.4. For the depicted change of the input signals a and b, a glitch occurs in the output signal d.

cell to the next. Second, combinational cells usually take outputs from cells in different stages of the circuit as input. For example, it can happen that the first input of a cell is generated by another combinational cell, while the second input is a direct input of the circuit.

The different arrival times of the input signals at a combinational cell lead to temporary states of the output of this cell. Such temporary states of the output of a combinational cell are called *glitches* or *dynamic hazards* [RCN03]. During such a glitch or dynamic hazard the output signal of the cell not always fully switches to the voltage level that corresponds to the logic value of the temporary state. The reason for this is that one of the inputs may change again while the cell is switching to the temporary state.

The number of glitches in digital circuits typically increases with the number of stages of the combinational circuit. Glitches propagate like an avalanche through such circuits. If there is a glitch in one of the early stages of the circuit, this single glitch leads to glitches at the outputs of all cells that are connected to this cell. In complex CMOS circuits, glitches can become a dominant factor in the total power consumption. The occurrence of glitches is data dependent.

> Glitches in CMOS circuits are data dependent and have a strong impact on the dynamic power consumption.

In order to illustrate the concept of glitches, we now discuss a simple example. Consider the small combinational circuit shown in Figure 3.4. This combinational circuit consists of an inverter (INV) and a NAND cell. The inputs of the circuit are a and b. The output of the inverter is denoted by c and the output of the circuit is denoted by d. Table 3.2 shows the truth table of the combinational circuit for settled input signals a and b.

Table 3.2.   Truth table of the combinational circuit shown in Figure 3.4.

a	b	$c = INV(b)$	$d = NAND(a, c)$
0	0	1	1
0	1	0	1
1	0	1	0
1	1	0	1

Now assume that the input signals of the combinational circuit switch from $a = 0$ and $b = 0$ to $a = 1$ and $b = 1$. For simplicity, we assume that the propagation delays of the wires are zero. The propagation delays of the inverter and the NAND cell are denoted by $t_{prop,INV}$ and $t_{prop,NAND}$, respectively. As shown in the waveform diagram in Figure 3.5, a glitch occurs at the output d of the combinational circuit. Although the output signal d should stay at 1 for the given change of the input signals, there is a short period of time where d temporarily switches to 0. Note that the length of the glitch depends on the propagation delay $t_{prop,INV}$ of the inverter.

3.2   Power Simulations and Power Models for Designers

Simulations of the power consumption of digital circuits are used during the design process in order to determine whether the circuits meet the requirements or not. Power simulations are crucial for designers in order to build reliable, power efficient, and, in the case of cryptographic devices, secure digital circuits. These simulations can be performed at different levels of accuracy. The higher the level of accuracy is, the more resources (simulation time, memory) are typically required. Hence, a reasonable trade-off between accuracy and the usage of resources has to be found. The most commonly used levels of accuracy for power simulations are the so-called analog level, the logic level, and the behavioral level.

3.2.1   Analog Level

The most precise way to simulate the power consumption of a digital circuit is to perform an analog simulation. The basis of such a simulation is a transistor netlist of the circuit that lists all transistors of the circuit and the connections between them. Furthermore, this netlist usually contains the parasitic elements of the circuit. Parasitic elements occur in digital circuits due to the way the circuits are manufactured. In particular, there are parasitic capacitances between the wires of a circuit, and there are also unwanted capacitances in the transistors. The size of the parasitic elements mainly depends on the used process technology.

The number of parasitic elements that occur in digital circuits is very large in practice. This is why certain simplifications are usually made in order to reduce the complexity of the simulation. A very common simplification is to model all parasitic capacitances of a cell or wire as a single capacitance at the output of a cell. As already mentioned, this model is called the *lumped-C model*. In the lumped-C model, it is also assumed that wires have a negligible resistance compared to the resistance between the drain and source terminals of MOS transistors. Due to these simplifications, simulations based on the lumped-C model not always precisely describe the instantaneous power consumption of a circuit. Clearly, the more precisely the parasitic elements of a circuit are modeled, the more precise is the simulation result.

Independent of the fact whether the used transistor netlist is based on certain simplifications or not, an analog simulation of a netlist is always performed in the same way. For the simulation, the analog circuit simulator takes the transistor netlist and then calculates the voltages and currents that occur in this circuit based on difference equations. This requires a lot of resources. Hence, analog simulations are typically only used for the most critical blocks of a CMOS circuit and not for the circuit as a whole.

Examples of analog circuit simulators are *SPICE* [Rab] from the University of California at Berkeley [Uni], *Spectre* from Cadence Design Systems [Cad], and *Nanosim* from Synopsys [Syn]. SPICE (Simulation Program with Integrated Circuit Emphasis) is the most famous analog circuit simulator and this is why analog simulations are often also called SPICE simulations. SPICE is the ancestor of many other simulators [Pes]. Analog circuit simulators differ mainly in terms of their field of application (analog circuits, digital circuits), their speed, and their accuracy.

> Analog simulations are based on transistor netlists that also contain parasitic elements of the circuit. The power consumption is calculated based on solving difference equations. The precision of the simulation essentially depends on the precision of the modeling of the circuit parasitics.

3.2.2 Logic Level

Power simulations at the logic level usually require less resources than analog circuit simulations at the price of a lower precision. The basis of a power simulation at the logic level is a netlist of the cells of the circuit. This netlist contains all cells of the circuit and the connections between them. Optionally, this netlist can include some information about signal delays that are caused by cells and wires. Furthermore, it can also include rise and fall times for the signals in the circuit. The process of introducing accurate signal delays as well as rise and fall times into a netlist is called back-annotation, see Section 2.2.2.

Based on a (back-annotated) netlist of a circuit, power simulations at the logic level are done in two steps. First, the transitions that occur in a digital circuit are simulated. The result of this simulation is a file that states for each cell of the circuit which transitions it performs at what times. Clearly, this result is the more precise, the better the back-annotation of the netlist is. If the netlist is not back-annotated at all, the rise and fall times as well as the propagation delays of the wires are usually set to 0. Furthermore, the propagation delays of the cells are set to some default value, e.g. 1 ns. These are assumptions, which are often quite inaccurate. For a precise simulation of the transitions occurring in a digital circuit, back-annotation is therefore crucial. Otherwise, important effects (in particular glitches) are not simulated accurately.

The second step of a power simulation at the logic level is to map the simulated transitions to a power trace. For this purpose, it is necessary to have power models that describe how the transitions at the outputs of the cells are related to the power consumption. Such power models are often included in modern standard-cell libraries. Usually, they are parameterized by the capacitive load of the output and by the transition times of the input and output signals. Besides the power models in standard-cell libraries, there is also a generic model that can be used to map simulated transitions to power traces. This is the so-called Hamming-distance (HD) model. The HD model is less precise than the power models of standard-cell libraries.

> Power simulations at the logic level are based on netlists of cells that ideally also contain back-annotated information about signal delays, rise times, and fall times. The precision of a logic simulation depends on the quality of the back-annotated information and on the power model used for the cells.

Hamming-Distance Model

The basic idea of the Hamming-distance model is to count the number of $0 \rightarrow 1$ and $1 \rightarrow 0$ transitions that occur in a digital circuit during a certain time interval. This number of transitions is then used to describe the power consumption of the circuit in this time interval. By cutting the entire simulation of a circuit into small intervals, a kind of power trace can be generated. This power trace does not contain actual power values in Watt but the number of transitions that occur in the corresponding time interval.

When using the Hamming-distance model to simulate the power consumption, the following assumptions are made. First, it is assumed that all $0 \rightarrow 1$ and $1 \rightarrow 0$ transitions in a digital circuit lead to the same power consumption. Second, all $0 \rightarrow 0$ and $1 \rightarrow 1$ transitions are also assumed to contribute equally to the power consumption. The Hamming-distance model hence does not consider differences in parasitic capacitances of wires or cells. It assumes that all cells contribute equally to the power consumption. It also completely ignores the static power consumption of the cells.

Due to its simplicity, the Hamming-distance model is commonly used for power simulations. Simulations based on this model provide a rough estimate of the power consumption that can be calculated relatively quickly. A formal definition of the Hamming distance is given in the following. The Hamming distance of two values v_0 and v_1 corresponds to the Hamming weight (HW) of $v_0 \oplus v_1$. The Hamming weight corresponds to the number of bits that are set to one, and hence, $HW(v_0 \oplus v_1)$ corresponds to the number of bits that differ in v_0 and v_1.

The Hamming-distance model assumes that all cells contribute to the power consumption equally and that there is no difference between $0 \rightarrow 1$ and $1 \rightarrow 0$ transitions. The Hamming distance between two values v_0 and v_1 can be calculated as follows.

$$HD(v_0, v_1) = HW(v_0 \oplus v_1) \tag{3.5}$$

3.2.3 Behavioral Level

Power simulations at the behavioral level are a very fast, but typically not very accurate method to simulate the power consumption of digital circuits. The basis of a power simulation at this level is a high-level description of the digital circuit. This high-level description contains the major components of the digital circuit (microcontrollers, memories, dedicated hardware modules, *etc.*) and some high-level power models of these components. These models are used during the simulation to map activities of components, such as the processing of data or the execution of instructions, to power consumption values.

Power simulations at the behavioral level are usually used to assess the average power consumption of complex circuits. In the context of power analysis attacks, only those behavioral-level power simulators are of interest that also include the data-dependent and the operation-dependent portions of the power consumption. Examples of behavioral-level power simulators are instruction set power simulators for microcontrollers like *SimplePower* [IKV01] and *Joule-Track* [SC01].

Power simulations at the behavioral level are based on high-level descriptions and high-level power models of digital circuits. When analyzing the security of digital circuits, only simulators are of interest that model the data dependency and the operation dependency of the power consumption.

Table 3.3. Simulation techniques and power models typically used by designers of cryptographic devices.

Simulation level	Analog	Logic
Power model	Difference equations	HD power model
Required knowledge about the circuit	Transistor netlist	Cell netlist
Resource usage	Very high	Medium
Signal transitions considered	Individually for each transistor	Equal for all cells
Glitches considered	Yes	Yes
Simulation result	Power consumption in Watt	Number of transitions

3.2.4 Comparison

Simulations at the analog level and at the logic level describe the power consumption of cryptographic devices sufficiently well to make statements on their resistance against power analysis attacks. Hence, these two techniques are commonly used by designers of cryptographic devices to analyze the devices during the design process.

In this book, we use simulations at the analog level in particular in those sections where we discuss logic styles to counteract power analysis attacks. In these sections, we simulate and compare logic cells that are implemented using different logic styles. These cells are small. Hence, it is possible to use the most precise type of simulation. When a larger part of a circuit or even an entire chip has to be simulated, analog simulations are usually too expensive in terms of the required resources. In these cases, we resort to simulations at the logic level based on back-annotated netlists. Usually, we use the Hamming-distance model for these simulations.

Table 3.3 compares the most important properties of the two types of simulations that we use in this book. Clearly, these two kinds of simulations are only suitable for designers of cryptographic devices. Attackers usually do not have access to transistor netlists or cell netlists of the attacked cryptographic devices.

3.3 Power Simulations and Power Models for Attackers

In power analysis attacks, it is often necessary to map data values that are processed by the attacked device to power consumption values. This is a kind of power simulation of the device. However, it is important to point out that the absolute values of the power consumption are not relevant in case of power analysis attacks. For an attack, only relative differences between simulated power consumption values are important.

Power simulations that are performed by attackers are typically much simpler than the simulations we have discussed in the previous section. This is due to the different requirements and due to the fact that attackers usually have only very limited knowledge about the attacked cryptographic device. In this section, we discuss different techniques of attackers to map data values to power consumption values. In particular, we introduce the Hamming-distance and the Hamming-weight model. These are two generic power models that are commonly used for power analysis attacks. We also briefly discuss some variants and alternatives of these models.

3.3.1 Hamming-Distance Model

The Hamming-distance model has already been discussed in the context of power simulations at the logic level, see Section 3.2.2. In such simulations, the HD model is used to map the transitions that occur at the outputs of cells of a netlist to power consumption values. An attacker usually does not have access to such a netlist and hence cannot perform such a simulation. However, in practice an attacker often has some clues about parts of this netlist. Therefore, the attacker can simulate the power consumption of these parts.

If the attacked device is for example a microcontroller, it is very likely that this microcontroller is built in a similar way as publicly known microcontrollers. Hence, it has registers, a data bus, some memory, an arithmetic logic unit (ALU), some communication interface, *etc.* These components have also some well-known properties. For example, a data bus is typically quite long and connected to many components. Therefore, the capacitive load of such a bus is quite big, and hence, it contributes significantly to the overall power consumption of the microcontroller. It typically also holds that there occur no glitches on a data bus because such a bus is usually driven directly by sequential cells. Furthermore, it can typically be assumed that the capacitive loads of all the individual wires of a bus are about equal.

Based on these observations, it is clear that the Hamming-distance model is very well suited to describe the power consumption of data buses. An attacker can map data values that are transferred over such a bus to power consumption values without the need to have a netlist of the device. The power consumption that is caused by changing the bus from a value v_0 to a value v_1 is proportional to $HD(v_0, v_1) = HW(v_0 \oplus v_1)$. Similar observations also hold for other kinds of buses, such as address buses.

Besides the power consumption of buses, also the power consumption of registers in hardware implementations of cryptographic algorithms can be described very well by the HD model. Registers are triggered by a clock signal, and hence, they change their value only once in each clock cycle. An attacker can simulate the power consumption of a register by calculating the Hamming distance of the values that are stored in consecutive clock cycles.

In general, the HD model can be used to simulate the power consumption of a part of a netlist whenever the attacker knows the consecutive data values that are processed by this part of the netlist. In case of combinational cells, the attacker usually does not know these data values due to the glitches that occur in the combinational circuit. The HD model is therefore mainly used for registers and buses.

Attackers commonly use the Hamming-distance model to describe the power consumption of buses and registers.

3.3.2   Hamming-Weight Model

The Hamming-weight model is simpler than the HD model. It is usually applied if the attacker has no information about the netlist at all, or if the attacker does not know consecutive data values for some known part of the netlist. The latter can happen if the attacker only knows one data value that is transferred over a bus. In order to apply the HD model, it is necessary to know either the preceding or the succeeding value of the bus. If neither of these values is known, the HD model cannot be used to simulate the power consumption of the bus.

In case of the Hamming-weight model, the attacker assumes that the power consumption is proportional to the number of bits that are set in the processed data value. The data values that are processed before and after this value are ignored. Therefore, this power model is in general not very well suited to describe the power consumption of a CMOS circuit. The power consumption of a CMOS circuit rather depends on the fact whether there occurs a transition in the circuit or not, and not on the processed value.

However, in practice the Hamming weight of a data value is usually not completely unrelated to the power consumption that is caused by the processing of this value. We now show this by looking at some concrete scenarios. For all these scenarios, we assume that some part of the attacked device first processes v_0, then v_1, and finally v_2. In all scenarios, the goal is to simulate the power consumption that is caused by the processing of v_1 without knowing v_0 or v_2. v_1 is involved in two transitions, *i.e.* $v_0 \rightarrow v_1$ and $v_1 \rightarrow v_2$. In the following scenarios, we just focus on the $v_0 \rightarrow v_1$ transition. All considerations we make for this transition are equally valid for the $v_1 \rightarrow v_2$ transition.

Bits of v_0 are Equal and Constant

The first scenario we consider is that all bits of v_0 are equal each time the $v_0 \rightarrow v_1$ transition occurs. An example of such a scenario is an n-bit bus that always processes the value $v_0 = 0$ (*i.e.* all n bits are set to 0) before processing the value v_1. In this case, the HW model is equivalent to the HD model, *i.e.* $HD(v_0, v_1) = HW(v_0 \oplus v_1) = HW(v_1)$.

In the case that all n bits of v_0 are set to 1, it holds that $HD(v_0, v_1) = HW(v_0 \oplus v_1) = n - HW(v_1)$. For power analysis attacks it does not matter whether the simulated power consumption is directly or inversely proportional to the actual power consumption. It is only important that it is proportional. Hence, the HW and the HD model are equivalent in terms of power analysis attacks, if all bits of v_0 are set to 0 or to 1 before the $v_0 \to v_1$ transition occurs.

Bits of v_0 are Constant

The second scenario is that all bits of v_0 are constant, but not set to the same value. This means v_0 is a constant value that is not known by the attacker. In this scenario, we cannot make such nice observations as in the previous one, unless we focus on only one bit of the transition $v_0 \to v_1$. If we just consider one bit, the HW and HD model are equivalent in terms of power analysis attacks by the same argument as in the previous scenario. The power consumption that is caused by a bit of v_1 is hence directly or inversely proportional to the value of this bit if v_0 is always set to the same value before the $v_0 \to v_1$ transition occurs.

> The power consumption that is caused by a bit v is directly or inversely proportional to the value of the bit if the cell that processes v always stores the same value before or after the processing of v.

In case not only one but all bits of v_0 and v_1 are considered, the HW model typically describes the power consumption of the transition $v_0 \to v_1$ not very well. However, clearly the more bits of v_0 are set to the same value, the more related is the HW of v_1 to the number of transitions that occur, *i.e.* the better is the power simulation of the attacker.

Bits of v_0 are Uniformly Distributed and Independent of v_1

The last scenario is the worst-case scenario for an attacker using the HW model. In this scenario, the bits of v_0 are not constant, but random for each simulation run of the attacker. It is clear that $HW(v_1)$ is independent of $HW(v_0 \oplus v_1)$ if v_0 is a random variable that is uniformly distributed and independent of v_1. The outputs of simulations based on the HW and the HD model are hence unrelated in this scenario.

Note that this does not imply that power simulations based on the HW model are useless. The power consumption of a cryptographic device is only roughly proportional to the number of transitions that occur in the device. The HD model amongst other things assumes that $0 \to 1$ and $1 \to 0$ transitions lead to the same power consumption, see Section 3.2.2. However, this assumption only holds to a certain degree. In practice, $0 \rightarrow 1$ and $1 \rightarrow 0$ transitions do not lead to exactly the same power consumption. For example, the power consumption for a $0 \rightarrow 1$ transition can be bigger than the one for a $1 \rightarrow 0$ transition. In this case, the power consumption is on average higher for values with a high Hamming weight than for values with a low Hamming weight.

The differences between $0 \rightarrow 1$ and $1 \rightarrow 0$ transitions exist due to the fact that the pull-up and the pull-down networks of CMOS cells have slightly different power consumption characteristics. Furthermore, the parasitic capacitances between the output of a cell and V_{DD} as well as the parasitic capacitances between the output of a cell and GND are in general not equal. However, the differences are typically small, and hence, the HD model is much more suitable than the HW model in this scenario. The HW model just relies on the small differences.

General Case

In practice, there exist many more scenarios than the three we have discussed so far. In particular, there are scenarios where v_0 and v_1 depend on each other, and there are also scenarios where the bits of v_0 are not uniformly distributed. Making a general statement about how well the HW model describes the power consumption in all these scenarios is difficult. However, due the fact that $0 \rightarrow 1$ and $1 \rightarrow 0$ transitions always lead to slightly different power consumptions, $HW(v_1)$ is typically at least somehow related to the actual power consumption. This relationship can of course become very weak. Attackers therefore use the HD model whenever possible.

> Attackers only resort to the Hamming-weight model if the Hamming-distance model cannot be applied.

3.3.3   Other Power Models

The HD and the HW model are the power models that are most commonly used by attackers to simulate the power consumption of a part of a cryptographic device. This is due to the fact that these two models are very simple and generic. Other power models that have been proposed so far are typically specific for a particular kind of device. Such device-specific models can for example be derived by extending the HD model, *i.e.* certain simplifying assumptions are not made any more.

The HD model, for example, assumes that all bits of a data value contribute equally to the power consumption, *i.e.* it is assumed that the output capacitances of the cells processing the different bits of the data value are equal. If an attacker knows that some cells of a circuit contribute more than others, this can of course

Table 3.4. Simulation techniques typically used by attackers of cryptographic devices.

Simulation level	Behavioral	Behavioral
Power model	HD power model	HW power model
Required knowledge about the circuit	Processed data and "guess" about architecture	Processed data
Resource usage	Very low	Very low
Signal transitions considered	Equal for all bits	Equal for all bits
Glitches considered	No	No
Simulation result	Number of transitions	Number of logic ones

be considered in the model by introducing different weights for the different bits of the processed data value. Another extension of the HD model is to introduce different weights for the different types of transitions that occur in the device. For example, a $0 \rightarrow 1$ transition can be weighted twice as much as a $1 \rightarrow 0$ transition. In the normal HD model, both transitions are weighted equally.

Besides deriving new models by extending the HD model, it is also possible to introduce power models that describe the power consumption of larger parts of a circuit. For example, multipliers with one operand at zero often require less power than in all other cases. Hence, a suitable power model for such a multiplier is that its power consumption is zero if one of the operands is zero, and the power consumption is one in all other cases.

3.3.4 Comparison

In general, there are many ways to map data values to power consumption values in the context of power analysis attacks. The more knowledge an attacker has about a cryptographic device, the better power models the attacker can derive. The quality of the power model has a strong impact on the effectiveness of an attack. Hence, finding suitable ways to simulate the power consumption of a part of the attacked device is an important task during an attack.

Table 3.4 compares the most important properties of power simulations based on the HW model and the HD model. Note that attackers can use the HD model only for parts of circuits, while designers can use this model for entire circuits, see Table 3.3.

3.4 Measurement Setups for Power Analysis Attacks

In this chapter we have so far mainly conducted theoretical discussions about the power consumption of cryptographic devices. Now, we switch to a more practical topic. For power analysis attacks, it is necessary to measure the power consumption of a cryptographic device while it executes a cryptographic algorithm. The current section provides a basic description of corresponding measurement setups and their most important components. Furthermore, we discuss the measurement setups that have been used for all examples provided in this book.

3.4.1 Typical Measurement Setups

Measurement setups for power analysis attacks usually consist of several components that interact with each other. A block diagram of a typical measurement setup is shown in Figure 3.6. The numbers in the figure indicate in which sequence the components interact with each other when a power trace is acquired. We now shortly describe the involved components and then we take a closer look at this sequence.

- **Cryptographic Device:** This is the device under attack. Cryptographic devices usually provide an interface to communicate with a PC. This interface can be used to send commands to the device that trigger the execution of a cryptographic algorithm. In this case, the PC sends some data to the device, and the device encrypts this data and sends the result back to the PC.

- **Clock Generator:** Cryptographic devices often need to be supplied with an external clock signal. Smart cards, for example, are supplied with a clock signal of up to 4 MHz.

- **Power Supply:** Cryptographic devices usually also need an external power supply. The power for a smart card is for example provided by a smart card reader. Such a reader can typically provide 5 V, 3 V, or 1.8 V to the inserted smart card.

- **Power Measurement Circuit or EM Probe:** The power consumption of a cryptographic device can be measured directly by inserting a power measurement circuit between the power supply and the cryptographic device or indirectly by an EM probe. We discuss these different methods in Section 3.4.2.

- **Digital Sampling Oscilloscope:** The power consumption signal that is provided by the measurement circuit or by the EM probe needs to be recorded. In most measurement setups this is done by a digital sampling oscilloscope. Modern oscilloscopes, as we describe them in detail in Section 3.4.3, can be controlled remotely by a PC via a general-purpose interface bus (GPIB) or an Ethernet interface. Based on this interface, the recorded traces can also be transferred to a PC.

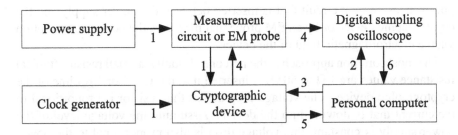

Figure 3.6. Block diagram of a typical measurement setup for power analysis attacks. The numbers indicate in which sequence the components interact with each other to acquire a power trace.

■ **Personal Computer:** The PC controls the whole measurement setup and stores the measured power traces. Every state-of-the-art PC has enough computing power to communicate with a cryptographic device and an oscilloscope. Hence, there are no special requirements for this component of the measurement setup.

In order to measure the power consumption of a cryptographic device while it executes a cryptographic algorithm, the listed components interact as follows. First, the cryptographic device is supplied with power and a clock signal (1). The device is thus operational and ready to receive commands. Next, the PC configures and arms the oscilloscope (2). In step (3), the PC sends commands to the cryptographic device that start the execution of a cryptographic algorithm. During the execution of this algorithm, the oscilloscope records the power consumption of the cryptographic device (4). The power consumption is measured by a power measurement circuit or an EM probe. Finally, the PC receives the output of the cryptographic algorithm from the device (5) and the recorded power trace from the oscilloscope (6). The steps (2) to (6) are repeated as often as it is necessary for the performed power analysis attack.

3.4.2 Power Measurement Circuits and EM Probes

Digital oscilloscopes are the preferred piece of equipment whenever it comes to recording electrical signals over time. However, oscilloscopes can only measure the voltage of a signal. It is not possible to measure other electrical properties of a signal directly with an oscilloscope. In order to measure properties like power or current with an oscilloscope, it is necessary to generate voltage signals that are proportional to these properties. These voltage signals can then be recorded with the oscilloscope.

In a measurement setup for power analysis attacks, there are essentially two ways to generate a voltage signal that is proportional to the power consumption of the cryptographic device. Such a voltage signal can be generated either by a power measurement circuit that is inserted between the power supply and the cryptographic device or by an EM probe that measures the power consumption via the electromagnetic field of the device.

The most common approach is the first one. Usually, a small resistor (typical resistance values are $1\,\Omega$ to $50\,\Omega$) is inserted into the *GND* or V_{DD} line of the cryptographic device. The voltage drop across this resistor is proportional to the current that is flowing into the device. Assuming the voltage level of the power supply is constant, this voltage drop is also proportional to the power consumption of the device. The insertion of a resistor into a power supply line of the cryptographic device is the simplest way of building a power measurement circuit.

There also exist proposals for more complex power measurement circuits. In [BGL$^+$06], for example, a power measurement circuit is presented that has been explicitly designed to enhance power analysis attacks. The basic idea of this circuit is to provide a highly stable supply voltage to the cryptographic device and to measure the power consumption using a transimpedance amplifier. However, building such active measurement circuits can be quite challenging in practice. These circuits contain several active components that can add noise to power traces and the circuits can also start to oscillate if they are not implemented carefully.

Another document that discusses power measurement circuits suitable for power analysis attacks is the IEC standard 61967 [Int03]. This standard defines how conducted and radiated emissions of an integrated circuit (IC) have to be measured in order to test the electromagnetic compatibility (EMC) of the IC. EMC testing and power analysis attacks have a lot in common. The power consumption signal of an IC that is exploited in power analysis attacks is viewed as a conducted or radiated emission in the context of EMC testing. Hence, measurement setups for EMC testing can also be used for power analysis attacks.

The IEC standard 61967 consists of six parts. The first part discusses general conditions and definitions for all measurement setups. The remaining five parts define measurement setups to analyze the conducted and radiated emissions of an IC. The setups are referred to as "TEM-cell method", "Surface scan method", "$1\,\Omega/150\,\Omega$ direct coupling method", "Workbench Faraday cage method", and "Magnetic probe method".

For measurement setups that measure the power consumption of the cryptographic device directly via a power measurement circuit, the "$1\,\Omega/150\,\Omega$ direct coupling method" is the most interesting one. This method uses a $1\,\Omega$ measurement resistor and a $49\,\Omega$ matching resistor to generate a voltage signal that can be recorded with a digital oscilloscope. A detailed specification of this setup can be found in [Int03].

The other methods that are presented in the IEC standard 61967 mainly describe measurement setups that measure the power consumption of the cryp-

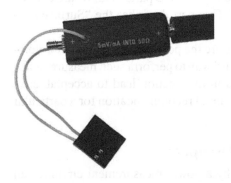

Figure 3.7. Contactless current probe to measure the current that is flowing in a wire.

Figure 3.8. H-field and E-field probes to measure the electromagnetic field of a cryptographic device.

tographic device indirectly via its electromagnetic field. For power analysis attacks, the "TEM-cell method", the "Surface scan method", and "Magnetic probe method" are the most relevant methods. The "TEM-cell method" measures the radiated emissions of the entire device in a shielded environment. The "Surface scan method" is based on small probes that can be used to analyze emissions of a specific part of an IC. The "Magnetic probe method" measures the magnetic field of wires that are connected to the cryptographic device. It therefore indirectly measures the current that is flowing in the wires. All these measurement methods are well suited to conduct power analysis attacks.

Besides the standardized methods to indirectly measure the power consumption of a cryptographic device, it is also possible to use simpler approaches to perform contactless measurements of the power consumption.

The simplest form of performing contactless measurements is to use an off-the-shelf contactless current probe. A contactless current probe (see Figure 3.7) measures the current flowing in a wire via the magnetic field that is caused by the current. For this purpose, the wire needs to be pulled through the probe as it is shown in Figure 3.7. Contactless current probes can be used for power analysis attacks. However, measurement setups based on current probes typically have a lower sensitivity than setups with power measurement circuits.

Another method to perform contactless measurements is to use H-field or E-field probes (see Figure 3.8). Based on H-field and E-field probes, the electromagnetic field of the attacked device can be measured. Depending on the size and the positioning of the probe, this field is proportional to the power consumption of a part or the entire device that is attacked. The probes that are shown in Figure 3.8 measure the power consumption of an entire IC. Suitable probes to measure the power consumption of specific parts of an IC are defined in part two of the IEC standard 61967. This part describes the "Surface scan method".

All in all, there are many ways to measure the power consumption of cryptographic devices. So far, there is no standard way to perform such measurements. Most methods that have been presented in this section lead to acceptable results. However, it is difficult to make a general recommendation for a particular method.

3.4.3   Digital Sampling Oscilloscopes

The voltage signal that is generated by a power measurement circuit or an EM probe is usually recorded with a digital sampling oscilloscope. Such an oscilloscope takes the analog voltage signal as input, converts it into a digital signal, and stores it in its memory. The analog-to-digital conversion that takes place in the oscilloscope is essentially characterized by three parameters. These parameters are the input bandwidth, the sampling rate, and the resolution of the oscilloscope.

■ **Input Bandwidth:** Every analog signal can be viewed as a sum of sinusoidal functions multiplied by some coefficients. These coefficients can be calculated by applying the Fourier transform to the analog signal, *i.e.* the Fourier transform reveals which frequency components are present in the signal, see for example [OSB99].

The bandwidth of a signal is defined as the distance between the highest and the lowest frequency component that is present in the signal. For an oscilloscope it is no problem to process signals with low frequency components. In fact, the minimum frequency of the recorded signal is typically 0 Hz. Hence, the input bandwidth of an oscilloscope is the maximum frequency that an input signal may have in order to be processed without distortion. Modern oscilloscopes have an input bandwidth of at least several hundred MHz. High-end oscilloscopes can also handle signals with frequency components in the GHz range. In typical measurement setups for power analysis attacks, the bandwidth of the power consumption signal is well below 1 GHz. Therefore, usually no distortion of the signal takes place in the oscilloscope. In the context of power analysis attacks, the input bandwidth of the oscilloscope is usually not a critical parameter.

■ **Sampling Rate:** The first step of an analog-to-digital conversion is to convert the input signal into a time-discrete signal. For this purpose, the amplitude of the input signal is sampled with a certain sampling rate. The sampling rate determines how many points of the analog signal are recorded per second. If the sampling rate is for example 1 GS/s, the amplitude of the input signal is recorded once every nanosecond.

According to the Nyquist-Shannon sampling theorem, the sampling rate needs to be more than twice as high as the highest frequency component of the input signal in order to avoid a loss of information, see for example [OSB99]. In the case of power analysis attacks, usually only noise leads to very high frequency components. As already mentioned, the power consumption signal itself is typically well below 1 GHz. Therefore, it is not necessary to always use the highest sampling rate of the oscilloscope. In practice, it is usually sufficient to focus on the most dominant frequency components of the power consumption signal. As a rule of thumb, we normally select a sampling rate that is a few times higher than the most dominant frequency component of the power consumption signal.

- **Resolution:** The second step of an analog-to-digital conversion is into convert the time-discrete signal to a time- and value-discrete signal. This means that the range of the sampled values is reduced from an infinite number of possible values to a finite number of possible values. The parameter that defines the number of possible values after the analog-to-digital conversion is called resolution. Most oscilloscopes have a resolution of 8 bits. This means that each sampled value is mapped to one out of $2^8 = 256$ possible output values. The error that is induced by the quantization of the sampled values is called quantization noise, see also Section 3.5.1.

3.4.4 Examples of Measurement Setups

In this book we use two cryptographic devices to present examples of power analysis attacks and countermeasures. The first device is a microcontroller that executes a software implementation of AES. The second device is a dedicated AES processor. We now present the measurement setups that have been used to conduct power analysis attacks on these two devices. Furthermore, we also discuss the basic properties of power traces that have been recorded with these setups.

Measurement Setup for the Microcontroller

Most power analysis attacks presented in this book target an 8-bit microcontroller executing an AES software implementation, see Appendix B.2 for details of the implementation. The microcontroller is small and simple. When we designed a measurement setup for this device, we decided to also keep the setup as simple as possible. Hence, the measurement setup is a straightforward design. Its simplicity allows us to describe power analysis attacks in a generic way. No special properties of the setup or the attacked device need to be considered in the discussion. Readers should be able to easily reproduce the presented results.

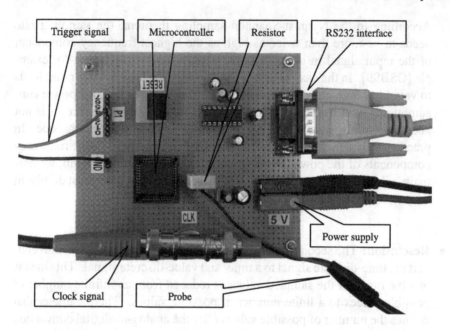

Figure 3.9. Picture of the measurement setup for the attacks on the 8-bit microcontroller.

Figure 3.9 shows a picture of the measurement setup for the 8-bit microcontroller. The controller is mounted on a prototyping board together with some basic components that are needed to run the device. The circuit is powered by a 5 V laboratory power supply and it is clocked with an external crystal oscillator running at 11 MHz. There are also several bypass capacitors on the board to stabilize the supply voltage. The microcontroller receives commands from a PC via an RS-232 interface. Essentially, the PC sends data blocks to the microcontroller. The microcontroller then encrypts these data blocks using AES and sends the results back to the PC.

In order to measure the power consumption of the microcontroller while it performs the AES encryption, a 1 Ω resistor has been inserted into the *GND* line of the power supply of the microcontroller. The output signal of this measurement circuit, *i.e.* the voltage drop across the resistor, is recorded with a digital oscilloscope. The oscilloscope is triggered by a signal on pin 4 of port 1 of the microcontroller. We intentionally set this trigger signal in our software implementation of AES in order to make power analysis attacks on this device a little bit easier than an attack on a device that has not been programmed by the attacker. In an attack on a device that has not been programmed by the attacker, typically the communication between the device and the PC has to be used as trigger signal. However, this communication often runs asynchronously to the clock signal of the device. This means that the distance between the trigger signal and the start of the AES encryption is not fixed. Therefore, it is necessary to correctly align the power traces before performing power analysis attacks. We discuss such alignment techniques in Section 8.2. In the current setup, it is not necessary to align the measured power traces because the trigger signal is synchronized to the clock signal of the microcontroller.

The oscilloscope we use for power analysis attacks on the microcontroller has an input bandwidth of 1 GHz and a resolution of 8 bits. The voltage drop across the resistor is sampled with 250 MS/s in most attacks. The power traces that are recorded with the oscilloscope are transferred to a standard PC via GPIB. Based on these traces, the PC subsequently performs different power analysis attacks.

Some examples of power traces that have been recorded with this setup have already been presented in Chapter 1. Figure 1.1 and Figure 1.4 show zoomed views of such power traces. In these zoomed views, it can be observed that there is one peak in each clock cycle in the power trace. The height of this peak is proportional to the switching activity of the device within this clock cycle.

Measurement Setup for the AES ASIC

The second device that we attack in the examples of this book is an AES ASIC implementation. We have designed and manufactured an AES encryption core using a 0.25 μm CMOS process technology, see Appendix B.3 for details of the architecture of the encryption core. In order to mount power analysis attacks on this device, we have built a dedicated printed circuit board. This PCB is shown in Figure 3.10.

The main components on the PCB are the AES ASIC and an FPGA. We use the FPGA to provide a simple interface between the ASIC and a PC. The communication between the PC and the FPGA is performed via an optically decoupled parallel interface. In order to encrypt a plaintext, the FPGA first receives a plaintext block via this interface from the PC and loads it into the ASIC. Subsequently, the ASIC performs an AES encryption and returns the corresponding ciphertext to the FPGA. The FPGA then forwards this ciphertext to the PC.

In contrast to the microcontroller we have described before, the ASIC has two separate V_{DD} lines. One V_{DD} line is connected to the encryption core and the second V_{DD} line powers the I/O cells that connect the encryption core to the pins of the package. The two corresponding GND lines are connected internally on the chip. Therefore, the power consumption of the encryption core cannot be measured separately based on inserting a resistor into the GND line. In the GND line it is only possible to measure the total power consumption of the encryption core and the I/O cells. This is why we have inserted a 1 Ω resistor into the V_{DD} line of the encryption core. The voltage drop across this resistor is measured using a differential probe.

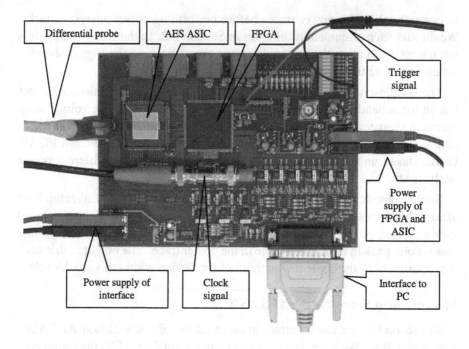

Figure 3.10.   Picture of the measurement setup for the attacks on the AES ASIC.

In the current setup, the power consumption signal of the encryption core is recorded using the same digital oscilloscope as in the previous setup. Just like in the setup for the microcontroller, a trigger signal is generated by the attacked device in order to facilitate attacks. This trigger signal indicates the start of an encryption in the AES ASIC.

Figure 3.11 shows the power consumption of the AES ASIC during one AES encryption. For this power measurement, the ASIC has been clocked with 2 MHz. Each AES encryption round takes eight clock cycles (see Appendix B.3 for details). Hence, there are eight peaks in the power trace for each round. The ten AES encryption rounds generate a repeating pattern in the power trace between 5 μs and 45 μs.

The ASIC implementation of AES performs encryptions in a much more parallel way than the software implementation on the microcontroller. As we will point out in the following chapters, it therefore takes significantly more effort to attack the ASIC than the software implementation. The effort that is needed to attack the software implementation is in fact quite low. Hence, attacks work very well even if they are conducted with a low-quality measurement setup. In case of the ASIC, the quality of the measurement setup is much more important. The following section discusses the most important quality criteria

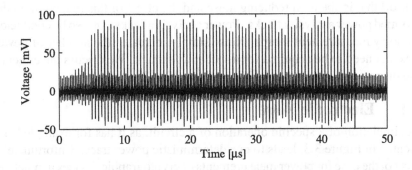

Figure 3.11. Power consumption of the AES ASIC during one AES encryption.

of measurement setups in general, and it also provides some guidelines for increasing the quality.

3.5 Quality Criteria for Measurement Setups

The power consumption of cryptographic devices is an analog high-frequency (HF) signal. A logic cell that is implemented in a current CMOS process technology switches its output typically in less than one nanosecond. Hence, the power consumption signals of the cells of a cryptographic device contain frequency components in the GHz range. Such HF signals are quite challenging to measure in practice because the signals are affected in many ways when they propagate from the cells to the digital oscilloscope. There are effects like thermal noise, reflections on wires, crosstalk, filtering, and all kinds of interactions with the environment. These effects occur in the cryptographic device and of course also in the measurement setup. The quality of a measurement setup is essentially determined by how well the setup copes with these HF effects.

In the context of power analysis attacks it is possible to model the impact of these effects on the measured power traces as different kinds of noise. The quality of a measurement setup can therefore be described relatively easily by the amount of noise that is present in the measured power traces. The two most important kinds of noise that are present in the measured power traces are electronic noise and switching noise.

> The two most important quality criteria for measurement setups are the amount of electronic noise and the amount of switching noise that are present in the recorded power traces.

In the following subsections we define and discuss the two kinds of noise. Furthermore, we provide several suggestions how to reduce the amount of noise in power traces. Typically, a reasonable compromise needs to be found between the effort that is spent on reducing noise and the effect of this reduction on the performed power analysis attacks. Power analysis attacks can also be conducted with noisy measurement setups. Noise simply increases the number of power traces that need to be acquired for a successful attack. We discuss the effect of noise on power analysis attacks in more detail in Chapter 4.

3.5.1   Electronic Noise

A simulation of a specific operation of a circuit, as it has for example been presented in Figure 3.3, leads to one deterministic power trace. Unfortunately, this is not the case for power measurements of cryptographic devices in practice. This is a consequence of electronic noise.

> When a measurement of a cryptographic device is repeated several times with constant input parameters, the resulting power traces are different. We refer to these fluctuations in the power traces as electronic noise.

Electronic noise is present in every measurement in practice and it is not possible to remove it completely. The sources of electronic noise are manifold. Some of these noise sources can be controlled rather easily while others are more difficult to cope with. We now give a brief overview of the most important sources of electronic noise in measurement setups for power analysis attacks. Additionally, we provide some hints how the influence of these noise sources on the power traces can be reduced.

- **Noise of the Power Supply:** All kinds of noise that are produced by the power supply of the attacked cryptographic device directly lead to noise in the measured power traces. Hence, a highly stable power supply should be used. The device under attack should never be supplied directly by a PC via the USB or PS/2 port. Instead, we recommend batteries or a laboratory power supply. The more stable the power supply is, the better is the setup. Details about the stability of power supply units can be found in the corresponding data sheets.

- **Noise of the Clock Generator:** A clock generator that is used for power analysis attacks needs to be stable in two ways. First of all, it is important that the clock frequency is highly stable. Only if the clock frequency is stable, it is possible to easily compare different parts of a power trace or to compare entire power traces. Misaligned power traces significantly increase the effort that is necessary for power analysis attacks. The second stability requirement for clock generators concerns the amplitude of the clock signal. The amplitude of the clock signal should also be free of noise as this noise finds its way into the power traces by various coupling effects. In order to reduce the amount of noise in the clock signal to a minimum, we recommend using a sinusoidal clock signal instead of a rectangular one. In practice, the clock signal is typically generated by a suitable crystal oscillator or a high-quality waveform generator.

- **Conducted Emissions:** The conducted emissions of all components that are directly connected to the device under attack also lead to noise in power traces. This is in particular true for components that are mounted on the same PCB as the attacked device. Therefore, we recommend using as few components on this PCB as possible. A common approach to reduce the number of components that are directly connected to the attacked device is to build a measurement setup based on two PCBs. The first PCB only contains the attacked device and a power measurement circuit. This so-called measurement board is then connected to an interface board that takes care of the communication with the PC. The measurement board and the interface board ideally are galvanically isolated by optocouplers or magnetic couplers in the communication lines. The sources of conducted emissions are therefore minimized. In case of our measurement setup for the AES ASIC, we have partly implemented this approach. The parallel interface is galvanically decoupled from the rest of the board.

- **Radiated Emissions:** In addition to conducted emissions, radiated emissions can also increase the noise in power traces. The impact of radiated emissions on a measurement setup can be reduced by shielding. A good approach to cope with radiated emissions is for example to put the PCB that features the attacked device into a Faraday cage. Additionally, also the communication and the measurement lines that are connected to the attacked device should be shielded or decoupled accordingly.

- **Quantization Noise:** Quantization noise is a consequence of the analog-to-digital conversion that is performed by the oscilloscope. The higher the resolution of the oscilloscope, the lower is the amount of quantization noise. For power analysis attacks, a resolution of 8 bits is usually sufficient. In this case, the effect of quantization noise typically is already much smaller than the effect of the other kinds of noise. However, there are also oscilloscopes with higher resolutions. It is important to point out, though, that a high resolution does not come for free in practice. It is usually necessary to make a trade-off between the resolution and sampling rate of the oscilloscope.

3.5.2 Switching Noise

During the computation of a cryptographic algorithm, the logic cells of an attacked device switch their output very frequently. This high switching activity leads to a high power consumption. In a power analysis attack, this power consumption is measured and analyzed in order to reveal the secret key that is used by the device. However, in most power analysis attacks it is actually not the total power consumption of the device that is relevant for the attack. Usually, it is the power consumption of a small part of the cryptographic device that allows the attacker to reveal the secret key. The power consumption of all other parts of the device is noise from the attacker's point of view. A perfect measurement setup would therefore be a setup that only measures the power consumption of the part of the cryptographic device that is relevant for the attack.

Measuring the power consumption of a specific part of a cryptographic device is a quite challenging task in practice. One way to perform such measurements is to use small EM probes that are precisely positioned above the relevant part of the cryptographic device, see Section 3.4.2. However, even if this technique is used, the power traces usually still contain contributions of cells that are not relevant for the attack. We refer to these contributions as switching noise.

> We refer to variations of power traces that are caused by cells that are not relevant for the attack as switching noise.

The amount of switching noise not only depends on the measurement setup. It also strongly depends on the architecture of the attacked device. Clearly, the more computations are performed in parallel by a device, the more switching noise is present in the total power consumption. This fact is often used by designers of cryptographic devices to increase the resistance of their devices against power analysis attacks, see also Chapter 7. Most measurement setups that are used in practice actually measure the total power consumption of a cryptographic device, and hence, using a highly parallel architecture can indeed help to make attacks more difficult.

We now elaborate on the amount of switching noise that occurs if an attacker measures the total power consumption of a device. Typically, the total power consumption is measured directly by inserting a measurement circuit between the attacked device and its power supply. In such a scenario, essentially two properties of the measurement setup have an impact on the amount of switching noise that is present in the power traces. The first parameter is the bandwidth of the connection between the logic cells in the cryptographic device and the oscilloscope. The second one is the clock frequency of the device.

Bandwidth

The bandwidth of a power consumption signal of a logic cell in a cryptographic device is typically in the GHz range. In order to perfectly measure the power consumption of all the logic cells, it would therefore be necessary to provide a path with such a high bandwidth from all the logic cells to the oscilloscope. However, this is not possible in practice. The paths from the cells to the oscilloscope contain a lot of parasitics that limit the bandwidth.

The first parasitics of these paths are located in the power supply grid of the chip. This grid consists of *GND* and V_{DD} wires that supply all the cells of the chip with energy. In order to provide a stable supply voltage to all cells, power supply grids are usually built in a way that there is a big capacitance between V_{DD} and *GND*. This capacitance acts as on-chip bypass capacitance and it is responsible for providing HF currents to the cells. While this capacitance is good for the functionality of the chip, it constitutes a big parasitic in the path of the power consumption signal.

The power supply grid of a chip is connected to the *GND* and the V_{DD} pins of the package. This connection is in most cases established via I/O cells and bonding wires. In particular the bonding wires add a parasitic inductance to the path of the power signal. This inductance and the on-chip bypass capacitance cannot be influenced by an attacker. The power consumption signal that can be measured at the pins of a cryptographic device is therefore always filtered. The exact bandwidth of the filter is determined by the size of the inductance, the bypass capacitance, and all the other smaller parasitics that are connected to the power supply grid of the IC. Typically, the bandwidth of this filter is significantly lower than the bandwidth of the original power consumption signals of the logic cells.

This low bandwidth implies that it is not possible to obtain distinct power signals for logic cells that switch within a few nanoseconds. The individual power consumption peaks of the logic cells are blurred by the filtering. One fixed moment of time in a measured power trace is hence influenced by switching activities of cells within a certain time interval. The smaller the bandwidth, the longer is this interval.

In most measurement setups that we have built so far, the bandwidth is so low that this interval is longer than a period of the maximum clock frequency of the chip. It is therefore not possible to measure distinct power signals within one clock cycle. We have only been able to measure the total power consumption of all cells switching within one clock cycle. In such a scenario, the switching noise corresponds to the power consumption of all cells that switch in the same clock cycle as the attacked cells. Based on our experience, we consider this to be the most common scenario in practice.

With our measurement setups it is only possible to measure the total power consumption of one clock cycle. The switching noise hence corresponds to the power consumption of all cells that switch in the same clock cycle as the attacked cells.

Clock Frequency

The clock frequency is another parameter that can affect the amount of switching noise in the power traces. If, for example, a high clock frequency is used for the attacked device, it can happen that the power consumption signals of consecutive clock cycles interfere with each other. We illustrate this effect in Figure 3.12.

The four traces in this figure show the power consumption of our AES ASIC during four consecutive clock cycles. The difference between the traces is that for each trace a different clock frequency has been used. For the first trace, the chip has been clocked with 2 MHz. In this case, there is a high peak in the power consumption every 0.5 μs. Such a high peak occurs every time the clock signal rises from low to high. The height of this peak is proportional to the switching activity that is triggered by the rising clock edge. Some smaller peaks occur at the moments of time when the clock signal falls from high to low. The power consumption at the falling clock edge is mainly caused by buffers of the clock tree in the chip.

When the clock frequency is increased, the distance between the peaks in the power trace is reduced. The second trace shows the power consumption of the chip when it is clocked with 10 MHz. At this clock frequency, the small and the big power consumption peaks, which previously occurred separately, start to overlap. The overlap is further increased when the clock frequency is set to 20 MHz. At this clock frequency, there is only one peak in each clock cycle. At 60 MHz, the peaks overlap even more and the power consumption stays at a high level.

The chip can be attacked at all clock frequencies up to 75 MHz, which is the maximum clock frequency of the chip. However, the higher the clock frequency, the higher is also the amount of switching noise in the measured power traces. This is due to the increasing overlap of the power consumption signals of consecutive clock cycles.

All attacks on the AES ASIC that we present in this book have been conducted with a clock frequency of 2 MHz. As a general rule to reduce switching noise in power traces, the clock frequency should be reduced to a level that leads to separate peaks in the power trace for every clock edge. Of course, this is only possible if the device allows such a low clock frequency.

3.6   Summary

Cryptographic devices are in most cases implemented using complementary CMOS. The power consumption of such devices can be divided into two parts: a static power consumption and a dynamic power consumption. Static power consumption occurs whenever a CMOS circuit is powered. The static power consumption depends on the sum of the leakage currents of all MOS transistors

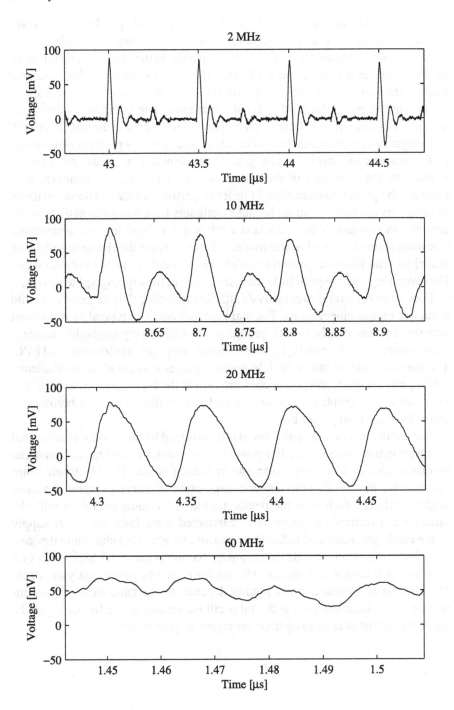

Figure 3.12. The power consumption of the AES ASIC during four clock cycles. A different clock frequency has been used for each of the four traces.

in the CMOS circuit and on the level of the supply voltage. Dynamic power consumption occurs whenever signals in the circuit change their values. The dynamic power consumption of a circuit depends on the switching activities of the logic cells of the circuit, on the level of the supply voltage, on the capacitive loads of the logic cells, on the duration of the signal transitions, and on the short-circuit current peaks during the signal transitions. The switching activities of the logic cells are strongly influenced by glitches that occur in the CMOS circuit.

Designers as well as attackers of cryptographic devices perform simulations of the power consumption. The goal of the simulations of the designers is to estimate the resistance of their devices. The motivation of attackers is to estimate the power consumption in order to perform attacks. The simulations of designers are quite accurate because designers have access to netlists of the devices. In contrast to this, attackers work at a very high level of abstraction. The most commonly used power models of attackers are the Hamming-distance model and the Hamming-weight model. These models are not very accurate. However, they only require little knowledge about the cryptographic device.

In order to perform power analysis attacks in practice, it is necessary to build a suitable measurement setup. The main components of a typical measurement setup are a power supply, a clock generator, the attacked cryptographic device, a measurement circuit or an EM probe, a digital sampling oscilloscope, and a PC. The measurement circuit or the EM probe provides a signal to the oscilloscope that is proportional to the power consumption of the cryptographic device. The PC controls the cryptographic device and the oscilloscope. Furthermore, it stores the measured power traces.

The quality of a measurement setup is determined by the amount of noise that is present in the power traces. For power analysis attacks, two kinds of noise can be distinguished: electronic noise and switching noise. The electronic noise accounts for the fact that the power consumption of a cryptographic device is slightly different each time the device performs the same operation with the same data. Electronic noise includes conducted noise from the power supply or the clock generator and radiated noise from neighboring electronic devices. Switching noise corresponds to the power consumption of all logic cells of a cryptographic device that are not relevant for a specific power analysis attack. The amount of switching noise mainly depends on the bandwidth of the connection of the logic cells to the digital oscilloscope and on the frequency of the clock signal that is used to operate the cryptographic device.

Chapter 4

STATISTICAL CHARACTERISTICS
OF POWER TRACES

After having discussed different measurement setups and their most important quality criteria in Chapter 3, we now analyze power traces from a statistical point of view. Power traces are vectors of voltage values that have been recorded with a digital sampling oscilloscope. The measured voltage values are proportional to the power consumption of a cryptographic device because the oscilloscope is connected to an appropriate measurement circuit or EM probe. The settings of the oscilloscope determine the length of the power traces and the number of points that are recorded per second.

In this chapter, we present statistical models for the individual points of power traces and for power traces as a whole. These models describe power traces in a way that power analysis attacks can be explained and analyzed relatively easily. We introduce these models by first presenting the most important components of power traces, *i.e.* we discuss which factors influence the power consumption most. Subsequently, we characterize these dependencies based on probability distributions and we introduce a notation for side-channel leakage. Using the conclusions gained from these discussions, we present different methods to compress power traces. In the last section of this chapter, we provide a brief introduction to confidence intervals and hypothesis testing.

4.1 Composition of Power Traces

Power analysis attacks exploit the fact that the power consumption of cryptographic devices depends on the operations they perform and on the data they process. These two dependencies are therefore the most interesting properties of the power consumption in the context of this book. This is also why we introduce a dedicated notation for the operation-dependent and the data-dependent components of power traces. For each single point of a power trace, we refer to the operation-dependent component of the point as P_{op}. With P_{data}, we refer to the data-dependent component of the point.

Besides P_{op} and P_{data}, each point of a power trace also depends on two additional factors. As already pointed out in Section 3.5.1, there is electronic noise in every power measurement in practice. When a power measurement of a fixed operation on some fixed data is repeated, the measurement is different for every repetition. We denote this noise component of the power consumption by $P_{el.\,noise}$. Besides the noise, each point of a power trace has also a constant component. Constant components are for example caused by leakage currents and by transistor switchings that occur independently of the performed operation and the processed data. We refer to this constant power consumption as P_{const}. The components P_{op}, P_{data}, $P_{el.\,noise}$, and P_{const} are additive. Therefore, it is possible to model each point of a power trace as the sum of these four components.

Each point of a power trace can be modeled as the sum of an operation-dependent component P_{op}, a data-dependent component P_{data}, electronic noise $P_{el.\,noise}$, and a constant component P_{const}.

$$P_{total} = P_{op} + P_{data} + P_{el.\,noise} + P_{const} \qquad (4.1)$$

Note that the four components P_{op}, P_{data}, $P_{el.\,noise}$, and P_{const} are functions of the time. For each point of a power trace, these components are potentially different. We do not write these components explicitly as a function of time as we usually only analyze single points based on this model.

In the context of power analysis attacks, the components P_{op}, P_{data}, and $P_{el.\,noise}$ are the most important ones. The component P_{const} is not relevant for power analysis attacks because it does not provide any exploitable information for an attacker. An attacker can only learn information about the key of a cryptographic device by analyzing P_{op} and P_{data}. This analysis becomes the more difficult, the bigger the noise component $P_{el.\,noise}$ is. In the following sections, we discuss the characteristics of the different additive components of the power consumption.

4.2   Characteristics of Single Points

We start the characterization of the different components of power traces by only looking at a single point of a power trace, *i.e.* we look at the power consumption of a cryptographic device at a fixed moment of time. For this fixed moment of time, we determine the probability distribution of $P_{el.\,noise}$, P_{data}, and P_{op}.

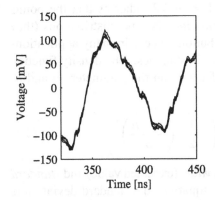

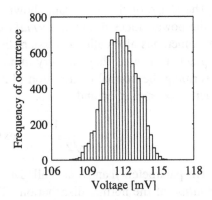

Figure 4.1. Power traces look very similar if the same data is processed.

Figure 4.2. Histogram of the power consumption at 362 ns of Figure 4.1.

4.2.1 Electronic Noise

In order to characterize the component $P_{el.\,noise}$, it is necessary to record the power consumption of the device while it performs some fixed operation on fixed data. As a concrete example, we now characterize the electronic noise of our microcontroller. For this purpose, we have recorded the power consumption of the microcontroller using the measurement setup described in Section 3.4.4 while it has repeatedly moved the data value 0 from its internal memory to a register—always the same instruction and the same data have been used. Figure 4.1 shows five power traces of this experiment. It is obvious that the five traces are very similar because the same instruction with the same data has been processed. However, there are differences in the power traces due to electronic noise.

In order to get a better idea of how the points of the power traces are distributed, we have performed the same experiment with 10 000 repetitions. From each of the 10 000 traces, we have only used the point at 362 ns. This is the time index of the highest peak in Figure 4.1. With these 10 000 points we have computed the histogram that is shown in Figure 4.2. In a histogram, the areas of the blocks represent the frequency of occurrence. In Figure 4.2, all blocks have the same width. Thus, the higher a block is in Figure 4.2, the more points have had the corresponding power consumption.

In this experiment, most points at 362 ns are around 112 mV and only very few points are below 109 mV or above 115 mV. If we would draw a histogram for the points at 400 ns, we would get a slightly different result. Figure 4.1 indicates that the power consumption at 400 ns is around −25 mV. Hence, we expect that most of the points would be in the range of −25 mV. However, the shape of the histogram would be the same.

The shape of the histogram shown in Figure 4.2 indicates that the points in the power traces follow a *normal distribution*. Normal distributions (they are sometimes also called Gaussian distributions) occur in many applications in practice. Hence, they are very important in statistics. The density function describing the normal distribution (4.2) depends on the parameters μ and σ, where $-\infty < \mu < \infty$ and $\sigma > 0$.

$$f(x) = \frac{1}{\sqrt{2 \cdot \pi \cdot \sigma}} \cdot \exp\left(-\frac{1}{2} \cdot \left(\frac{x - \mu}{\sigma}\right)^2\right) \tag{4.2}$$

The parameters μ and σ are called *mean value* (expected value) and *standard deviation* of the normal distribution. The square of the standard deviation is also called *variance*.

$$\mu = E(X) \tag{4.3}$$
$$\sigma^2 = Var(X) = E((X - E(X))^2) \tag{4.4}$$

We write $X \sim \mathcal{N}(\mu, \sigma)$ for a variable X that is normally distributed with mean value μ and standard deviation σ. If $\mu = 0$ and $\sigma = 1$, we refer to the distribution as standard normal distribution. The cumulative distribution function of the standard normal distribution is typically denoted by $\Phi(x)$. The mean value and the standard deviation fully define the normal distribution.

Mean and Variance

In our experiment, the variable X defines the power consumption at 362 ns. We refer to the realizations of this experiment with x. In a normal distribution, the mean value (expected value) is the most likely outcome of an experiment. In other words, it is the outcome that occurs the most often. In addition, most outcomes of an experiment are very close to the mean value. Hence, we can estimate $\mu = E(X)$ with the *average* value $\bar{x}$, see (4.5). In our experiment, the average $\bar{x}$ is 111.86 mV. We can also relate the average to the histogram that is shown in Figure 4.2. It is the center of this histogram.

$$\bar{x} = \frac{1}{n} \cdot \sum_{i=1}^{n} x_i \tag{4.5}$$

We can estimate the standard deviation $\sigma = \sqrt{Var(X)}$ with the square root of the *empirical variance* s^2, see (4.6).

$$s^2 = \frac{1}{n - 1} \cdot \sum_{i=1}^{n} (x_i - \bar{x})^2 \tag{4.6}$$

The standard deviation essentially reflects the width of the distribution. The wider the distribution of the power consumption, the higher is the standard

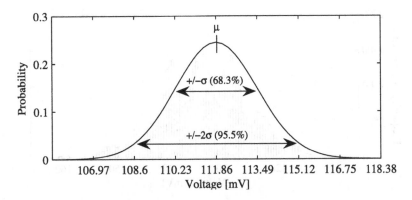

Figure 4.3. The normal distribution $\mathcal{N}(111.86, 1.63)$ models the power consumption at 362 ns.

deviation. An important property of the standard deviation is that 68.3% of all x_i are within one standard deviation around the average, 95.5% are within two standard deviations, and 99.99% are within four standard deviations, see Figure 4.3.

In our experiment, s is 1.63 mV. Consequently, the power consumption at 362 ns can be described by a normal distribution with $\mu = 111.86$ mV and $\sigma = 1.63$ mV: $X \sim \mathcal{N}(111.86, 1.63)$. This normal distribution is shown in Figure 4.3. It matches the histogram in Figure 4.2 very well.

We now assign the properties of this distribution to the components of the power consumption that have been introduced in Section 4.1. As pointed out in this section, all constant components of the power consumption are modeled as P_{const}. Hence, it holds that $E(P_{op}) = E(P_{data}) = E(P_{el.\,noise}) = 0$ and that $Var(P_{const}) = 0$. In the current experiment, always the same operation on the same data value has been performed. Hence, $Var(P_{data})$ and $Var(P_{el.\,noise})$ are also zero. The variance of the distribution that is shown in Figure 4.3 therefore corresponds to the variance of $P_{el.\,noise}$. This means that the electronic noise in the power traces of our microcontroller is normally distributed with $\mu = 0$ mV and $\sigma = 1.63$ mV. In fact, the electronic noise in the power traces of most cryptographic devices is normally distributed. The standard deviation is of course specific for each device.

> The distribution of the electronic noise $P_{el.\,noise}$ is a normal distribution:
> $P_{el.\,noise} \sim \mathcal{N}(0, \sigma)$.

4.2.2 Data Dependency

We now focus on the data dependency of power traces. Our goal is to determine the distribution of the power consumption of cryptographic devices

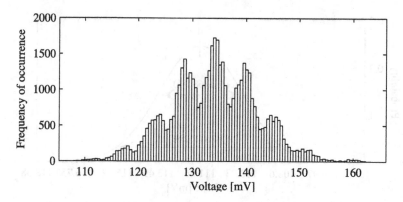

Figure 4.4. Histogram of the power consumption at 362 ns if different data values are transferred from the internal memory to a register.

when they process different data. Of course, the distribution of P_{data} not only depends on the cryptographic device, but also on the distribution of the data that is processed. In all our discussions, we assume that the processed data is uniformly distributed, *i.e.* we assume that each data value occurs with the same probability.

In order to get an idea of how the distribution of the power consumption looks like when our microcontroller processes uniformly distributed data, we have performed an experiment that is similar to the one in Section 4.2.1. We have again measured the power consumption of the microcontroller while it has moved data from its internal memory to a register. However, instead of keeping the data constant, we have now run through all 256 data values. We have recorded the power consumption of the microcontroller 200 times for each of the 256 data values. Thus, we have measured $256 \cdot 200 = 51\,200$ power traces in total.

The power traces look similar to the ones shown in Figure 4.1, except for one important difference. The power traces of the current experiment vary more than the ones shown in Figure 4.1. This is because the data is not kept constant. In order to compare the distribution of the power traces of the current experiment with the one from the previous experiment, we have again generated a histogram of the power consumption at 362 ns, see Figure 4.4.

The histogram shows that the power consumption of the microcontroller indeed varies more than before. The histogram also indicates that the power consumption is no longer normally distributed. However, when looking closer at Figure 4.4 it seems that the new distribution is composed out of nine normal distributions with different amplitudes. The first distribution seems to have its mean at about 112 mV, the second one at about 118 mV, and the third one at about 123 mV, *etc.*

Table 4.1. Probability distribution for the Hamming weight of a uniformly distributed 8-bit value.

HW	0	1	2	3	4	5	6	7	8
Prob.	0.004	0.031	0.109	0.219	0.273	0.219	0.109	0.031	0.004

In order to analyze this property of the histogram, we have generated separate histograms of the power consumption for all data values. We have generated a histogram for the case that the microcontroller processes a 0 (this histogram has already been shown in Figure 4.2), one for the case that it processes a 1, a 2, *etc.* The analysis of these histograms shows that for all data values with the same Hamming weight, the power consumption of the microcontroller is distributed in approximately the same way. For each Hamming weight there is a normal distribution with a different mean—the standard deviation of the normal distributions is approximately the same. There are nine possible Hamming weights for an 8-bit value: $0, \ldots, 8$. Hence, there are nine normal distributions with different means.

The Hamming weights of the 8-bit values that are processed by the microcontroller are binomially distributed. Table 4.1 shows the probabilities for the Hamming weights of a uniformly distributed 8-bit value. Hamming weight 4 has the highest probability of occurrence and the Hamming weights 0 and 8 have the lowest one.

The shape of the histogram in Figure 4.4 can therefore be explained as follows. The histogram is the result of the superposition of two effects. First, the power consumption of the microcontroller is proportional to the Hamming weight of the data that it moves from the memory to the register. This data-dependent component P_{data} of the power consumption is binomially distributed. Second, no matter which data the microcontroller processes, there is always a certain amount of electronic noise in the power traces. This noise is largely independent of the data that is being processed, and it is normally distributed as already pointed out in Section 4.2.1.

From Section 4.2.1, we also already know that the standard deviation of the electronic noise of the microcontroller is $1.63\,\mathrm{mV}$. In order to accurately model the total power consumption of the microcontroller in the current experiment, we therefore only need to additionally characterize the component P_{data} of the power consumption. However, it is not possible to measure P_{data} separately because the electronic noise cannot be turned off.

In order to characterize P_{data}, it is necessary to remove the electronic noise from the power traces in a different way. As discussed in Section 4.6.1, the variance of $Var(P_{el.\,noise})$ can be reduced by calculating the mean of several power traces. In the current experiment, we have therefore calculated the mean

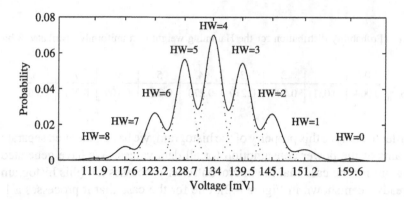

Figure 4.5. The distribution of the power consumption when the microcontroller transfers different data from the internal memory to a register.

values of the power consumption for the nine different Hamming weights. These nine mean values can be estimated based on the 51 200 power traces we have recorded. The estimated mean values are 111.9, 117.6, 123.2, 128.7, 134, 139.5, 145.1, 151.2, and 159.6. Since we model all constant components of the power consumption as P_{const}, we need to remove the mean of these values, *i.e.* $E(P_{data})$ needs to be zero. The voltage levels for the nine Hamming weights are hence -22.67, -16.92, -11.35, -5.86, -0.49, 4.96, 10.53, 16.68, and 25.12. The Hamming weights are binomially distributed. Therefore, P_{data} is binomially distributed as well.

By superposing the distributions of P_{const}, P_{data}, and $P_{el.\,noise}$ it is possible to plot the distribution of the total power consumption. This distribution is shown in Figure 4.5. In addition to the distribution of the total power consumption, this figure also shows the nine normal distributions for the different Hamming weights (dotted lines). It can be observed that the plotted distributions match the histogram of Figure 4.4 very well. Hence, we have found a quite accurate model for the power consumption of the microcontroller when it transfers different data from its memory to a register using the same instruction. Notice that the power consumption is actually inversely proportional to the Hamming weight of the processed data, *i.e.* Hamming weight 0 leads to the highest power consumption.

Of course, the microcontroller has many more instructions and it can also perform many more operations. We have generated histograms of the power consumption for other operations of the microcontroller in order to find out how the power consumption depends on the processed data in these cases. The experiments have shown that the power consumption of the microcontroller is inversely proportional to the Hamming weight of the processed data. The model fits quite well—however, it is not perfect. As it can be observed in Figure 4.5, the distance between the nine mean values for the different Hamming weights is not exactly the same. Nevertheless, the Hamming-weight model describes the microcontroller sufficiently well to analyze most attacks and countermeasures.

> The component P_{data} of the power consumption of our microcontroller is approximately inversely proportional to the Hamming weight of the data that our microcontroller processes.

Data Dependency of Cryptographic Devices in General

It is difficult to make a general statement about the data dependency of the power consumption of cryptographic devices. After the discussion about the different power models in Section 3.3, it is clear that P_{data} does not always depend on the Hamming weight of the processed data. The microcontroller that we use in the context of this book leaks the Hamming weight of the intermediate values because it *precharges* the bus lines. This means, before a value is sent over the bus, the bus lines are set to logical one. Many other microcontrollers do not leak the Hamming weight, but the Hamming distance or some property of the processed data. In the worst case, each data value leads to a different power consumption.

The fact which model fits best for a given cryptographic device can always be determined using the same strategy. First, multiple power traces need to be recorded for each data value. Subsequently, the mean of these power traces needs to be calculated for every data value in order to remove the electronic noise. The resulting mean values and their frequency of occurrence constitute the data-dependent component P_{data} of the power consumption.

Although it is not possible to make a general statement about the way P_{data} depends on the processed data, it is possible to make a statement on the shape of the distribution of P_{data} for most cryptographic devices. The component P_{data} can usually be approximated by a normal distribution. This means that the power consumption of a cryptographic device at a fixed moment of time is more or less normally distributed, when it processes uniformly distributed data values at this time. This approximation is valid in most cases no matter whether the power consumption depends on the Hamming weight, the Hamming distance, or some other property of the processed data. The approximation becomes better, the more data bits the device processes in parallel. Even for our 8-bit microcontroller such an approximation is possible. The histogram shown in Figure 4.4 can quite well be approximated by one big normal distribution.

> For most cryptographic devices, it is valid to approximate the distribution of the data-dependent component P_{data} of the power consumption by a normal distribution if the processed data is uniformly distributed.

4.2.3   Operation Dependency

The power consumption of a cryptographic device at some fixed moment of time not only depends on the processed data, but also on the operation that is performed. The operation-dependent component P_{op} can be determined by measuring the power consumption of a cryptographic device while it performs different operations with the same data. Just like in the case of P_{data}, multiple measurements need to be made for every operation and subsequently the electronic noise needs to be removed by calculating the mean of the traces for each operation. The distribution of the calculated mean values constitutes the distribution of P_{op}. Usually, this distribution can be approximated quite well by a normal distribution.

> The distribution of the operation-dependent component P_{op} can usually be approximated by a normal distribution.

In case of our microcontroller, the components P_{op} and P_{data} are largely independent. The power consumption is inversely proportional to the Hamming weight of the processed data in case of all operations. However, the data dependency is located at different voltage levels for different operations. This is the only effect of P_{op}. However, in practice not every device behaves like this. In fact, there are many devices where P_{op} and P_{data} are not independent. In case of such a device, a change of an operation also leads to a change of P_{data}. This is in particular true for hardware implementations of cryptographic algorithms. For such devices, P_{data} needs to be characterized separately for every operation.

4.3   Leakage of Single Points

In the previous section, we have discussed the distributions of $P_{el.\,noise}$, P_{data}, and P_{op}. These components of the power consumption are the most relevant ones for power analysis attacks because they determine how much information an attacker can learn from the power traces. In this section, we introduce a metric to quantify this information leakage for each point of a power trace. Our metric is based on a signal-to-noise ratio (SNR).

4.3.1   Signal and Noise

Different types of power analysis attacks often exploit very different properties of P_{op} and P_{data}. In fact, it is not unusual that there exist two or even more power analysis attacks that exploit different properties of the same point of a power trace. This is the reason why it is difficult to talk about leakage of a point in general. Rather than talking about leakage in general, we talk about a particular leakage for every attack scenario. Consequently, we also introduce our metric in the context of a given attack scenario. We essentially ask the following two questions to quantify leakage: Which information is the attacker looking for and what information do the points of a power trace give to this attacker?

In order to answer these two questions, we introduce a dedicated notation for the component of the power consumption that is exploitable in a given attack scenario. We refer to this component as the exploitable power consumption P_{exp}. The component P_{exp} corresponds to P_{data}, P_{op}, or a combination of parts of both. SPA attacks, for example, often exploit combinations of P_{op} and P_{data}. DPA attacks typically only exploit a small part of P_{data}. As already defined in Section 3.5.2, we refer to the part of the power consumption that is not exploitable in a given attack scenario and that is not electronic noise, as switching noise $P_{sw.\,noise}$. The relationship between the different components of the power consumption is given in (4.7).

$$P_{op} + P_{data} = P_{exp} + P_{sw.\,noise} \tag{4.7}$$

In DPA attacks, $Var(P_{op})$ is zero. The attacker records power traces of a cryptographic device while it performs the same operation with different data. Hence, there is no operation dependency. In SPA attacks, P_{op} and P_{data} are exploitable. The remaining parts of P_{op} and P_{data} correspond to $P_{sw.\,noise}$. Based on (4.7), it is possible to rewrite (4.1). Hence, we obtain a way to model the power consumption of a point of a power trace in a given attack scenario.

In the context of a given attack scenario, the power consumption of a point of a power trace can be modeled as the sum of the exploitable power consumption P_{exp}, the switching noise $P_{sw.\,noise}$, the electronic noise $P_{el.\,noise}$, and the constant component P_{const}.

$$P_{total} = P_{exp} + P_{sw.\,noise} + P_{el.\,noise} + P_{const} \tag{4.8}$$

It is important to understand that the components P_{exp}, $P_{sw.\,noise}$, $P_{el.\,noise}$, and P_{const} are independent of each other. Therefore, all parts of the power consumption that depend on the information that the attacker is looking for, need to be modeled as P_{exp} and not as $P_{sw.\,noise}$. Assume for example that our microcontroller processes an 8-bit value where each bit is uniformly distributed. Furthermore, assume that the value of bit two is the complement of the value of bit one in each experiment. Hence, the two bits depend on each other. All other bits of the processed values are independently distributed.

Now consider an attacker who is interested in the power consumption of bit one. In this case, P_{exp} needs to be defined as the power consumption that is caused by the bits one and two, while $P_{sw.\,noise}$ corresponds to the power consumption of the other six bits. It is not possible to model the power

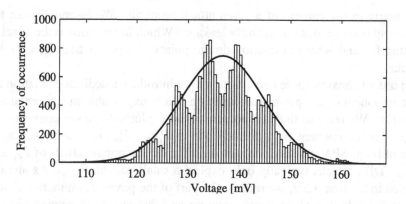

Figure 4.6. Histogram of the total noise ($P_{sw.\,noise} + P_{el.\,noise}$) if the exploitable signal is the LSB of the byte that the microcontroller processes. This noise is approximately normally distributed.

consumption that is caused by bit one as P_{exp} because bit one and two depend on each other. The power consumption of both bits constitutes the exploitable information. This is independent of the fact how much of the information is actually exploited in an attack.

We next look in more detail at an example where the microcontroller processes uniformly distributed data, *i.e.* each bit is independently and uniformly distributed. For this example, we use the power measurements described in Section 4.2.2. In this section, we have characterized P_{data}. Notice that this characterization has been done independently of potential attack scenarios.

Assume again that the attacker is only interested in the power consumption of the least significant bit (LSB) of the data that the microcontroller processes. In this scenario, P_{exp} is the power consumption that is caused by the processing of the LSB. The other seven bits that are processed by the microcontroller are switching noise. They are independent of the LSB.

In this attack scenario, the total noise ($P_{sw.\,noise} + P_{el.\,noise}$) can be characterized by keeping the LSB constant and by randomly varying all other bits. We have obtained such power traces by taking a corresponding subset of the power traces we have previously analyzed in Section 4.2.2. In this section, we have measured 51 200 power traces while the microcontroller has transferred uniformly distributed byte values. By selecting those 25 600 power traces where the LSB was 1, we have obtained the desired power traces. Figure 4.6 shows a histogram of these traces at 362 ns. This is the same moment of time we have considered for the histograms in Section 4.2.

The histogram of the total noise is the result of the superposition of two effects. First, there is the electronic noise we have already characterized in Section 4.2.1. Second, there are seven bits that switch randomly. This switching noise is binomially distributed because the power consumption of the micro-controller depends on the Hamming weight of the processed data. The total distribution of the noise has eight peaks as a consequence of this binomial distribution. However, as shown in Figure 4.6, it is possible to approximate the distribution of the total noise by a normal distribution. The standard deviation of this distribution is 7.54 mV. Based on the definition of P_{exp} and $P_{sw.\,noise}$ we now introduce a signal-to-noise ratio that quantifies the leakage of points of power traces.

4.3.2 Signal-to-Noise Ratio

Signal-to-noise ratios are commonly used in electrical engineering and signal processing. An SNR is the ratio between the signal and the noise component of a measurement. The general definition of an SNR in a digital environment is given in (4.9).

$$SNR = \frac{Var(Signal)}{Var(Noise)} \tag{4.9}$$

In case of a power analysis attack, the *signal* corresponds to P_{exp}. This is the component of the power consumption that is exploitable. The exploitable power consumption P_{exp} is the only component that contains relevant information for an attacker in a given attack scenario. The *noise* component is given by the sum of $P_{sw.\,noise}$ and $P_{el.\,noise}$. This is the total noise the attacker is confronted with.

In the context of a given attack scenario, the signal-to-noise ratio of a point of a power trace is given by the following equation.

$$SNR = \frac{Var(P_{exp})}{Var(P_{sw.\,noise} + P_{el.\,noise})} \tag{4.10}$$

The SNR quantifies how much information is leaking from a point of a power trace. The higher the SNR, the higher is the leakage.

The variance $Var(P_{exp})$ quantifies how much a point of a power trace varies because of the exploitable signal. The variance $Var(P_{sw.\,noise} + P_{el.\,noise})$ quantifies how much a point varies because of noise. It is quite clear that the higher the SNR, the easier it is to detect P_{exp} in the noise. In order to illustrate this fact, we now compare two attack scenarios for our microcontroller.

In the first attack scenario, we exploit the fact that the power consumption of the microcontroller depends on the 8-bit values it processes. In the second scenario, we only exploit the fact that the power consumption of the microcontroller depends on the LSB. All other bits are independently distributed, and hence, they are switching noise.

Table 4.2. Variance of the components of the power consumption according to the models discussed in (4.1) and (4.8).

Component	Variance	
	8-bit scenario	1-bit scenario
P_{data}	61.12	61.12
P_{op}	0.00	0.00
$P_{el.\,noise}$	2.67	2.67
P_{exp}	61.12	6.87
$P_{sw.\,noise} + P_{el.\,noise}$	2.67	56.85

We start our analysis of the SNR in the two attack scenarios by first only looking at one particular point of the power consumption. In fact, we look again at the power consumption of our microcontroller at 362 ns while it moves uniformly distributed data from its internal memory to a register. In Section 4.2.2, we have already measured $256 \cdot 200 = 51\,200$ power traces while the microcontroller performs this operation. Hence, we can reuse some results and traces from Section 4.2.2.

One of these results is that the standard deviation of the electronic noise at 362 ns is 1.63 mV. Hence, the variance $Var(P_{el.\,noise})$ is $2.67\,(\mathrm{mV})^2$. We have also learned that the power consumption at this moment of time is inversely proportional to the Hamming weight of the byte the microcontroller processes. In fact, we have also determined the mean power consumption for each of the nine possible Hamming weights. Based on these mean values, we now calculate the variance of P_{data}. For this purpose, we proceed as follows.

The 51 200 power traces that we have measured in Section 4.2.2 correspond to the total power consumption, *i.e.* these traces also contain electronic noise. In order to calculate the variance of P_{data} at 362 ns, it is necessary to replace each of the 51 200 power consumption values that have been measured at this position by the average power consumption value for the corresponding Hamming weight. For example, there are 200 traces that correspond to the processing of a value with Hamming weight 0. All these values need to be replaced by 25.12. As determined in Section 4.2.2, this is the data-dependent component P_{data} for the processing of a value with Hamming weight 0. This value has been calculated by removing the electronic noise and P_{const} from the total power consumption. By replacing all of the 51 200 power consumption values by the corresponding values for P_{data}, we obtain $Var(P_{data}) = 61.12\,(\mathrm{mV})^2$. The first three lines of Table 4.2 show the variance of P_{data}, P_{op}, and $P_{el.\,noise}$ at 362 ns. Because the operation is fixed in both attack scenarios, the variance of the operation-dependent part of the power consumption P_{op} is zero.

In order to determine the SNR in our two attack scenarios we need to switch from the power model with P_{data}, P_{op}, and $P_{el.\,noise}$ to the power model stated in (4.8). This means we need to determine P_{exp} and $P_{sw.\,noise}$ for the two attack scenarios. This procedure is straightforward for the first attack scenario where we exploit the power consumption of all 8 bits the microcontroller processes. In this case we have $P_{exp} = P_{data}$. This also implies that $Var(P_{sw.\,noise}) = 0$. Hence, the total noise $Var(P_{sw.\,noise} + P_{el.\,noise}) = Var(P_{el.\,noise})$. The column labeled "8-bit scenario" in Table 4.2 shows the corresponding variances in the lines four and five. Filling these variances into (4.10), we obtain $SNR = 22.89$.

In case of the second attack scenario, the exploitable power consumption P_{exp} is the power consumption that is caused by the processing of the LSB. We calculate the variance of P_{exp} again based on the power traces recorded in Section 4.2.2. The first step of this calculation is to determine the mean of the power consumption for the case that $LSB = 0$ and for the case that $LSB = 1$. These mean values turn out to be 131.49 mV and 136.73 mV. In contrast to the mean values for the Hamming weights, these two mean values occur equally likely. Therefore, the two squared differences occurring during the calculation of $Var(P_{exp})$ are both weighted in the same way. The calculation of $Var(P_{exp})$ based on the 51 200 power traces leads to 6.87 $(mV)^2$.

The total noise that occurs in the attack on the LSB has already been analyzed in Section 4.3.1. We have determined that the standard deviation of $P_{sw.\,noise} + P_{el.\,noise} = 7.54$ mV. Hence, the variance is $7.54^2 = 56.85\,(mV)^2$. Given the variance of the signal and the noise, it turns out that $SNR = 0.12$ in the second attack scenario. An important observation in this scenario is that $Var(P_{sw.\,noise})$ is about seven times higher than $Var(P_{exp})$. This is due to the fact that only one out of eight bits is attacked.

Table 4.2 shows all the variances for the different attack scenarios. Note that the sum of the variances is always the same no matter whether the power consumption is modeled as the sum of P_{data}, P_{op}, and $P_{el.\,noise}$ or as the sum of P_{exp}, $P_{sw.\,noise}$, and $P_{el.\,noise}$. The different models just explain the variance of the power consumption in different ways. Clearly, the SNR is much higher in the "8-bit scenario" than the one in the "1-bit scenario".

Signal-to-Noise Ratios of Neighboring Points

So far, we have only analyzed the SNR at one point of the power consumption. In both attack scenarios, we have looked at the power consumption at 362 ns. It has turned out that the SNR at this moment of time is 22.89 and 0.12, respectively. We now analyze the SNR of neighboring points in both attack scenarios in order to illustrate which points of a power trace usually provide the best SNR.

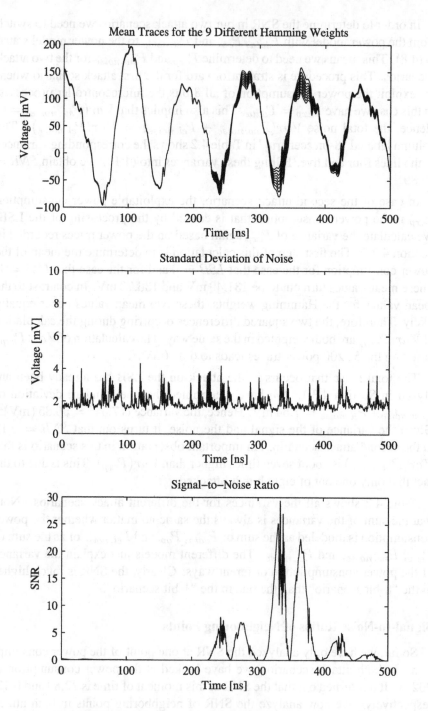

Figure 4.7. The signal levels, the standard deviation of the noise, and the SNR when attacking a uniformly distributed 8-bit data value on our microcontroller.

We start with the "8-bit scenario". The first plot of Figure 4.7 shows the power consumption of our microcontroller while it transfers uniformly distributed data values from its internal memory to a register. The nine traces show the mean power consumption for the different Hamming weights of the processed data. It can be observed that the nine traces not only differ at 362 ns, but that they actually differ at many points between 200 ns and 500 ns. There is a signal that the attacker can exploit at each moment of time where the mean traces do not match exactly. At all these points $Var(P_{exp}) \neq 0$.

The second plot shows the standard deviation of the power traces for the case that the microcontroller processes some fixed data, *i.e.* this plot shows the standard deviation of the electronic noise $P_{el.\,noise}$. An important observation here is that the noise is not the same at all moments of time. In general, the noise is low at those moments of time when there are no fast changes of the power consumption. The noise is typically higher if the power consumption changes quickly. This is due to the fact that even a small jitter in the clock signal leads to a big difference for points that are sampled at a moment of time when the power consumption changes quickly.

The third plot finally shows the SNR of all points between 0 ns and 500 ns. The shape of this SNR plot is actually very typical for many cryptographic devices. The SNR usually reaches its maximum at the positive and at the negative peaks of the power consumption. In between it drops to zero. It is also very common that the exploitable signal is present in several peaks of the power consumption. In the current example, the SNR is very good at the negative peak of the power consumption at 330 ns and at the positive peak at 362 ns. The SNR is also non-zero in some peaks before and after these moments of time. However, the SNR is much lower there.

We now do the same analysis for the "1-bit scenario". The first plot of Figure 4.8 shows the mean power consumption of the microcontroller when it processes data with $LSB = 0$ and $LSB = 1$, respectively. The difference between these two traces is significantly smaller than the maximum difference we have observed in the "8-bit scenario". Hence, also $Var(P_{exp})$ is significantly smaller. However, like in the "8-bit scenario", there are many points between 200 ns and 500 ns where $Var(P_{exp}) \neq 0$.

The second plot of Figure 4.8 shows the standard deviation of the noise in the "1-bit scenario". In contrast to before, $Var(P_{sw.\,noise})$ is not equal to zero in the current scenario. Hence, the noise we have to consider now is $Var(P_{sw.\,noise} + P_{el.\,noise})$. The fact that $Var(P_{sw.\,noise}) \neq 0$ leads to some peaks of the standard deviation between 200 and 500 ns. At these moments of time, not only the attacked LSB is processed, but also the seven other bits that switch randomly. Hence, a lot of switching noise occurs at these moments of time. Every time $Var(P_{exp})$ is high, also $Var(P_{sw.\,noise})$ is high. Besides the

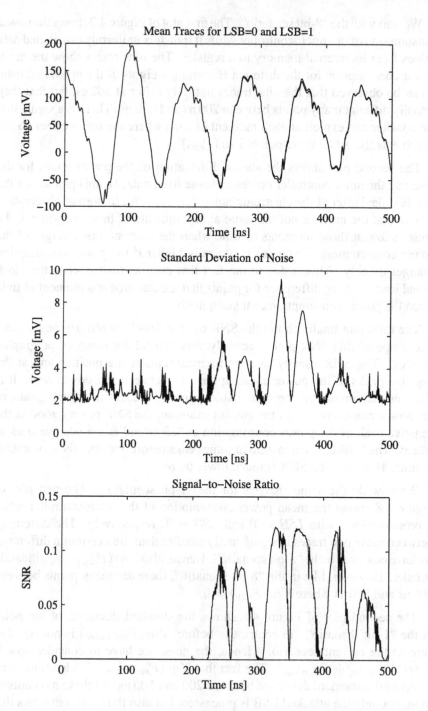

Figure 4.8. The signal levels, the standard deviation of the noise, and the SNR when attacking a uniformly distributed 1-bit data value on our microcontroller.

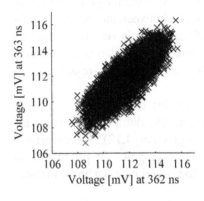

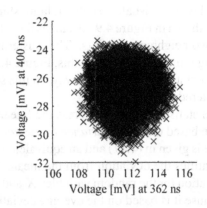

Figure 4.9. Scatter Plot: The power consumption at 362 ns is correlated to the power consumption at 363 ns.

Figure 4.10. Scatter Plot: The power consumption at 362 ns is largely uncorrelated to the power consumption at 400 ns.

fact that $Var(P_{exp})$ is lower in the "1-bit scenario" than in the "8-bit scenario", this is the main reason why the SNR is significantly lower in the "1-bit scenario".

The SNR for all points between 0 ns and 500 ns in the "1-bit scenario" are shown in the third plot of Figure 4.8. The peaks in this plot are wider and significantly smaller than before. However, the general observations we have previously made for the "8-bit scenario" also hold in this attack scenario. The SNR again reaches its maximum at the positions, where there are positive or negative peaks in the power consumption. These points of the power traces leak the most information.

4.4 Characteristics of Power Traces

In the previous sections, we have looked in detail at the characteristics of single points of power traces. We have discussed different models for the components of the power consumption at fixed moments of time and we have finally presented how the leakage of single points can be quantified by an SNR. In all our discussions, we looked at the points of the power traces independently. This is different in the current section. This section introduces statistical methods to describe the relationship between two or more points within a power trace. Based on these methods we then derive a model for entire power traces.

4.4.1 Correlation

In the previous sections, we have looked at the electronic noise of different points of a power trace independently. However, electronic noise usually does not change significantly from one point of a power trace to the next one. Hence, the electronic noise that is present in neighboring points is typically related. A good way to visualize such a relationship is to draw a scatter plot. The scatter plot shown in Figure 4.9 visualizes the relationship between the electronic noise of two neighboring points. The clustering of the points is very tight around a straight line. In contrast to this, Figure 4.10 visualizes the relationship between two points that are farther away. In this scatter plot, the points are only loosely clustered.

In statistics, we can express the linear relationship between two points of a trace based on the *covariance* or the *correlation*. The definition of the covariance is given in (4.11) and an equivalent form is given in (4.12). The covariance quantifies the deviation from the mean. It is the average of the product of the deviation for the random variables X and Y. The covariance is a linear measure because it is based on the average deviation.

The equivalent form (4.12) shows that the covariance is also related to the concept of statistical dependence. If X and Y are statistically independent, then $E(XY) = E(X) \cdot E(Y)$, and therefore $Cov(X, Y) = 0$. For normally distributed X and Y also the converse holds. This means, if $Cov(X, Y) = 0$, then X and Y are independent.

$$\begin{aligned} Cov(X,Y) &= E((X - E(X)) \cdot (Y - E(Y))) &(4.11)\\ &= E(XY) - E(X) \cdot E(Y) &(4.12) \end{aligned}$$

As in case of μ and σ, the covariance is typically not known and needs to be estimated. The estimator c of the covariance is given in (4.13).

$$c = \frac{1}{n-1} \cdot \sum_{i=1}^{n} (x_i - \bar{x}) \cdot (y_i - \bar{y}) \qquad (4.13)$$

A related, but more commonly used method to measure a linear relationship between two values is the correlation coefficient $\rho(X, Y)$. The correlation coefficient is defined in terms of the covariance as can be seen in (4.14). It is a dimensionless quantity and it can only take values between plus and minus one: $-1 \leq \rho \leq 1$.

> The correlation coefficient measures the linear relationship between two variables. It is always between -1 and 1.

Also ρ is typically unknown and needs to be estimated. The estimator r is defined in (4.15).

$$\rho(X,Y) = \frac{Cov(X,Y)}{\sqrt{Var(X) \cdot Var(Y)}} \qquad (4.14)$$

$$r = \frac{\sum_{i=1}^{n}(x_i - \bar{x}) \cdot (y_i - \bar{y})}{\sqrt{\sum_{i=1}^{n}(x_i - \bar{x})^2 \cdot \sum_{i=1}^{n}(y_i - \bar{y})^2}} \qquad (4.15)$$

The correlation coefficient for the electronic noise of the two neighboring points at time index 362 ns and 363 ns is $r = 0.82$. The correlation coefficient between the noise at time 362 ns and 400 ns is $r = 0.12$. As expected, the closer the points in the scatter plot are concentrated around the straight line, the higher is the correlation coefficient.

4.4.2   Multivariate-Gaussian Model

In Section 4.2.1, we have determined that the electronic noise of each point of a power trace is normally distributed. The corresponding parameters μ and σ can be estimated with $\bar{x}$ and s. Unfortunately, this model of the power consumption does not take into account the correlation between neighboring points. In order to also consider the correlation between the points, it is necessary to model a power trace t as a *multivariate normal distribution*. The multivariate normal distribution is a generalization of the normal distribution to higher dimensions. It can be described by a covariance matrix $\mathbf{C}$ and a mean vector $\mathbf{m}$. The probability density function of the multivariate normal distribution is given in (4.16).

$$f(\mathbf{x}) = \frac{1}{\sqrt{(2 \cdot \pi)^n \cdot det(\mathbf{C})}} \cdot exp\left(-\frac{1}{2} \cdot (\mathbf{x} - \mathbf{m})' \cdot \mathbf{C}^{-1} \cdot (\mathbf{x} - \mathbf{m})\right) \quad (4.16)$$

The covariance matrix $\mathbf{C}$ contains the covariances $c_{ij} = Cov(X_i, X_j)$ of the points at time index i and j. The mean vector $\mathbf{m}$ lists the mean values $m_i = E(X_i)$ for all points in the trace. When filling $\mathbf{C}$ and m into (4.16), the value of the probability density function for the vector $\mathbf{x}$ is returned. Note that the vector-matrix product in the exponent leads to a scalar.

$$\mathbf{C} = \begin{pmatrix} c_{11} & c_{12} & \cdots \\ c_{21} & c_{22} & \cdots \\ \vdots & \vdots & \ddots \end{pmatrix} \quad (4.17)$$

$$\mathbf{m} = (m_1, m_2, \ldots)' \quad (4.18)$$

In order to illustrate the basic idea of a multivariate normal distribution, we now provide a concrete example. Like in the case of a normal distribution, we typically have to estimate the covariance matrix and the mean vector in practice. For the sake of readability, we do not introduce separate names for the estimators of $\mathbf{C}$ and m. Instead, we reuse these two letters also for their estimators. We have characterized the electronic noise of our microcontroller in a short time interval starting at 362 ns while it transfers a fixed data value from its memory to a register. The corresponding covariance matrix and the mean vector are:

$$
\mathbf{C} \;=\;
\begin{pmatrix}
2.67 & 1.50 & 1.36 & 1.19 & 1.10 \\
1.50 & 2.82 & 1.61 & 1.38 & 1.32 \\
1.36 & 1.61 & 2.77 & 1.55 & 1.45 \\
1.19 & 1.37 & 1.55 & 2.88 & 1.56 \\
1.10 & 1.32 & 1.45 & 1.56 & 2.78
\end{pmatrix}
\tag{4.19}
$$

$$
\mathbf{m} \;=\; (111.86, 111.57, 110.33, 108.99, 107.56)'
\tag{4.20}
$$

This example shows important properties of the covariance matrix. First, the covariance matrix is symmetric. This is because $Cov(X_i, X_j) = Cov(X_j, X_i)$. Second, the main diagonal of the matrix contains the variance of the considered points. For example, the variance of the electronic noise at $362\,\mathrm{ns}$ is $1.63^2 = 2.67\,(\mathrm{mV})^2$. Third, the determinant of the matrix is a positive number. This guarantees that the square root can be calculated.

In our example, the covariance matrix $\mathbf{C}$ essentially characterizes $P_{el.\,noise}$. However, the multivariate normal distribution can also be used to characterize other components of the power consumption. One of the problems when using the multivariate normal distribution is that the computational effort to calculate $\mathbf{C}$ grows quadratically with the number of considered points. Therefore, in practice either only small parts of power traces are characterized in such a detailed way or the power traces need to be compressed before the characterization.

4.5   Compression of Power Traces

In practice, compression techniques are often used to reduce the complexity of power analysis attacks. Compression techniques are motivated by the fact that there is usually a lot of redundancy in power traces. Already in the context of the discussion of switching noise in Section 3.5, we have provided arguments why it is usually not possible to measure the power consumption of individual logic cells separately. With our measurement setups, we are only able to measure the power consumption that is caused by the overall switching activity of logic cells within one clock cycle. We are not able to measure distinct and independent power consumption peaks within one clock cycle.

However, the oscilloscope usually records the power consumption several times during one clock cycle, see Section 3.4.3. Hence, each clock cycle is represented by several measurement points. This redundancy is typically not desired. Long power traces with many redundant points make power analysis attacks computationally much more expensive than it is actually necessary. This is why often compression techniques are applied before characterizing or attacking cryptographic devices. With compression we mean that the length of the power traces is reduced without a significant loss of leaking information, *i.e.* mainly the redundancy is removed.

In order to compress power traces, it is necessary to know which points of the traces actually carry relevant information for an attacker. The discussion of leakage in Section 4.3 has already indicated that such information is mainly present in the peaks of the power consumption of our microcontroller. In the following section, we confirm this observation also for our AES ASIC implementation. However, instead of analyzing the SNR, we now analyze which points of the power traces of the AES ASIC are correlated most to the number of transitions that occur inside the chip. This is another technique to determine the points of power traces that have the highest information leakage.

4.5.1 Relevant Points of Power Traces

In order to determine the points of the power traces of our AES ASIC that have the highest information leakage, we have first recorded its power consumption while it has encrypted 1 000 random plaintexts. This has been done using the measurement setup described in Section 3.4.4. Each plaintext has been sent 100 times to the device and the power consumption has been recorded for each of these 100 encryption runs. Subsequently, the average of the 100 power traces has been calculated in order to obtain average power traces with almost no electronic noise. In this way, we have calculated an average power trace for each of the 1 000 random plaintexts.

In the second step of our analysis, we have simulated the power consumption of the AES ASIC based on a back-annotated netlist of the chip. We have used the same 1 000 plaintexts for our simulations and for the actual measurements. For each plaintext, we have counted the number of transitions that occur in the chip during each clock cycle of an encryption run. Hence, we have obtained a simulated power trace for each of the 1 000 plaintexts. Simulations at the cell level are unfortunately not as precise as simulations at the transistor level, see Section 3.2. However, simulations of the entire chip at the transistor level would have been too resource intensive.

After having obtained measured and simulated power traces of the AES ASIC for the same set of plaintexts, we have analyzed how these power traces are related to each other. The goal of this analysis was to find out which points of the measured power traces are correlated to the transition counts that we have simulated. Figure 4.11 shows a representative result of this analysis.

The upper plot of the figure shows four clock cycles of a measured power trace. There occurs a damped oscillation with a high amplitude for each positive edge of the clock and a smaller damped oscillation for every negative edge. The lower plot shows the correlation between the 1 000 measured power traces and the number of simulated transitions for the second clock cycle. This means that we have taken the 1 000 simulated transition counts for the second clock cycle and we have correlated them with the measured power traces at all possible moments of time. As expected, there is no correlation during the first clock

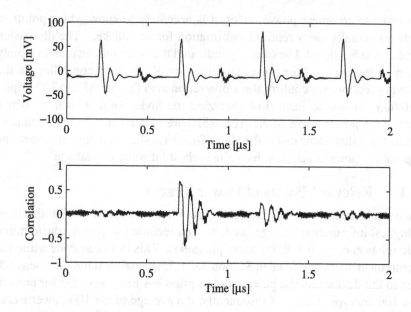

Figure 4.11. The upper plot shows a power trace of the AES ASIC during four clock cycles. The lower plot shows the correlation between the measured power traces and the simulated number of transitions that occur during the second clock cycle.

cycle. During the second clock cycle, there is a significant correlation with a peak value of more than 0.7. Hence, the amplitude of the damped oscillation that occurs upon the rising clock edge for the second clock cycle (see upper plot) is roughly proportional to the simulated number of transitions. During the clock cycles three and four, there is again almost no correlation.

The correlation between the simulated and the measured traces is lower than 1 during the second clock cycle because the simulation is not perfect. In particular, we have not considered parasitic capacitances in our simulation based on the back-annotated netlist, see Section 3.2. Therefore, it is clear that the simulation does not match the measured traces perfectly. However, 0.7 is already a quite strong correlation that clearly indicates at which moments of time the power consumption depends on the number of transitions.

We have repeated our analysis also for all other clock cycles that occur during an encryption, *i.e.* we correlated the corresponding simulated number of transitions with the measured power traces. The result of the analysis has always been the same. The amplitude of the damped oscillations in the measured power traces has been roughly proportional to the total number of transitions occurring in the corresponding clock cycles. The number of transitions that occur during a certain clock cycle depends on the operation that is performed and on the data that is being processed. The peaks of the damped oscillation therefore carry all the information that is necessary to perform a power analysis attack.

4.5.2 Examples

Based on the observation that the peaks of the recorded power traces are the most relevant points for power analysis attacks, different proposals for compression techniques have been made. In the context of this book, we focus on the two most common ones. The first technique we present is to extract the maximum peak of every clock cycle. The second technique is to integrate the power consumption signal.

Maximum Extraction

The lower plot of Figure 4.11 shows the correlation between the simulated number of transitions for the second clock cycle and the measured traces. The highest correlation occurs exactly at the position where the power consumption of the second clock cycle reaches its maximum. The amplitude of this peak in the power consumption is proportional to the energy that is dissipated by the chip during this clock cycle.

It is therefore reasonable to choose this point as representative point for the entire clock cycle. A power trace can be compressed significantly by simply extracting the maximum peak values of the recorded clock cycles. In order to verify that this compression technique actually works, we have compressed the 1 000 power traces that we have measured in Section 4.5.1. Each of these power traces corresponds to one full AES encryption and it consists of 80 clock cycles.

After having extracted the 80 maxima of each of these power traces, we have correlated the compressed power traces with the simulated ones. Figure 4.12 shows the correlation between the extracted peaks from the power consumption and the simulated number of transitions for each clock cycle. For all clock cycles, there is quite a high correlation. The compressed power traces essentially contain all the information about the switching activities that take place during an encryption.

Integration

An alternative to the maximum extraction is the integration of the power traces during each clock cycle or during smaller time intervals. In contrast to the previous approach, this approach uses all points of the damped oscillations in the power traces and not only the peak values. As a consequence, this approach is typically more robust than the previous one.

When integrating a power trace in a given time interval, the signal as well as the noise of the points in this interval are summed up. This also affects the

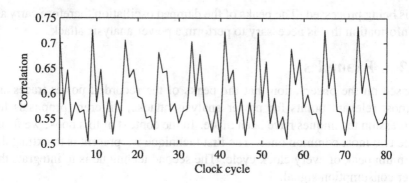

Figure 4.12. Correlation between the simulated number of transitions and the maximum peaks of the compressed power trace.

signal-to-noise ratio. The SNR of the sum of the points depends on P_{exp} and on $P_{sw.\,noise} + P_{el.\,noise}$ of the points as well as on the correlation between the signal and the noise components. In practice, it can happen that the SNR of the sum is higher than the one of the individual points, but it can also happen that the SNR is decreased by the integration.

Whether the SNR increases or decreases essentially depends on the size of the time interval that is used for the integration. The SNR typically increases, if there are many points with a strong signal component in the integrated time interval. The SNR is lower than the one of a single point if this point is combined with points that leak no or almost no information. Therefore, choosing a suitable time interval for the integration is crucial for the success of this compression technique. In practice, usually time intervals up to the length of a single clock cycle are used.

Depending on whether or not a preprocessing step is done before the integration, we distinguish the following methods to compress power traces.

Raw Integration: In this approach, the sum of all points of each time interval is computed. These sums are then used as compressed power trace.

Sum of Absolute Values: In this approach, the sum of the absolute values of all points in each time interval is computed. These sums of absolute values are then used as compressed power trace.

Sum of Squares: In this approach, the sum of the squares of all points in each time interval is computed. These sums of squared values are then used as compressed power trace.

4.6   Confidence Intervals and Hypothesis Testing

At the beginning of this chapter, we have verified in an experiment that the electronic noise of a point in a power trace can be characterized by a normal distribution. A normal distribution is defined by its mean μ and its variance σ^2. As we typically do not know these parameters in practice, we have to estimate them. The best estimators are the average $\bar{x}$ and the empirical variance s^2.

In this section, we discuss the properties of mean and variance in more detail. The motivation for this further study is the importance of statistics in power analysis attacks. Remember how the attack in Chapter 1 has worked. In this attack, we have compared two mean values. The difference between these mean values has allowed us to determine the key. We have measured a certain number of traces to calculate the mean values. In practice, the number of traces often determines how practical an attack is. The fewer traces an attacker needs to acquire, the more practical an attack is. Therefore, it is important to have tools that allow assessing the number of traces. Statistics is the tool that is most suited for this purpose. For this reason, we now discuss different statistical concepts that are relevant in the context of power analysis attacks. More extensive discussions of these concepts can be found in textbooks on statistics such as [FPP97], and [Ric94].

4.6.1 Sampling Distribution

We focus on a single point in a trace. We have already determined that the electronic noise of a single point can be characterized by a normal distribution. For some fixed value (and operation), the power consumption is fixed, except for the variance that is introduced by electronic noise. Hence, if we measure the power consumption for a fixed instruction and a fixed value several times, we get slightly different results as Figure 4.1 shows.

Suppose that we would repeat the experiment of Section 4.2.1 that has led to Figure 4.1. This means, we would make another set of measurements for the same instruction on the same data, and look at the average $\bar{x}$ and the empirical standard deviation s at time index 362 ns. Probably, we would get similar values, but it is unlikely that we would get exactly the same values. The average value and the standard deviation would change slightly for every experiment. This means, we can view the average value and the empirical standard deviation as random variables. In other words, the average value can be described as random variable $\bar{X}$ and the standard deviation can be described as random variable S. Consequently, we can characterize the average value and the empirical standard deviation in the same way as we have characterized the power consumption at 362 ns—they both have a mean value and a standard deviation. Summarizing, the mean value and the empirical standard deviation can be characterized by a probability distribution as well. This important concept is called *sampling distribution*.

> The sampling distribution determines how well a certain parameter can be estimated.

In case of the average value $\bar{X}$, the sampling distribution describes how well the average value estimates μ. The sampling distribution of $\bar{X}$ is a normal distribution: $\bar{X} \sim \mathcal{N}(\mu, \sigma/\sqrt{n})$. For normally distributed values it is known that the average is a very good estimator for μ, see (4.21). The more traces, *i.e.* points x, are used to compute $\bar{X}$, the better is the estimation. To be more precise, if n traces are used, the variance of the average decreases by a factor of n, see (4.22). Hence, the average value gets closer to the mean value the more measurements are used. This has the important consequence that we can use the average to decrease the amount of electronic noise in measurements. The average of n traces has a variance that is n times smaller than the variance of a single trace. Consequently, it contains n times less electronic noise.

$$E(\bar{X}) \;=\; \mu \tag{4.21}$$

$$Var(\bar{X}) \;=\; \frac{\sigma^2}{n} \tag{4.22}$$

Similar observations hold for the sampling distribution of S^2. However, the sampling distribution of S^2 is a chi-square distribution. The mean value of S^2 is σ^2 as given in (4.23). Also, the empirical variance S^2 gets closer to σ^2, the more measurements are used as can be seen in (4.24).

$$E(S^2) \;=\; \sigma^2 \tag{4.23}$$

$$Var(S^2) \;=\; \frac{2 \cdot \sigma^4}{n - 1} \tag{4.24}$$

4.6.2 Confidence Intervals

In the previous section, we have determined that the estimators for the mean and the variance, the average $\bar{x}$ and the empirical variance s^2, can be seen as random variables $\bar{X}$ and S^2. They both have a certain distribution that can be characterized by a mean and a variance. In case of the average, the distribution has turned out to be a normal distribution. Because we rather work with $\bar{X}$ than with S in the context of this book, we limit the following discussion to $\bar{X}$.

We have concluded from $\bar{X} \sim \mathcal{N}(\mu, \sigma/\sqrt{n})$ that the approximation of μ by $\bar{X}$ gets the better the more traces are used. However, it is still unclear how good a given approximation is or how many traces are needed for a good approximation. So-called *confidence intervals* are a statistical concept that allows telling how good a given approximation is.

> Confidence intervals quantify how close a certain approximation is to the real parameter.

When we say that we have a 0.99 confidence interval for μ, then we mean that our construction delivers an interval that contains μ with probability 0.99.

Hypotheses tests are closely connected to confidence intervals. In a hypothesis test, we test whether a certain hypothesis that we make is true or not. For instance, we can test whether μ is equal to a certain value μ_0 or not. In this case, we formulate two hypotheses. The first hypothesis is that $\mu = \mu_0$ whereas the other hypothesis is $\mu \neq \mu_0$. One of the two hypotheses is called the null hypothesis H_0. The alternative hypothesis is called H_1. For instance, we can define the null hypothesis to be H_0: $\mu = \mu_0$ and the alternative hypothesis to be H_1: $\mu \neq \mu_0$.

Remember that a confidence interval contains all the values that are reasonably close to a certain parameter. Hence, if we want to know whether a certain parameter is equal to a certain value, we have to determine whether this value lies in the confidence interval of the parameter or not. In other words, a hypothesis test accepts all those values μ_0 of the null hypothesis, which lie in the confidence interval of μ.

> Confidence intervals consist of those values for which the null hypothesis is accepted.

In the next section, we show in detail how to calculate a confidence interval for the parameter μ of the normal distribution and we look at the corresponding hypothesis test. The sections thereafter provide the same discussions for the difference of two means, for the correlation coefficient, and for the difference of two correlation coefficients. However, compared to the next section, the sections thereafter are kept much shorter.

4.6.3 Confidence Interval and Hypothesis Test for μ

Before we can derive a confidence interval for the mean, we need to introduce the concept of quantiles. The *quantile* z_α of the standard normal distribution has the property that $p(Z \leq z_\alpha) = \alpha$. Because $p(Z \leq z_\alpha) = \Phi(z_\alpha)$ we have that $\Phi(z_\alpha) = \alpha$. This property is also shown in Figure 4.13. Hence, the quantile z_α can be derived by applying the inverse of the standard normal distribution function to α. A property that is frequently used is that $z_\alpha = -z_{1-\alpha}$. Table 4.3 lists some quantiles. Now we can derive a confidence interval for μ.

Confidence Interval

We know already that the distribution of $\bar{X}$ is a normal distribution: $\bar{X} \sim \mathcal{N}(\mu, \sigma/\sqrt{n})$. Hence, the distribution of the variable $Z = (\bar{X} - \mu) \cdot \sqrt{n}/\sigma$ is a standard normal distribution. By the definition of z_α (see also Figure 4.13), we have:

$$p(z_{\alpha/2} \leq Z \leq z_{1-\alpha/2}) = 1 - \alpha \tag{4.25}$$

The value α is often called *error probability*. The interval $\left[z_{\alpha/2}, z_{1-\alpha/2}\right]$ is the confidence interval for Z. However, we are interested in the confidence interval

Table 4.3.   Quantiles z_α of the $\mathcal{N}(0,1)$ distribution for some values of α.

α	0.800	0.850	0.900	0.950	0.975	0.990	0.995	0.999	0.9999
z_α	0.842	1.036	1.282	1.645	1.960	2.326	2.576	3.090	3.7190

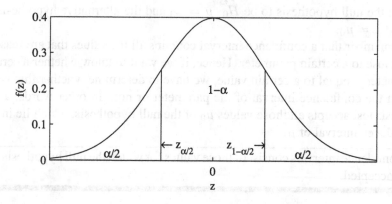

Figure 4.13.   A standard normal density function showing the quantiles $z_{\alpha/2}$ and $z_{1-\alpha/2}$.

for μ. Consequently, we substitute $Z = (\bar{X} - \mu) \cdot \sqrt{n}/\sigma$ in (4.25) and rewrite the inequalities:

$$p(\bar{X} - \frac{\sigma}{\sqrt{n}} \cdot z_{1-\alpha/2} \leq \mu \leq \bar{X} + \frac{\sigma}{\sqrt{n}} \cdot z_{1-\alpha/2}) = 1 - \alpha$$

Now the inequalities are written with respect to μ. Consequently, the mean μ is contained in the interval:

$$\left[\bar{X} - \frac{\sigma}{\sqrt{n}} \cdot z_{1-\alpha/2}, \; \bar{X} + \frac{\sigma}{\sqrt{n}} \cdot z_{1-\alpha/2}\right] \tag{4.26}$$

The interval $\left[\bar{X} - \frac{\sigma}{\sqrt{n}} \cdot z_{1-\alpha/2}, \; \bar{X} + \frac{\sigma}{\sqrt{n}} \cdot z_{1-\alpha/2}\right]$ is a $(1 - \alpha)$ confidence interval for μ.

The interval in (4.26) has upper and lower bounds unequal to infinity; it is called a two-sided confidence interval for μ. By working out $p(Z \geq z_\alpha) = 1 - \alpha$ and $p(Z \leq z_{1-\alpha}) = 1 - \alpha$ we can also determine one-sided confidence intervals:

$$\left(-\infty, \bar{X} + \frac{\sigma}{\sqrt{n}} \cdot z_{1-\alpha}\right] \text{ and } \left[\bar{X} - \frac{\sigma}{\sqrt{n}} \cdot z_{1-\alpha}, \infty\right)$$

We can also write the interval with respect to $\bar{X}$ by rewriting (4.25) once more:

$$p(\mu - \frac{\sigma}{\sqrt{n}} \cdot z_{1-\alpha/2} \leq \bar{X} \leq \mu + \frac{\sigma}{\sqrt{n}} \cdot z_{1-\alpha/2}) = 1 - \alpha$$

Hence the interval, written with respect to $\bar{X}$, is:

$$\left[\mu - \frac{\sigma}{\sqrt{n}} \cdot z_{1-\alpha/2}, \ \mu + \frac{\sigma}{\sqrt{n}} \cdot z_{1-\alpha/2}\right]$$

Obviously, we can write the confidence intervals that we have derived for μ also with respect to Z, and $\bar{X}$.

We have implicitly assumed that the standard deviation σ of $\bar{X}$ is known in our current discussion. If this parameter needs to be estimated as well, *i.e.* if only S is known and not σ, then, the variable $T = (\bar{X} - \mu) \cdot \sqrt{n}/S$ is known to be t-distributed. Hence, the confidence interval is then given by the quantiles of the t distribution. However, in practice if there are sufficiently many traces available ($n \geq 30$), s can be used instead of σ and the standard normal distribution can be used instead of the t-distribution.

Hypothesis Test

Suppose that we want to test whether μ equals a certain value μ_0 or not. Then, the null hypothesis is H_0: $\mu = \mu_0$ and the alternative hypothesis is H_1: $\mu \neq \mu_0$. A corresponding test can now be formulated easily. Since we do not know μ, we estimate it by $\bar{X}$. If $|\bar{X} - \mu_0|$ is much larger than some constant c that we define on beforehand, we decide against H_0, otherwise we accept H_0. The constant c defines the so-called *critical region*.

Ideally, our test would always give the correct result. This means, that if μ is indeed equal to μ_0, then the test would accept the null hypothesis, otherwise it would reject it. However, since we can only estimate μ with $\bar{X}$ there is always a chance that we make an error. Our goal is therefore to minimize such an error. In particular, we would like to minimize the chance that we reject H_0 although it is correct. In other words, we would like to choose the constant c such that the probability of incorrectly deciding against H_0 is small:

$$p(|\bar{X} - \mu_0| > c) = \alpha \qquad (4.27)$$

Similarly to the calculations that we have made to derive the confidence intervals in the previous section, we can work out (4.27) in order to determine c.

$$\begin{aligned} p(|\bar{X} - \mu_0| > c) &= p(|\bar{X} - \mu_0| \cdot \sqrt{n}/\sigma > c \cdot \sqrt{n}/\sigma) = \alpha \\ p(|Z| > c \cdot \sqrt{n}/\sigma) &= \alpha \\ 2p(Z > c \cdot \sqrt{n}/\sigma) &= \alpha \\ c &= \frac{\sigma}{\sqrt{n}} \cdot z_{1-\alpha/2} \end{aligned}$$

This means, if we choose $c = \sigma \cdot z_{1-\alpha/2}/\sqrt{n}$, then the probability to incorrectly decide against H_0 is α. We can also write down the critical regions with respect to Z. Furthermore, by rewriting we can also derive the critical regions for μ_0 and $\bar{X}$.

$$
Z \;:\; \left(-\infty,\, -z_{1-\alpha/2}\right) \cup \left(z_{1-\alpha/2},\, \infty\right)
$$

$$
\mu_0 \;:\; \left(-\infty,\, \bar{X} - \frac{\sigma}{\sqrt{n}} \cdot z_{1-\alpha/2}\right) \cup \left(\bar{X} + \frac{\sigma}{\sqrt{n}} \cdot z_{1-\alpha/2},\, \infty\right)
$$

$$
\bar{X} \;:\; \left(-\infty,\, \mu_0 - \frac{\sigma}{\sqrt{n}} \cdot z_{1-\alpha/2}\right) \cup \left(\mu_0 + \frac{\sigma}{\sqrt{n}} \cdot z_{1-\alpha/2},\, \infty\right)
$$

Apparently, the critical regions are the complementary sets to the confidence intervals that we have derived before. This shows the duality between confidence intervals and hypothesis tests: A confidence interval consists of those values for which the null hypothesis is accepted.

Number of Traces

An important aspect in practice is the number of traces that have to be measured in order to get reliable results. Confidence intervals (or critical regions) allow determining the number of traces. From the previous discussions we know already that the limit of the confidence interval is given by $c = \sigma \cdot z_{1-\alpha/2}/\sqrt{n}$. From this equation we can determine the number of traces n that are necessary to ensure with probability $1 - \alpha$ to be "close" (within a range of c) to μ.

The number of traces n to estimate the mean μ of the normal distribution $\mathcal{N}(\mu, \sigma)$ with confidence $1 - \alpha$ and precision c is:

$$
n = \frac{\sigma^2}{c^2} \cdot z_{1-\alpha/2}^2 \tag{4.28}
$$

A related, but even more important, question in the context of power analysis attacks is how many traces are needed in order to distinguish a certain parameter from zero based on its estimator. Hence, in this case we are interested in the special case that $p(\bar{X} < 0) = \alpha$ given $\mu = \mu_0$ with $\mu_0 < 0$ (the same calculation can be done for $\mu > 0$ and $p(\bar{X} > 0)$). Then the probability $p(\bar{X} < 0)$ should be high because $\mu_0 < 0$. Hence, we want to have that $p((\bar{X} - \mu) \cdot \sigma/\sqrt{n} < -\mu \cdot \sigma/\sqrt{n}) = 1 - \alpha$. This allows deriving the number of traces n: $p((\bar{X} - \mu) \cdot \sigma/\sqrt{n} < -\mu \cdot \sigma/\sqrt{n}) = \Phi(-\mu \cdot \sigma/\sqrt{n}) = 1 - \alpha$.

The number of traces n that is necessary to assert with confidence $1 - \alpha$ that the mean of the normal distribution $\mathcal{N}(\mu, \sigma)$ is non-zero is given by:

$$
n = \frac{\sigma^2}{\mu^2} \cdot z_{1-\alpha}^2 \tag{4.29}
$$

Example

We have previously determined that the distribution of the power consumption at 362 ns is a normal distribution with an estimated $\mu = \bar{x} = 111.86 \, \text{mV}$ and $\sigma = s = 1.63 \, \text{mV}$. Based on the 10 000 traces, we can also determine a confidence interval for μ. For example, a confidence interval with probability 0.99 for μ is $\left[111.86 - \frac{2.576 \cdot 1.63}{\sqrt{10\,000}}, \, 111.86 + \frac{2.576 \cdot 1.63}{\sqrt{10\,000}} \right] = [111.82 \, \text{mV}, 111.90 \, \text{mV}]$. Note that $z_{1-\alpha/2} = z_{0.995} = 2.576$ can be derived from Table 4.3.

Next we test the hypothesis that $\mu_0 = 112 \, \text{mV}$. This means, we test H_0: $\mu = 112 \, \text{mV}$ against H_1: $\mu \neq 112 \, \text{mV}$. The error probability is assumed to be 0.01. Then, the critical region for μ_0 is $(-\infty, 111.82 \, \text{mV}) \cup (111.90 \, \text{mV}, \infty)$. Because $\mu_0 = 112 \, \text{mV}$ is in the critical region, we decide against H_0, *i.e.* we decide that $\mu \neq 112 \, \text{mV}$.

Last, we determine the number of traces. The number of traces n to estimate μ with precision $c = 0.01$ is 176 306. The number of traces n to distinguish the mean of the distribution $\mathcal{N}(1.86, 1.63)$ from zero with confidence 0.99 is $n = 4$. Note that in practice we would estimate μ and σ.

4.6.4 Confidence Interval and Hypothesis Test for $\mu_X - \mu_Y$

Comparing two sets of traces is another important task in the context of power analysis attacks. Suppose that our set X consists of n traces ($X = \{X_1, X_2, \ldots, X_n\}$) and our set Y consists of m traces ($Y = \{Y_1, Y_2, \ldots, Y_m\}$). Remember that we only look at one fixed point of each trace. We suspect the two sets to be from different normal distributions. More precisely, we suspect the traces to have different means but the same variance: $X_i \sim \mathcal{N}(\mu_X, \sigma)$ and $Y_j \sim \mathcal{N}(\mu_Y, \sigma)$. The distribution of the difference of two traces is a normal distribution. Its mean value is given by the difference of the means of two normal distributions and its variance given by the sum of the variances of the two normal distributions.

Confidence Interval

We are mostly interested in the difference of means of two sample sets. The distribution of the difference of means is then a normal distribution:

$$\bar{X} - \bar{Y} \sim \mathcal{N} \left(\mu_X - \mu_Y, \, \sigma \sqrt{\frac{m+n}{m \cdot n}} \right)$$

Thus, the distribution of the variable

$$Z = \frac{\bar{X} - \bar{Y} - (\mu_X - \mu_Y)}{\sigma \cdot \sqrt{\frac{m+n}{m \cdot n}}}$$

is a standard normal distribution. Like in Section 4.6.3, we can derive the two-sided confidence interval for $\mu_X - \mu_Y$ for known σ by rewriting the inequalities in $p(z_{\alpha/2} \leq Z \leq z_{1-\alpha/2})$. The two-sided confidence interval is then given by:

$$\left[\bar{X} - \bar{Y} - \sigma \cdot \sqrt{\frac{m+n}{m \cdot n}} \cdot z_{1-\alpha/2}, \; \bar{X} - \bar{Y} + \sigma \cdot \sqrt{\frac{m+n}{m \cdot n}} \cdot z_{1-\alpha/2} \right] \quad (4.30)$$

If the standard deviation is not known and has to be estimated, the distribution of $\bar{X} - \bar{Y}$ becomes a t-distribution. As mentioned in the previous section, the t-distribution is well approximated by the standard normal distribution for a reasonable (≥ 30) amount of traces. The standard deviation $s_{\bar{X}-\bar{Y}}$ is estimated by the square root of the so-called pooled variance s_p^2.

$$s_p^2 = \frac{(n-1) \cdot s_X^2 + (m-1) \cdot s_Y^2}{m+n-2} \quad (4.31)$$

$$s_{\bar{X}-\bar{Y}} = s_p \cdot \sqrt{\frac{m+n}{m \cdot n}} \quad (4.32)$$

Since in power analysis attacks both sets have typically about $n/2$ traces, we can substitute $n/2$ for m and n and simplify the previous formulas. We use the simplification from now on.

Hypothesis Test

In the same way as for the testing of a hypothesis about μ, we can test a hypothesis about $\mu_X - \mu_Y$. Hence, a two-sided test compares H_0: $\mu_X - \mu_Y = 0$ versus H_1: $\mu_X - \mu_Y \neq 0$. This means we test if $Z = (\bar{X} - \bar{Y})/(\sigma \cdot \sqrt{4/n})$ is in the critical region. We reject H_0 if Z is in the critical region. Remember that the critical region is the complement of the two-sided confidence interval for $\bar{X} - \bar{Y}$.

Number of Traces

We can derive the number of traces in the same way as we have derived it for μ. This means we use the limit c of the confidence interval, which is given by $c = \sigma \cdot \sqrt{4/n} \cdot z_{1-\alpha/2}$, in order to determine the number of traces.

The number of traces n to estimate the difference of means $\mu_X - \mu_Y$ of the normal distribution $\mathcal{N}(\mu_X - \mu_Y, 2 \cdot \sigma/\sqrt{n})$ with confidence $1 - \alpha$ and precision c is:

$$n = 4 \cdot \frac{\sigma^2}{c^2} \cdot z_{1-\alpha/2}^2 \quad (4.33)$$

Like in discussion with respect to μ before, the most important question is actually how many traces are needed to distinguish $\bar{X}$ from $\bar{Y}$. Hence, the farther away their difference distribution is from zero, the easier it is to distinguish them. Without loss of generality, we assume that $\mu_X - \mu_Y < 0$ (if not, we just switch X and Y). We want the probability that the difference distribution Z is smaller than zero to be large: $p(Z < 0) = 1 - \alpha$. We know that $p(Z < 0)$ equals $\Phi(-(\mu_X - \mu_Y) \cdot \sqrt{n}/(2/\sqrt{\sigma}))$ and therefore we have $\Phi(-(\mu_X - \mu_Y) \cdot \sqrt{n}/(2/\sqrt{\sigma})) = 1 - \alpha$.

Rewriting this probability allows us to determine n.

The number of traces n that is necessary to assert with a confidence of $1 - \alpha$ that the two normal distributions $X \sim \mathcal{N}(\mu_X, \sigma/\sqrt{n/2})$ and $Y \sim \mathcal{N}(\mu_Y, \sigma/\sqrt{n/2})$ are different is given by:

$$n = \frac{4 \cdot \sigma^2}{(\mu_X - \mu_Y)^2} \cdot z_{1-\alpha}^2 \qquad (4.34)$$

In the case that the two distributions have different variances, the previous formulas need to be adapted. However, for most practical applications, the given formulas suffice.

Example

Now we discuss the result of the DPA attack, which we have performed in Section 1.3, in the context of confidence intervals and hypothesis tests. In this attack, we have used the difference of means that is defined by some intermediate value v (the MSB of the output of SubBytes) to determine the key. It has turned out that for key 119, the difference of means is significantly different from zero at several times. We focus now on the largest difference that occurs and compute the confidence interval for it. The largest difference of means is $\bar{x} - \bar{y} = 7.86\,\text{mV}$. Its estimated standard deviation is $s_{\bar{X}-\bar{Y}} = 0.60\,\text{mV}$. Hence, a confidence interval with probability 0.99 is $[6.31\,\text{mV}, 9.41\,\text{mV}]$. In order to get a confidence interval with $c = 1$, about $n = 4\,823$ traces would be required according to (4.33).

As explained before, the most interesting question in the context of power analysis attacks is the determination of the number of traces for a successful attack. In order to assert with confidence 0.999 that the two distributions can be distinguished from each other, we need $n = 112$ traces according to (4.34).

4.6.5 Confidence Interval and Hypothesis Test for ρ

The sampling distribution of the correlation coefficient r is complicated. However, if a sufficiently large number of traces (≥ 30) can be measured, a transformation that is due to Fisher can be used to map the random variable R to a random variable Z_1 that has a normal distribution, see (4.35). The mean of Z_1 is then given by μ, see (4.36), and the variance is σ^2, see (4.37). The value n denotes the number of traces. Note that the fraction $\rho/(2 \cdot (n - 1))$

approaches zero for large n and can thus be omitted from the formula.

$$Z_1 = \frac{1}{2} \ln \frac{1+R}{1-R} \tag{4.35}$$

$$\mu = \frac{1}{2} \ln \frac{1+\rho}{1-\rho} + \frac{\rho}{2 \cdot (n-1)} \tag{4.36}$$

$$\sigma^2 = \frac{1}{n-3} \tag{4.37}$$

Confidence Interval

Because of Fisher's transformation, we have that $Z = (Z_1 - \mu)/\sigma \sim \mathcal{N}(0, 1)$. Hence, we can derive the confidence interval for μ by working out $p(z_{\alpha/2} \leq Z \leq z_{1-\alpha/2})$. The confidence interval for μ is given in (4.38).

$$\left[\frac{1}{2} \cdot \ln \frac{1+R}{1-R} - \frac{z_{1-\alpha/2}}{\sqrt{n-3}}, \frac{1}{2} \cdot \ln \frac{1+R}{1-R} + \frac{z_{1-\alpha/2}}{\sqrt{n-3}} \right] \tag{4.38}$$

Hypothesis Test

Suppose we have a set of traces and we want to test a hypothesis about the correlation coefficient. Testing $H_0: \rho = \rho_0$ against $H_1: \rho \neq \rho_0$ simply requires checking whether $\mu_0 = 1/2 \cdot \ln ((1 + \rho_0)/(1 - \rho_0))$ is in the confidence interval (4.38) that is constructed based on the traces.

Number of Traces

Like before, we can use the limit c of the confidence interval to determine the number of traces. However, now we also have to apply the Fisher transformation to c: $c_F = 1/2 \cdot \ln ((1 + c)/(1 - c))$. The number of traces n that is needed to estimate the mean of the normal distribution $\mathcal{N}(1/2 \cdot \ln ((1 + \rho)/(1 - \rho)), 1/\sqrt{n-3})$ with a confidence level of $1 - \alpha$ and precision c is:

$$n = 3 + 4 \cdot \frac{z_{1-\alpha/2}^2}{\ln^2 \frac{1+c}{1-c}} \tag{4.39}$$

The number of traces n that is necessary to assert with confidence $1 - \alpha$ that the mean of the normal distribution $\mathcal{N}(1/2 \cdot \ln ((1 + \rho)/(1 - \rho)), 1/\sqrt{n-3})$ is non-zero is given by:

$$n = 3 + 4 \cdot \frac{z_{1-\alpha/2}^2}{\ln^2 \frac{1+\rho}{1-\rho}} \tag{4.40}$$

Example

We take the estimation $r = 0.82$ for the two correlated points $362 \, \mathrm{ns}$ and $363 \, \mathrm{ns}$ that we have analyzed in Section 4.4.1. The confidence interval for

μ, given that $n = 10\,000$ and $\alpha = 0.01$, is then $[1.13, 1.18]$. In order to approximate an arbitrary ρ with precision $c = 0.01$ we need about $n = 66\,356$ samples.

4.6.6 Confidence Interval and Hypothesis Test for $\rho_0 - \rho_1$

In practice, we are often interested whether two correlation coefficients are different. As explained in the previous section, we can map the estimator R for ρ to a variable Z that has a standard normal distribution by the Fisher transformation.

$$R_0 \mapsto Z_0 = \frac{1}{2} \cdot \ln \frac{1 + R_0}{1 - R_0}$$

$$\mu_0 = E(Z_0) = \frac{1}{2} \cdot \ln \frac{1 + \rho_0}{1 - \rho_0} + \frac{\rho_0}{2 \cdot (n - 1)}$$

$$\sigma^2 = Var(Z_0) = \frac{1}{n - 3}$$

$$R_1 \mapsto Z_1 = \frac{1}{2} \cdot \ln \frac{1 + R_1}{1 - R_1}$$

$$\mu_1 = E(Z_1) = \frac{1}{2} \cdot \ln \frac{1 + \rho_1}{1 - \rho_1} + \frac{\rho_1}{2 \cdot (n - 1)}$$

$$\sigma^2 = Var(Z_1) = \frac{1}{n - 3}$$

Confidence Interval

The distribution of the difference $Z_0 - Z_1$ is a normal distribution $Z_0 - Z_1 \sim \mathcal{N}(\mu_0 - \mu_1, \sqrt{2} \cdot \sigma)$. Hence, $((Z_0 - Z_1) - (\mu_0 - \mu_1))/(\sqrt{2} \cdot \sigma)$ has a standard normal distribution. The two-sided confidence interval for $\mu_0 - \mu_1$ is

$$\left[(Z_0 - Z_1) - z_{1-\alpha/2} \cdot \sqrt{\frac{2}{n - 3}}, \ (Z_0 - Z_1) + z_{1-\alpha/2} \cdot \sqrt{\frac{2}{n - 3}} \right]. \quad (4.41)$$

Hypothesis Test

Remember that with the Fisher transformation we can map ρ to a variable that has a normal distribution, see (4.35) to (4.37). Hence, testing the difference of two correlation coefficients works in the same way as testing the difference of means of two normal distributions.

Number of Traces

The number of traces n to estimate the difference of means $\mu_0 - \mu_1$ of the normal distribution $\mathcal{N}(\mu_0 - \mu_1, \sqrt{2} \cdot \sigma)$ with confidence $1 - \alpha$ and precision c

is:

$$n = 3 + 8 \cdot \frac{z_{1-\alpha/2}^2}{\ln^2 \frac{1+c}{1-c}} \tag{4.42}$$

Like in the previous sections, the most important practical question is how many samples are needed in order to distinguish the two distributions that arise from ρ_0 and ρ_1. Also like before, we can assert the number of traces by looking at how much the difference distribution deviates from zero.

The number of traces n that is necessary to assert with a confidence of $1 - \alpha$ that the two normal distributions $Z_0 \sim \mathcal{N}(\mu_0, \sqrt{1/(n-3)})$ and $Z_1 \sim \mathcal{N}(\mu_1, \sqrt{1/(n-3)})$ are different is given by:

$$n = 3 + 8 \cdot \frac{z_{1-\alpha}^2}{\left(\ln \frac{1+\rho_0}{1-\rho_0} - \ln \frac{1+\rho_1}{1-\rho_1} \right)^2} \tag{4.43}$$

Example

The special case that $\rho_0 = 0$, is especially relevant in practice. The rule for calculating the number of traces simplifies in this case. It is given in (4.44). For example, $2\,751$ traces are necessary to distinguish $\rho = 0.1$ from $\rho = 0$ with $\alpha = 0.0001$.

The number of traces n that is necessary to assert with confidence $1 - \alpha$ that the two normal distributions $Z_0 \sim \mathcal{N}(0, \sqrt{1/(n-3)})$ and $Z_1 \sim \mathcal{N}(\mu, \sqrt{1/(n-3)})$ are different is given by:

$$n = 3 + 8 \cdot \frac{z_{1-\alpha}^2}{\ln^2 \frac{1+\rho}{1-\rho}} \tag{4.44}$$

4.7   Summary

The power consumption of cryptographic devices consists of different components. Each point of a power trace can be modeled as the sum of an operation-dependent component P_{op}, a data-dependent component P_{data}, electronic noise $P_{el.\,noise}$, and a constant component P_{const}. These components can usually be approximated by normal distributions. In case of our microcontroller, P_{data} is binomially distributed. The power consumption of the microcontroller is inversely proportional to the Hamming weight of the data it processes.

It is also possible to model the power consumption in the context of a given attack scenario. We have defined the exploitable power consumption P_{exp} as the power consumption that is caused by the processing of the information the attacker is looking for. The power consumption that does not contain exploitable information in the considered attack scenario is referred to as $P_{sw.\,noise}$. It holds that $P_{exp} + P_{sw.\,noise} = P_{data} + P_{op}$. Furthermore, it is important to point out that P_{exp} and $P_{sw.\,noise}$ are independent components. Therefore, it is necessary to model the power consumption that is caused by data that depends on the information the attacker is looking for as P_{exp}.

The fact that P_{exp}, $P_{sw.\,noise}$, and $P_{el.\,noise}$ are independent of each other can be used to define a signal-to-noise ratio. The signal corresponds to P_{exp} and the noise corresponds to $P_{sw.\,noise} + P_{el.\,noise}$. This SNR can be used to quantify the leakage of a cryptographic device in a given attack scenario. An analysis of the SNR of different points of power traces of our microcontroller has turned out that the SNR reaches its maximum at the peaks of the power traces. This observation is the motivation to compress power traces. The goal of compression methods for power traces is to remove the redundancy from power traces. As the amplitude of the peaks usually contains all relevant information, the extraction of maxima and the integration of parts of power traces are suitable compression methods.

The linear relationship between different points of power traces can be measured using the covariance or the correlation coefficient. The correlation coefficient has the advantage that it leads to a dimensionless result between -1 and 1. Hence, it is easy to compare correlation coefficients. The linear relationship between all points of a power trace is usually described by a covariance matrix. The covariance matrix and the corresponding mean trace can be used to model a power trace as a multivariate normal distribution. This way of modeling power traces is usually only used for small parts of power traces as the covariance matrix grows quadratically with the number of points in the traces.

In the last section of this chapter, we have discussed statistical concepts that are important for power analysis attacks. In particular, we have discussed properties of estimators for the parameters of a normal distribution and for the correlation coefficient. These estimators are distributed according to sampling distributions that determine how well the parameters can be estimated. An important observation is that the standard deviation is reduced by $\sqrt{n}$ when taking the mean of n power traces. Based on the sampling distribution, we have also explained confidence intervals and hypothesis tests. Confidence intervals are intervals that contain the estimated parameters with a given probability. Hypotheses tests can be used to determine with a given probability whether a certain parameter has a certain property or not. Based on these concepts, we have provided formulas that allow deriving the number of power traces that are needed to get reliable statistical results.

Chapter 5

SIMPLE POWER ANALYSIS

Simple power analysis (SPA) attacks are characterized by Kocher *et al.* in [KJJ99] in the following way: "SPA is a technique that involves directly interpreting power consumption measurements collected during cryptographic operations." In other words, the attacker tries to derive the key more or less directly from a given trace. This can make SPA attacks quite challenging in practice. Often, they require detailed knowledge about the implementation of the cryptographic algorithm that is executed by the device under attack. Furthermore, if only one power trace is available, usually complex statistical methods have to be used in order to extract the signal.

SPA attacks are useful in practice if only one or very few traces are available for a given set of inputs. Consider for example a scenario where a consumer uses a smart card to pay for gas at a gas station. The customer has to refill the gas tank of the car on a regular basis and always buys a similar amount of gas. A malicious smart card reader could record the power consumption of the card. In this way, the attacker could gather a couple of traces for similar plaintexts.

In this chapter, we discuss different types of SPA attacks, like the visual inspection of power traces, template-based SPA attacks and collision attacks. The examples for attacks that we provide in this chapter have been produced with the AES software implementation that we describe in Appendix B and the measurement setup that we describe in Section 3.4.4.

5.1   General Description

The goal of SPA attacks is to reveal the key when given only a small number of power traces (for a small number of plaintexts). In the most extreme case, this means that the attacker attempts to reveal the key based on one single power trace. In order to distinguish between the extreme and the normal SPA assumption, we distinguish between *single-shot SPA attacks* and *multiple-shot*

SPA attacks. In single-shot SPA attacks, only one power trace can be recorded. In multiple-shot SPA attacks, multiple power traces can be recorded.

In multiple-shot SPA attacks, either we can measure the power consumption for the same plaintext multiple times, or we can even supply different plaintexts. The advantage of having several traces for one plaintext is that we can use them to reduce the noise by computing the mean of the traces.

Despite the differences in taking a single measurement or taking multiple measurements, the principle of SPA attacks is always the same. The attacker needs to be able to monitor the power consumption of the device under attack. In the attacked device, the key must have (directly or indirectly) a significant impact on the power consumption.

SPA attacks exploit key-dependent differences (patterns) within a trace. They use only one trace or very few traces.

5.2   Visual Inspections of Power Traces

Every algorithm that runs on a cryptographic device is executed in a sequential manner. The operations that are defined by the algorithm are translated into instructions that are supported by the device. For example, AES consists of ten different steps, which are called rounds. Each round consists of four round transformations, which are called AddRoundKey, SubBytes, ShiftRows, and MixColumns. Each round transformation works on one or several bytes of the AES state, see Appendix B.

Suppose that AES is implemented in software on a microcontroller. In this case, the round functions are implemented using the instructions of the microcontroller. Microcontrollers have an instruction set that typically consists of arithmetic instructions (such as addition), logical instructions (such as exclusive-or), data transfer instructions (such as move), and branching instructions (such as jump). Each instruction works on a number of bytes and involves different components of the microcontroller, such as the arithmetic-logic unit, memory (external or internal RAM or ROM), or some peripheral (such as a communication port). These are physically separate components of the microcontroller and they differ in functionality and implementation. Therefore, they have a characteristic power consumption, which leads to a characteristic pattern in the power trace. For instance, a move instruction that operates on data that is stored in internal memory needs fewer clock cycles than a move instruction that operates on values that are located in external memory. Furthermore, the external buses often cause a higher power consumption than the internal buses. These facts make it possible to distinguish instructions in a power trace.

The possibility to distinguish instructions within a trace can lead to a serious security problem if the sequence of instructions directly depends on the key.

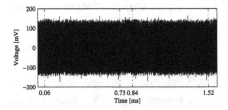

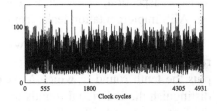

Figure 5.1. Two rounds of AES. The power trace has not been compressed.

Figure 5.2. One round of AES. The power trace has been compressed.

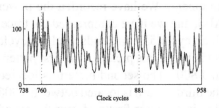

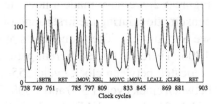

Figure 5.3. The sequence of AddRoundKey, SubBytes, and ShiftRows operations.

Figure 5.4. The annotated sequence of AddRoundKey, SubBytes, and ShiftRows operations.

```
            LCALL SET_ROUND_TRIGGER
            MOV A,ASM_input + 0      ; load a0
            XRL A,ASM_key + 0        ; add k0
            MOVC A,@A + DPTR         ; S-box look-up
            MOV ASM_input, A         ; store a0
            LCALL CLEAR_ROUND_TRIGGER
```

Figure 5.5. The sequence of assembly instructions that corresponds to Figure 5.4.

Hence, if a certain instruction only occurs if a key bit is 1 and if another instruction only occurs if a key bit is 0, then it is possible to derive the key from the power traces by looking at the sequence of instructions in the power trace. Especially for implementations of public-key cryptosystems, this has turned out to be a problem.

5.2.1   Example for Software

We now perform a visual inspection of a power trace that we have acquired during the execution of our AES implementation on the microcontroller. As pointed out before, an SPA attack that is based on a visual inspection of the power trace succeeds if the sequence of instructions depends on the key. In our AES implementation, this is not the case. Hence, we cannot reveal the key by a visual inspection. However, we can still investigate which information about our AES implementation leaks from the power trace.

Figure 1.1 in Section 1.3 shows a power trace of a full AES encryption run on our microcontroller. It is possible to see the nine normal AES rounds plus the tenth round in which MixColumns is skipped. Hence, it is easy to distinguish the individual rounds in the power traces. Figure 5.1 zooms in on the first two rounds. The first round is located between 0.06 ms and 0.73 ms, the second round is located between 0.84 ms and 1.52 ms. There are three tiny peaks in the upper half of the picture that facilitate recognizing the rounds. For Figure 5.2, we have made a new measurement. We have recorded the power consumption during the first round of AES, and we have compressed the trace, see Section 4.5. The round starts at clock cycle 555 and ends at clock cycle 4 305. There are three different patterns visible in this picture. The first pattern is located between clock cycle 555 and 1 800. The second pattern is located between clock cycle 1 800 and 4 305. The third pattern is located between 4 305 and 4 931. A typical AES implementation in software first performs the round functions that operate on individual bytes and then performs MixColumns, which operates on a column (4 bytes) of the AES state. If the key generation is done on-the-fly, it occurs in between the rounds. Consequently, the first pattern in Figure 5.2 corresponds to the sequence of AddRoundKey, SubBytes, and ShiftRows operations that are applied to the 16 bytes of the AES state. The second pattern corresponds to MixColumns and the third pattern corresponds to the generation of the first round key.

Within the first pattern, one can count 16 tiny peaks (on the bottom of the trace) that are closely located next to each other. Each peak corresponds to a sequence of AddRoundKey, SubBytes, and ShiftRows. In Figure 5.3, we zoom in on Figure 5.2 and show the power consumption of the first two peaks only. The first peak is located between clock cycle 760 and 881. The shape of the trace shortly after clock cycle 760 is very similar to the shape of the trace after clock cycle 881. By visually inspecting a power trace, it is possible to detect patterns like this and to make guesses about their meaning. We now use the sequence of executed assembly instructions in order to do a more detailed analysis. The sequence of assembly instructions is displayed in Figure 5.5.

First, the function SET_ROUND_TRIGGER is called to set a trigger signal. The SET_ROUND_TRIGGER function executes a SETB (set bit) and a RET (return) instruction. Then one byte of the AES state is processed. Afterwards, the trigger signal is cleared by calling the CLEAR_ROUND_TRIGGER function, which executes a CLRB (clear bit) and a RET instruction. On our microcontroller, different instructions require different numbers of clock cycles for their execution. SETB, MOV, and XRL require 12 clock cycles whereas the other instructions require 24 clock cycles.

In Figure 5.4, we have zoomed in on Figure 5.3 a little more and we have assigned the instructions from our assembly code to the patterns in the power trace. Now we can see that different instructions indeed lead to different patterns in the power traces. Furthermore, we can see that a MOVC looks different from a MOV instruction. Hence, different types of move instructions can be distinguished from each other.

5.3 Template Attacks

Template attacks exploit that the power consumption also depends on the data that is being processed. In template attacks, we characterize power traces by a multivariate normal distribution, see Section 4.4.2. In contrast to other types of power analysis attacks, template attacks usually consist of two phases: A first phase, in which the characterization takes place, and a second phase, in which the characterization is used for an attack.

5.3.1 General Description

According to Chapter 4, power traces can be characterized by a multivariate normal distribution, which is fully defined by a mean vector and a covariance matrix $(\mathbf{m}, \mathbf{C})$. We refer to this pair $(\mathbf{m}, \mathbf{C})$ as *template* from now on. In a template attack, we assume that we can characterize the device under attack. This means, we can determine templates for certain sequences of instructions. For example, we might possess another device of the same type as the attacked one that we can fully control. On this device, we execute these sequences of instructions with different data d_i and keys k_j in order to record the resulting power consumption. Then, we group together the traces that correspond to a pair of (d_i, k_j), and estimate the mean vector and the covariance matrix of the multivariate normal distribution. As a result, we obtain a template for every pair of data and key (d_i, k_j): $h_{d_i,k_j} = (\mathbf{m}, \mathbf{C})_{d_i,k_j}$.

> A template h is a pair that consists of a mean vector $\mathbf{m}$ and a covariance matrix $\mathbf{C}$.

Later on, we use the characterization together with a power trace from the device under attack to determine the key. This means, we evaluate the probability density function of the multivariate normal distribution with $(\mathbf{m}, \mathbf{C})_{d_i,k_j}$ and the power trace of the device under attack. In other words, given a power trace $\mathbf{t}$ of the device under attack, and a template $h_{d_i,k_j} = (\mathbf{m}, \mathbf{C})_{d_i,k_j}$, we compute the probability:

$$p(\mathbf{t}; (\mathbf{m}, \mathbf{C})_{d_i,k_j}) = \frac{exp\left(-\frac{1}{2} \cdot (\mathbf{t} - \mathbf{m})' \cdot \mathbf{C}^{-1} \cdot (\mathbf{t} - \mathbf{m})\right)}{\sqrt{(2 \cdot \pi)^T \cdot det(\mathbf{C})}} \tag{5.1}$$

We do this for every template. As a result, we get the probabilities $p(\mathbf{t}; (\mathbf{m}, \mathbf{C})_{d_1,k_1}), \ldots, p(\mathbf{t}; (\mathbf{m}, \mathbf{C})_{d_D,k_K})$. The probabilities measure how well the templates fit to a given trace. Intuitively, the highest probability should indicate the correct template. Because each template is associated with a key, we also get an indication for the correct key.

This intuition is also supported by the statistical literature, see [Kay98]. If all keys are equally likely, then the decision rule which minimizes the probability for a wrong decision is to decide for h_{d_i,k_j} if

$$p(\mathbf{t}; h_{d_i,k_j}) > p(\mathbf{t}; h_{d_i,k_l}) \; \forall l \neq j. \tag{5.2}$$

This is also called a maximum-likelihood (ML) decision rule.

5.3.2   Template Building Phase

In the previous section, we have explained that in order to *characterize* a device, we execute a certain sequence of instructions for different pairs (d_i, k_j) and record the power consumption. Then, we group together the traces that correspond to the same pair and estimate the mean vector and the covariance matrix of the multivariate normal distribution. Note that the size of the covariance matrix grows quadratically with the number of points in the trace. Clearly, one needs to find a strategy to determine the *interesting points*. We denote the number of interesting points by N_{IP}.

> The interesting points are those points that contain the most information about the characterized instruction(s).

In practice, there are different ways to build templates. For instance, the attacker can build templates for a specific instruction such as the MOV instruction, or the attacker can build templates for a longer sequence of instructions, such as the sequence of assembly instructions that we have given in Figure 5.5. Which strategy is best typically depends on what is known about the device that is attacked. In the following, we discuss some strategies for building templates.

Templates for Pairs of Data and Key

The first strategy, which is also the one that we have mentioned in the previous section, is to build templates for each pair of (d_i, k_j). The interesting points of a trace, which are used to build the templates, are therefore all points that correlate to (d_i, k_j). This implies that all instructions that involve d_i, k_j, and functions of (d_i, k_j) lead to interesting points.

For example, we can build templates for the sequence of instructions in our AES assembly implementation that implement AddRoundKey, SubBytes, and ShiftRows. Or, we can build templates for one round of the key schedule in our AES assembly implementation.

Templates for Intermediate Values

The second strategy is to build templates for some suitable function $f(d_i, k_j)$. The interesting points of a trace, which are used to build the templates, are therefore all points that correlate to the instructions that involve $f(d_i, k_j)$.

For example, suppose that we build templates for the MOV instruction in our AES assembly implementation that moves the S-box output from the accumulator back into the state register. This means, we build templates that allow us to deduce k_j given $S(d_i \oplus k_j)$. Instead of building 256^2 templates, one for each pair (d_i, k_j), we can simply build 256 templates $h_{v_{ij}}$, with $v_{ij} = S(d_i \oplus k_j)$, *i.e.* one for each output of the S-box. Note that the 256 templates can be assigned to the 256^2 pairs (d_i, k_j).

Templates with Power Models

For both previously described strategies, it is possible to take knowledge about the power consumption characteristics into account. For example, if a device leaks the Hamming weight of the data, then moving the value 1 will virtually lead to the same power consumption as moving the value 2. Consequently, the template that is associated with the value 1 will match as well to a trace in which the value 1 is moved, as the template that is associated with the value 2. This means, it does not make sense to build different templates for values with identical Hamming weights. Assume that we want to build templates for the S-box output. Then, we can simply build nine templates, one for each Hamming weight of the S-box output. This means we build $h_{HW(v_{ij})}$ with $v_{ij} = S(d_i \oplus k_j)$. Note that the nine templates can be assigned to the 256^2 pairs (d_i, k_j).

We typically follow this last approach when we build templates for our AES software implementation. Note that if a device only leaks the Hamming weight, it is typically not possible to determine the key from one single trace in this way.

5.3.3 Template Matching Phase

In practical template attacks, there are difficulties that can occur during the template matching phase. They are related to the covariance matrix. First, the size of the covariance matrix depends on the number of interesting points. Clearly, the number of interesting points must be chosen carefully. Second, the covariance matrix tends to be badly conditioned. This means, we run into numerical problems during the inversion, which needs to be done in (5.1). In addition, the values that are calculated in the exponent tend to be very small, which often leads to more numerical problems.

In principle, template matching means to evaluate (5.1) for a given trace. According to (5.2), the template that leads to the highest probability indicates the correct key.

> The template with the highest probability indicates the correct key:
>
> $$p(\mathbf{t}; \hbar_{d_i,k_j}) > p(\mathbf{t}; \hbar_{d_i,k_l}) \quad \forall l \neq j \qquad (5.3)$$

In order to avoid the exponentiation, one typically applies the logarithm to (5.1). Then, the template that leads to the smallest absolute value of the logarithm of the probability indicates the correct key:

$$\ln p(\mathbf{t}; (\mathbf{m}, \mathbf{C})) = -\frac{1}{2}(\ln((2 \cdot \pi)^{N_{IP}} \cdot det(\mathbf{C})) +$$
$$(\mathbf{t} - \mathbf{m})' \cdot \mathbf{C}^{-1} \cdot (\mathbf{t} - \mathbf{m})) \qquad (5.4)$$

$$|\ln p(\mathbf{t}; \hbar_{d_i,k_j})| < |\ln p(\mathbf{t}; \hbar_{d_i,k_l})| \quad \forall l \neq j \qquad (5.5)$$

In order to avoid problems with the inversion of the covariance matrix, we can set the covariance matrix equal to the identity matrix. This essentially means that we do not take the covariances between the points into account. A template that only consists of a mean vector is called a *reduced template*.

> Reduced templates only consist of a mean vector.

Setting the covariance matrix equal to the identity matrix simplifies the multivariate normal distribution:

$$p(\mathbf{t}; \mathbf{m}) = \frac{1}{\sqrt{(2 \cdot \pi)^{N_{IP}}}} \cdot exp\left(-\frac{1}{2} \cdot (\mathbf{t} - \mathbf{m})' \cdot (\mathbf{t} - \mathbf{m})\right) \qquad (5.6)$$

In order to avoid numerical problems in the exponentiation, we can apply the logarithm again. Then, the logarithm of the probability is simply calculated as

$$\ln p(\mathbf{t}; \mathbf{m}) = -\frac{1}{2}(\ln(2 \cdot \pi)^{N_{IP}} + (\mathbf{t} - \mathbf{m})' \cdot (\mathbf{t} - \mathbf{m})). \qquad (5.7)$$

As before, the template that leads to the smallest absolute value of the logarithm indicates the correct key, see (5.5). This method, which uses reduced templates, is also called *least-square* (LSQ) test in the literature. Since the only relevant term in (5.7) is the square of the difference of t and m, the decision rule for reduced templates simplifies to (5.8).

> The reduced template that leads to the smallest square difference indicates the correct key:
>
> $$(\mathbf{t} - \mathbf{m}_{d_i,k_j})' \cdot (\mathbf{t} - \mathbf{m}_{d_i,k_j}) < (\mathbf{t} - \mathbf{m}_{d_i,k_l})' \cdot (\mathbf{t} - \mathbf{m}_{d_i,k_l}) \quad \forall l \neq j \quad (5.8)$$

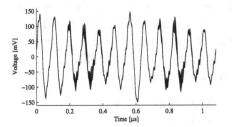 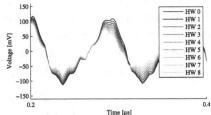

Figure 5.6. Power traces of MOV instructions with different Hamming weights.

Figure 5.7. Zoomed in on the MOV instructions with different Hamming weights.

5.3.4 Example for a MOV Instruction

We have already seen in the example of Section 5.2 that different instructions, which are executed on a microcontroller, lead to different power consumption patterns. In other words, the shape of the power trace depends on the instruction that is being executed. In the previous section, we have introduced templates that allow exploiting the data dependency in power traces. In this example, we investigate how to apply the concept of template attacks to our microcontroller.

Figure 5.6 shows nine power traces plotted on top of each other. Each trace shows the power consumption during the execution of a MOV instruction on our microcontroller. Each MOV instruction moves a byte with a certain Hamming weight from a register to the accumulator. In Figure 5.6, the distance between the nine traces is largest between $0.2\,\mu s$ and $0.4\,\mu s$, and between $0.7\,\mu s$ and $0.9\,\mu s$. Figure 5.7 zooms in on the traces between $0.2\,\mu s$ and $0.4\,\mu s$. The legend shows which trace corresponds to which Hamming weight. First, one can see that the power traces can be clearly distinguished from each other. Second, one can see that they are equidistant to each other. Third, one can see that processing a byte with Hamming weight 0 causes the highest power consumption, whereas processing a byte with Hamming weight 8 causes the lowest power consumption. This behavior is due to the fact that our microcontroller has a precharged bus.

Figure 5.7 proves that the height of several points within the power trace of a MOV instruction is inversely proportional to the Hamming weight of the processed data. Consequently, we can build templates that allow us to classify MOV instructions according to the Hamming weight. We can use the templates during an attack to deduce the Hamming weight of the processed data. Because there are nine different Hamming weights, we build nine templates. Remember that a template consists of a mean vector and a covariance matrix. The size of the covariance matrix grows quadratically in the number of points. As pointed out before, we have to identify the interesting points in order to minimize the size of the template. This is simple because we can identify the interesting points directly in Figure 5.6. All the points in which the power traces differ significantly can be used as interesting points. We took in total five of those points to build the templates. The template $h_0 = (m_0, C_0)$ for the MOV instruction that operates on data with Hamming weight 0 is given below.

$$C_0 = \begin{pmatrix} 1.77 & 1.03 & 0.29 & 0.22 & 0.34 \\ 1.03 & 2.71 & 0.50 & 0.54 & 0.75 \\ 0.29 & 0.50 & 1.88 & 0.17 & 0.22 \\ 0.22 & 0.54 & 0.17 & 1.32 & 0.21 \\ 0.34 & 0.75 & 0.22 & 0.21 & 1.56 \end{pmatrix} \tag{5.9}$$

$$m_0 = (-81.52, -78.65, 70.93, 74.01, 49.21)' \tag{5.10}$$

Now we check how well this template matches to some other power traces that we have recorded. Below, there is a matrix that consists of the interesting points of nine traces that we acquired during the execution of the MOV instruction:

$$T = \begin{pmatrix} -81.20 & -77.70 & 71.20 & 74.30 & 49.70 \\ -78.80 & -74.10 & 68.60 & 66.60 & 45.10 \\ -76.30 & -71.20 & 67.10 & 63.40 & 44.30 \\ -74.20 & -68.50 & 65.60 & 59.30 & 42.80 \\ -72.40 & -64.20 & 63.50 & 56.00 & 41.30 \\ -71.90 & -62.70 & 59.00 & 50.10 & 37.70 \\ -70.80 & -60.00 & 57.70 & 44.90 & 36.00 \\ -65.50 & -53.50 & 58.20 & 44.60 & 37.80 \\ -60.60 & -45.90 & 56.70 & 40.20 & 37.00 \end{pmatrix} \tag{5.11}$$

The first power trace t'_1, *i.e.* the first row of T, corresponds to the execution of the MOV instruction with data of Hamming weight 0, the second power trace t'_2 corresponds to data with Hamming weight 1, *etc.* Now, we match our template to these nine traces by computing (5.4) for each of the nine traces. The results are:

$$\ln p(t'_1; h_0) = -5.98$$
$$\ln p(t'_2; h_0) = -49.54$$
$$\ln p(t'_3; h_0) = -99.36$$
$$\ln p(t'_4; h_0) = -182.09$$
$$\ln p(t'_5; h_0) = -295.03$$
$$\ln p(t'_6; h_0) = -486.59$$
$$\ln p(t'_7; h_0) = -681.45$$
$$\ln p(t'_8; h_0) = -796.20$$
$$\ln p(t'_9; h_0) = -1\,131.03$$

The template h_0 fits best to t'_1, since it leads to the lowest absolute value. Trace t'_1 is indeed the trace where the MOV instruction with data of Hamming weight 0 has been executed.

5.3.5 Example for the AES Key Schedule

We now discuss how the ability to determine the Hamming weights of intermediate values that occur in AES allows revealing the AES key. We assume that the attacker has the following capabilities. First, the attacker has one plaintext and the corresponding ciphertext from the device under attack. Second, the attacker is able to identify the parts that correspond to the AES key expansion in the power trace, and to characterize the power consumption. Third, the attacker is able to extract the Hamming weights of the round-key bytes from the relevant parts of the trace.

The first assumption can typically be fulfilled easily. The second and third assumptions go along with the assumptions that we have made for SPA attacks. In particular, the second assumption can be fulfilled by using the template approach. The attacker characterizes the device under attack, *i.e.* templates are built that allow deducing the Hamming weights of the key bytes.

Of course, just knowing the Hamming weights of the round-key bytes does not simply lead to the key. The number of 8-bit values that have Hamming weight w is $\binom{8}{w}$. This implies that the number of possible keys is considerably reduced. However, it is still too large for a practical attack.

For AES, an efficient way to exploit the knowledge about the Hamming weights of the bytes of the round keys has been presented in [Man03a]. The idea here is the following: The attacker uses the dependencies between the bytes of the round keys within the AES key schedule to reduce the number of possible key values. The key is then determined by using the known plaintext-ciphertext pair.

The AES key expansion is described in Figure B.8 of Appendix B. According to it, eleven round keys are generated. The first round key is just the cipher key and it is used in the very first AddRoundKey operation. The second round key, however, depends in a rather simple way on the first round key. For instance, the first byte $W[4]_1$ of $W[4]$ depends on the first byte $W[0]_1$ of $W[0]$ and the second byte $W[3]_2$ of $W[3]$. However, $W[3]$ is sent through the AES S-box before it is used. Hence, by running through all values for $W[0]_1$ and $W[3]_2$ and by comparing their Hamming weights and the Hamming weights of their related intermediate results with the observed Hamming weights, one can exclude many possible values for $W[0]_1$ and $W[3]_2$.

According to this idea, we have produced Table 5.1. The first column in Table 5.1 lists the four intermediate values for which we can derive the Hamming weights from the trace. The second column lists the correspondingly observed Hamming weights. Each of the remaining four columns lists a combination

Table 5.1.   Observing the Hamming weights of intermediate values allows reducing the brute-force search space drastically. The only pairs (W[0]$_1$,W[3]$_2$) that lead to the observed Hamming weights are $\{(0, 5), (0, 10), (0, 24), (0, 192)\}$.

Intermediate value	HW	Combination of keys			
W[0]$_1$	0	0	0	0	0
W[3]$_2$	2	5	10	24	192
SubBytes(W[3]$_2$)	5	107	103	173	186
W[4]$_1$	5	107	103	173	186

of values for W[0]$_1$, W[3]$_2$, and the corresponding intermediate results that leads to the observed Hamming weights. Only those four combinations (out of 256^2 combinations) result in the observed Hamming weights. Hence the search space for a brute-force search for the two key bytes W[0]$_1$ and W[3]$_2$ has been reduced from 65 536 to 4.

Of course, we can do the same for the other bytes of W[0], W[3], and W[4] as well. We can also extend this idea to other round keys. For instance, the first byte of W[8] depends on the first byte of W[4] and the second byte of W[7], which depends on its four predecessors. Hence, we could simultaneously run through the second byte of W[1], W[2], W[3], and W[4] and the first byte of W[4] and compute with them all intermediate values according to the AES key schedule. Then we compare the Hamming weights of the intermediate values with the observed Hamming weights. This allows us to exclude most of the possible values of the key. There are no theoretical results available that allow telling exactly how much this method reduces the search space. However, in [Man03a] several results from practical experiments have been reported. For instance, if about 81 Hamming weights are used in such an attack, the average number of keys that remain for a brute-force search is 11.

5.4   Collision Attacks

Suppose we are able to observe the power consumption for two encryption runs with two different inputs d_i and d_i^*, and the unknown key k_j. In a collision attack we exploit the fact that in two encryption runs, a certain intermediate value $v_{i,j}$ can be equal: $v_{i,j} = f(d_i, k_j) = f(d_i^*, k_j)$. If an intermediate value in one encryption run is equal to the corresponding value in the other encryption run, we say that the intermediate value collides. The important observation is that for two inputs d_i and d_i^* a collision cannot occur for all key values, but only for a certain subset of keys. Hence, each collision allows reducing the search space for the key. If several collisions can be observed, the key can even be uniquely identified.

As we have explained in Section 5.2, every algorithm is executed in a number of steps. Hence, the intermediate value that collides is processed several times within the device under attack. First, it is processed when it is computed. Then, it is typically stored somewhere in memory, and later on, it is fetched from memory again. Depending on the structure of the cipher, it might also be subject to different transformations that lead to other intermediate values that then collide as well. Summarizing, a collision in one intermediate value of an algorithm leads to several colliding intermediate values during the execution on the device under attack. This facilitates detecting collisions in practice.

In order to detect a collision in practice, we first need to identify the part in the power trace in which the colliding intermediate value is processed. Then, given two power traces, we have to decide whether those two parts are equal or not. There are several ways to make this decision:

Template creation on beforehand: Assume that we can characterize the device before the attack. This means, we build templates for the part of the trace in which the intermediate value collides. If we assume to work with bytes, we have to build 256 templates. Later on, in the attack, we match the templates with the two traces. A collision has occurred if the same template fits best to both traces.

Template creation on the fly: Assume that we just get the two power traces, and that we know in which part of the trace the (colliding) intermediate value is processed. Then we could extract these parts (or some of the points) from both traces, consider them as reduced templates, and match them with each other. Because we use reduced templates only, we have to use the LSQ test, see (5.7).

Other ideas are to compute the correlation coefficient between the points in the traces, or to use pattern-matching techniques. The template-based technique without characterization has been used in [SLFP04] in order to mount a collision attack on AES.

5.4.1 Example for Software

We assume that the attacker has the following capabilities. First, the attacker may choose the plaintexts for encryption. Second, the attacker is able to identify the parts in the power trace that correspond to the AES round function. Third, the attacker is able to extract the Hamming weights of the bytes of the round key from the relevant parts of the trace.

In the round function of AES, a collision can occur after the first Mix-Columns operation, see [SLFP04]. Hence, we look at the first byte of the state after the MixColumns operation for the plaintexts d and d^*:

$$b_0 = 02 \cdot S(d_0 \oplus k_0) \oplus 03 \cdot S(d_1 \oplus k_1) \oplus S(d_2 \oplus k_2) \oplus S(d_3 \oplus k_3)$$
$$b_0^* = 02 \cdot S(d_0^* \oplus k_0) \oplus 03 \cdot S(d_1^* \oplus k_1) \oplus S(d_2^* \oplus k_2) \oplus S(d_3^* \oplus k_3)$$

Suppose that we may choose the two plaintexts d and d^* such that $d_0 = d_0^* = 0$, $d_1 = d_1^* = 0$, $d_2 = d_3 = a$, and $d_2^* = d_3^* = b$. Then, if b_0 and b_0^* collide, it holds that

$$b_0 \oplus b_0^* = S(a \oplus k_2) \oplus S(a \oplus k_3) \oplus S(b \oplus k_2) \oplus S(b \oplus k_3) = 0. \quad (5.12)$$

We know a and b, and thus we can run through all possible values for k_2 and k_3 and determine which of them satisfy (5.12). Because only a subset of the 2^{16} values satisfies (5.12), the search space for a brute-force key search is reduced. The same idea can also be applied to the other bytes of the state after the MixColumns operation. By adaptively choosing the plaintexts, we can even determine the key bytes uniquely, see [SLFP04] for a detailed description.

5.5   Notes and Further Reading

Basic SPA Techniques. Kocher *et al.* [KJJ99] were the first to discuss the application of power analysis to implementations of cryptographic algorithms. They were also the first to point out that the dependency of the power traces on the executed instructions can lead to a serious security problem. As an example for potentially vulnerable cryptosystems, they named the Data Encryption Standard (DES). Typical implementations of DES have key-dependent conditional branches in the key schedule and in the permutations. They pointed out that conditional branching in general is a risky operation. In addition, they highlighted that exponentiations (that make use of repeated multiplication and squaring) leak the entire key if the multiplication and squaring operations have different power consumption patterns. Furthermore, one of the techniques that they frequently use is called *trace-pair analysis*. In this technique, the attacker compares two power traces and looks for differences. Once such differences are found, the attacker needs to find out whether or not they are systematic in the sense that particular message differences lead to particular differences in the power trace. If such a relationship can be found, information about the internal state and thereby the key can be derived. Unfortunately, there are not many technical papers available from Cryptography Research Inc. Besides [Jaf06a], we have no further references about this technique.

Messerges *et al.* [MDS99a] discussed the application of SPA to an implementation of DES on an 8-bit microcontroller. They assumed that the attacker can observe the Hamming weights of several operands during the DES key schedule. They showed that knowing the Hamming weights of all eight DES key bytes reduces the brute-force search space from 2^{56} to approx. 2^{38} keys. They also pointed out that the knowledge of more intermediate values (that depend on the key bytes) allows reducing the search space even further. Messerges *et al.* further mentioned that their attack can be adapted to work with the Hamming-distance model.

Mayer-Sommer [MS00] discussed the leakage of the MOV instruction on a specific microcontroller that leaks the Hamming weight of the operands. She concluded that if the microcontroller runs at a low frequency, but with a high supply voltage, the Hamming weight can be obtained by an SPA attack. In addition, she observed that the leakage is mainly caused by the transfer of the data over the bus. Bertoni *et al.* [BZB+05] sketched an attack on AES that is based on recognizing cache misses in power traces.

Mangard [Man03a] discussed the application of an SPA attack to the AES key schedule, see also Section 5.3.5.

Profiling. In several of the previously discussed examples, we have assumed that the attacker is able to identify certain parts in a power trace. In other words, we have assumed that the attacker can assign certain parts of a cryptographic algorithm to certain parts of a power trace. This is called profiling. We have shown for an AES implementation on a microcontroller that under the assumption that the attacker knows the details of the implementation, a visual inspection of a power trace allows doing this profiling. However, one can argue that an attacker might not have such detailed information. Instead, the attacker could only have the possibility to use one or several devices, which are identical to the device under attack, to do the profiling.

Biham and Shamir [BS99] were the first to describe a profiling method that allows determining the parts of a power trace that correspond to the key scheduling. Their profiling method requires measuring the execution of the encryption function on several devices (each having a unique key) using different input data. First, they concentrate on the evaluation of the measurements taken from only one device. They determine the data-dependent parts of the power trace by studying the variance of the trace. Parts of the power trace with a high variance depend on the data, which varies from execution to execution, whereas the parts of the trace with a low variance depend on the key, which is constant for all executions on a device. Second, they compare the data-independent parts of the traces from several devices. Small variations in these data-independent parts are related to the key, because the key is different on different devices.

Fahn and Pearson [FP99] also discussed a scenario in which the attacker first can do some profiling, and then later on during the attack uses the profile to extract the key. They called their concept inferential power analysis. In contrast to Biham and Shamir, they do not require several devices during profiling. They extend the profiling and actually characterize the device. Fahn and Pearson reveal the location of the manipulation of the key bits by comparing the power consumption of the different rounds of the key schedule instead of comparing the power consumption of different executions of the key schedule. After they have determined the precise locations of the processing of the bits of the key, they also investigate which power level corresponds to which logic value of the bit of the key. In other words, they first profile the device and then build templates for the key bits.

Template and Collision Attacks.   Chari *et al.* [CRR03] were the first to observe that using multivariate statistics allows stronger attacks. They invented the name *template attacks* and showed how to apply this concept to an implementation of RC4. They also discussed how their technique is linked to standard signal processing techniques. In particular, they pointed out that the template approach can be seen as optimal in the sense that the probability for a wrong decision is minimized.

Rechberger and Oswald [RO04] discussed the practical aspects of template attacks. They pointed out that a simple, yet efficient way to find the interesting points is to mount a DPA attack on the device. They also provided some insights into the way the number of interesting points influences the probability of success of the attack.

Agrawal *et al.* [ARRS05] showed that template attacks can also be used to defeat masking schemes. We discuss this approach in Section 10.

The original idea of collision attacks came from Wiemers [Wie01], but Schramm *et al.* [SWP03] were the first to discuss collision attacks in the open scientific literature. They pointed out that the ability to identify a collision in the power trace allows breaking a DES implementation with a very small number of traces. In their attack, they need to do some profiling on beforehand to determine the parts in which the colliding intermediate value potentially occurs. Later on in the attack, they have to identify the collision from a single trace. Schramm *et al.* [SLFP04] showed that this principle can be applied to AES. In this article, they used reduced templates to identify collisions. Ledig *et al.* [LMV04] extended the work by Schramm *et al.* by using so-called almost collisions, and by providing a general framework for collision attacks on ciphers with a Feistel structure.

SPA on Asymmetric Cryptosystems.   Kocher *et al.* [KJJ99] pointed out that an implementation of a modular exponentiation, which uses the square-and-multiply algorithm, is susceptible to SPA attacks if square and multiply operations can be distinguished in the power trace. Modular exponentiation is used in many public-key encryption and signature schemes. For example, RSA decryption and RSA signature generation both require the execution of this modular exponentiation with the private key. Hence, any straightforward implementation of RSA is susceptible to SPA attacks.

Even if RSA decryption is implemented efficiently by using the Chinese Remainder Theorem (CRT), SPA attacks are possible. Trace-pair analysis can be naturally applied to many of such implementations because often the implementations first check whether the input message is larger than the modulus p (or q) or not. If the message is larger, a reduction is performed. This conditional reduction can be observed in the power trace and it depends on the message. With trace-pair analysis, one can find this systematic dependence by choosing messages of different size and looking whether they lead to a reduction or not. In this way, one can successively choose messages that come closer to the unknown modulus. Essentially, in each step of the analysis, one reveals one more bit of the modulus, which eventually allows recovering the modulus.

Novak [Nov02] showed how to exploit the conditional addition in Garner's algorithm (a popular version of the CRT), in order to determine a (small) interval in which the prime p must be in. As soon as p has been determined, the other RSA prime factor q can be determined easily, and the private key can be computed. The attack of Novak is a chosen-plaintext attack. Fouque *et al.* [FMP03] discussed an extension of Novak's idea in which they need only known messages. However, they assume that the RSA primes p and q are slightly unbalanced, *i.e.* they are of different bit-lengths. Similarly to Novak, they use SPA to determine those inputs which cause a conditional addition in Garner's algorithm. However, they use this information differently. They try to find a number of messages with this property, and then use a lattice reduction technique to determine the RSA primes.

Coron [Cor99] recognized that also elliptic curve cryptography (ECC) involve operations that are susceptible to SPA. For instance, the elliptic curve digital signature algorithm (ECDSA) requires executing a scalar multiplication of an elliptic curve point. This scalar multiplication is typically implemented as a sequence of point doubling and point addition operations. In a straightforward implementation, the sequence of point doubling and point addition operations corresponds to the sequence of zero bits and one bits in the ephemeral key. Örs *et al.* [OOP03] used such an implementation to demonstrate that power analysis attacks on FPGAs are feasible in practice. The knowledge of the ephemeral key allows extracting the private key easily.

For all types of attacks on public-key cryptosystems, one fact is of particular importance. In most public-key cryptosystems, knowing only a small fraction of the ephemeral key or the private key typically allows determining the entire key with low computational effort. Consequently, SPA attacks are a particular threat even if they only reveal a small fraction of the key. The rest of the bits can be found by other cryptanalytic techniques. A good overview of such techniques to break RSA can be found in May's PhD thesis [May03]. There are several articles that deal with techniques to break (EC)DSA, see [NS02], [NS03], [HGS01], and [RS01].

Philosophy: Building Templates.   An important practical aspect for template attacks is the building of the templates. Recall that we have hypothesized in the general description for template attacks that the attacker has full control over a device that is similar to the device under attack. With such a similar device, it is easy to build templates. However, one can also imagine other scenarios for building templates.

One possibility would be that the attacker breaks a device with another (non-template) power analysis attack. Then, using the derived key, the attacker can characterize the device and use this information to attack other devices of the same type.

Another possibility is that the attacker has a similar device but not the program code for the algorithm that runs on the targeted device. In this case, the attacker can build templates for different instructions (move, exclusive-or, *etc.*) that are likely to occur in the implementation on the targeted device. Then, it is a matter of mapping the templates to the appropriate parts of the power traces of the targeted device. In case of a software implementation with a nice power consumption trace, as shown in Example 5.2, this is not so difficult.

Yet another possibility is that the attacker characterizes the device under attack based on known plaintexts (or ciphertexts). Typically, the plaintext is moved at the beginning of the first encryption round into the accumulator. Hence, it is possible to characterize the move instruction of the device based on the plaintext. The move instruction is typically used for many operations of block ciphers. Therefore, it is possible to use the templates that are built based on the known plaintexts to attack other parts of the cipher.

Chapter 6

DIFFERENTIAL POWER ANALYSIS

Differential power analysis (DPA) attacks are the most popular type of power analysis attacks. This is due to the fact that DPA attacks do not require detailed knowledge about the attacked device. Furthermore, they can reveal the secret key of a device even if the recorded power traces are extremely noisy.

In contrast to SPA attacks, DPA attacks require a large number of power traces. It is therefore usually necessary to physically possess a cryptographic device for some time in order to mount a DPA attack on it. Consider for example an owner of an electronic purse. This person can record a large number of power traces by transferring small amounts of money to and from the purse. These traces could then be used to reveal the cryptographic key that is used by the purse.

In this chapter, we provide a comprehensive introduction to DPA attacks. We discuss and compare different kinds of DPA attacks and we also illustrate them based on several examples. For this purpose, we use the software and the hardware implementation of AES that are described in Appendix B. We also elaborate on issues like the simulation of DPA attacks and the calculation of the number of traces that are needed to perform successful DPA attacks.

6.1 General Description

The goal of DPA attacks is to reveal secret keys of cryptographic devices based on a large number of power traces that have been recorded while the devices encrypt or decrypt different data blocks. The main advantage of DPA attacks compared to SPA attacks is that no detailed knowledge about the cryptographic device is necessary. In fact, it is usually sufficient to know the cryptographic algorithm that is executed by the device.

Another important difference between the two kinds of attacks is that the recorded traces are analyzed in a different way. In SPA attacks, the power consumption of a device is mainly analyzed along the time axis. The attacker tries to find patterns or tries to match templates in a single trace. In case of DPA attacks, the shape of the traces along the time axis is not so important. DPA attacks analyze how the power consumption at fixed moments of time depends on the processed data. Hence, DPA attacks focus exclusively on the data dependency of the power traces.

> DPA attacks exploit the data dependency of the power consumption of cryptographic devices. They use a large number of power traces to analyze the power consumption at a fixed moment of time as a function of the processed data.

We now discuss in detail how such an analysis reveals secret keys of cryptographic devices. In contrast to SPA attacks, there exists a general attack strategy that is used by all DPA attacks. This strategy consists of five steps.

Step 1: Choosing an Intermediate Result of the Executed Algorithm. The first step of a DPA attack is to choose an intermediate result of the cryptographic algorithm that is executed by the attacked device. This intermediate result needs to be a function $f(d, k)$, where d is a known non-constant data value and k is a small part of the key. Intermediate results that fulfill this condition can be used to reveal k. In most attack scenarios, d is either the plaintext or the ciphertext.

Step 2: Measuring the Power Consumption. The second step of a DPA attack is to measure the power consumption of the cryptographic device while it encrypts or decrypts D different data blocks. For each of these encryption or decryption runs, the attacker needs to know the corresponding data value d that is involved in the calculation of the intermediate result chosen in step 1. We write these known data values as vector $\mathbf{d} = (d_1, \ldots, d_D)'$, where d_i denotes the data value in the ith encryption or decryption run.

During each of these runs the attacker records a power trace. We refer to the power trace that corresponds to data block d_i as $\mathbf{t}_i' = (t_{i,1}, \ldots, t_{i,T})$, where T denotes the length of the trace. The attacker measures a trace for each of the D data blocks, and hence, the traces can be written as matrix $\mathbf{T}$ of size $D \times T$. It is important for DPA attacks that the measured traces are correctly aligned. This means that the power consumption values of each column $\mathbf{t}_j$ of the matrix $\mathbf{T}$ need to be caused by the same operation. In order to obtain aligned power traces, the trigger signal for the oscilloscope needs to be generated in such a way that the oscilloscope records the power consumption of exactly the same sequence of operations during each encryption or decryption run. In case such a trigger signal is not available, the power traces need to be aligned using the techniques described in Section 8.2.2.

Step 3: Calculating Hypothetical Intermediate Values. The next step of the attack is to calculate a *hypothetical intermediate value* for every possible choice of k. We write these possible choices as vector $\mathbf{k} = (k_1, \ldots, k_K)$, where K denotes the total number of possible choices for k. In the context of DPA attacks, we usually refer to the elements of this vector as key hypotheses. Given the data vector $\mathbf{d}$ and the key hypotheses $\mathbf{k}$, an attacker can easily calculate hypothetical intermediate values $f(d, k)$ for all D encryption runs and for all K key hypotheses. This calculation (6.1) results in a matrix $\mathbf{V}$ of size $D \times K$. The first part of Figure 6.1 illustrates this calculation step.

$$v_{i,j} = f(d_i, k_j) \quad i = 1, \ldots, D \quad j = 1, \ldots, K \tag{6.1}$$

Column j of $\mathbf{V}$ contains the intermediate results that have been calculated based on the key hypothesis k_j. It is clear that one column of $\mathbf{V}$ contains those intermediate values that have been calculated in the device during the D encryption or decryption runs. Remember, $\mathbf{k}$ contains all possible choices for k. Hence, the value that is used in the device is an element of $\mathbf{k}$. We refer to the index of this element as ck. Hence, k_{ck} refers to the key of the device. The goal of DPA attacks is to find out which column of $\mathbf{V}$ has been processed during the D encryption or decryption runs. As soon as we know which column of $\mathbf{V}$ has been processed in the attacked device, we immediately also know k_{ck}.

Step 4: Mapping Intermediate Values to Power Consumption Values. The next step of a DPA attack is to map the hypothetical intermediate values $\mathbf{V}$ to a matrix $\mathbf{H}$ of *hypothetical power consumption values*, see Figure 6.1. For this purpose, the attacker uses the simulation techniques we have discussed in Section 3.3. Using one of these techniques, the power consumption of the device for each hypothetical intermediate value $v_{i,j}$ is simulated in order to obtain a hypothetical power consumption value $h_{i,j}$.

The quality of the simulation strongly depends on the knowledge of the attacker about the analyzed device. The better the simulation of the attacker matches the actual power consumption characteristics of the device, the more effective is the DPA attack. The most commonly used power models to map $\mathbf{V}$ to $\mathbf{H}$ are the Hamming-distance and the Hamming-weight model. However, as pointed out in Section 3.3, there are also many other ways to map data values to power consumption values.

Step 5: Comparing the Hypothetical Power Consumption Values with the Power Traces. After having mapped $\mathbf{V}$ to $\mathbf{H}$, the final step of a DPA attack can be performed. In this step, each column $\mathbf{h}_i$ of the matrix $\mathbf{H}$ is compared with each column $\mathbf{t}_j$ of the matrix $\mathbf{T}$. This means that the attacker compares the hypothetical power consumption values of each key hypothesis with the recorded traces at every position. The result of this comparison is a matrix $\mathbf{R}$

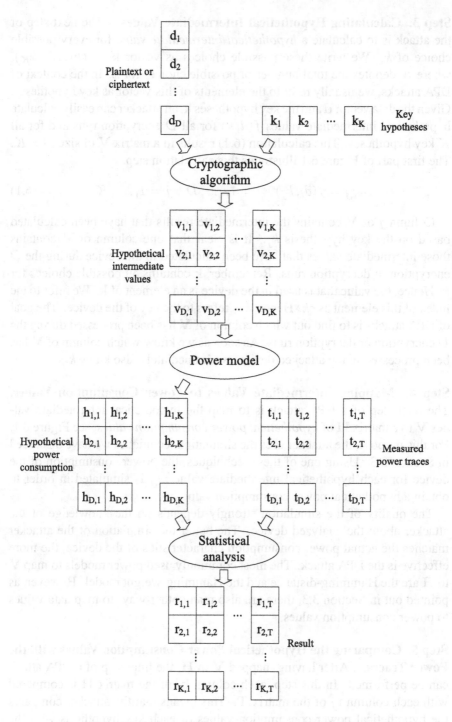

Figure 6.1.   Block diagram illustrating the steps 3 to 5 of a DPA attack.

of size $K \times T$, where each element $r_{i,j}$ contains the result of the comparison between the columns $\mathbf{h}_i$ and $\mathbf{t}_j$. The comparison is done based on algorithms we discuss later in this chapter. All algorithms have the property that the value $r_{i,j}$ is the higher, the better the columns $\mathbf{h}_i$ and $\mathbf{t}_j$ match. The key of the attacked device can hence be revealed based on the following observation.

The power traces correspond to the power consumption of the device while it executes a cryptographic algorithm using different data inputs. The intermediate result that has been chosen in step 1 is a part of this algorithm. Hence, the device needs to calculate the intermediate values $\mathbf{v}_{ck}$ during the different executions of the algorithm. Consequently, also the recorded traces depend on these intermediate values at some position. We refer to this position of the power traces as ct, *i.e.* the column $\mathbf{t}_{ct}$ contains the power consumption values that depend on the intermediate values $\mathbf{v}_{ck}$.

The hypothetical power consumption values $\mathbf{h}_{ck}$ have been simulated by the attacker based on the values $\mathbf{v}_{ck}$. Therefore, the columns $\mathbf{h}_{ck}$ and $\mathbf{t}_{ct}$ are strongly related. In fact, these two columns lead to the highest value in $\mathbf{R}$, *i.e.* the highest value of the matrix $\mathbf{R}$ is the value $r_{ck,ct}$. All other values of $\mathbf{R}$ are low because the other columns of $\mathbf{H}$ and $\mathbf{T}$ are not strongly related. An attacker can hence reveal the index for the correct key ck and the moment of time ct by simply looking for the highest value in the matrix $\mathbf{R}$. The indices of this value are then the result of the DPA attack.

The indices of the highest values of the matrix $\mathbf{R}$ reveal the positions at which the chosen intermediate result has been processed and the key that is used by the device.

It is important to point out though that it can also happen in practice that all values of $\mathbf{R}$ are approximately the same. In this case, the attacker has usually not measured enough power traces to estimate the relationship between the columns of $\mathbf{H}$ and $\mathbf{T}$. The more traces an attacker measures, the more elements are in the columns of $\mathbf{H}$ and $\mathbf{T}$, and the more precisely the attacker can determine the relationship between the columns. This also implies that the more measurements are made, the smaller relationships between the columns can be determined.

6.2 Attacks Based on the Correlation Coefficient

The correlation coefficient is the most common way to determine linear relationships between data. Therefore, it is also an excellent choice when it comes to performing DPA attacks. There exists a well-established theory for the correlation coefficient that can be used to model statistical properties of DPA attacks. Furthermore, this theory also makes comparisons of different attacks quite easy.

We have already discussed the basics of the correlation coefficient in Chapter 4. Section 4.4.1 has provided a definition of the correlation coefficient (4.14) as well as a formula to estimate its value (4.15). The sampling distribution of the estimator has then been discussed in Section 4.6.5.

In DPA attacks, the correlation coefficient is used to determine the linear relationship between the columns $\mathbf{h}_i$ and $\mathbf{t}_j$ for $i = 1, \ldots, K$ and $j = 1, \ldots, T$. This results in a matrix $\mathbf{R}$ of estimated correlation coefficients. We estimate each value $r_{i,j}$ based on the D elements of the columns $\mathbf{h}_i$ and $\mathbf{t}_j$. Using the notation of the previous section, we can therefore rewrite (4.15) as (6.2). In (6.2), the values $\bar{h}_i$ and $\bar{t}_j$ denote the mean values of the columns $\mathbf{h}_i$ and $\mathbf{t}_j$.

$$
r_{i,j} = \frac{\sum_{d=1}^{D}(h_{d,i} - \bar{h}_i) \cdot (t_{d,j} - \bar{t}_j)}{\sqrt{\sum_{d=1}^{D}(h_{d,i} - \bar{h}_i)^2 \cdot \sum_{d=1}^{D}(t_{d,j} - \bar{t}_j)^2}} \tag{6.2}
$$

In order to illustrate how DPA attacks based on the correlation coefficient work, we now discuss several examples of such attacks. We first show attacks on a software implementation of AES, and subsequently we look at a hardware implementation.

6.2.1   Examples for Software

The target of the first DPA attack is the AES software implementation that is described in Appendix B.2. We have executed this implementation on our microcontroller. The power consumption of the microcontroller has been measured using the setup presented in Section 3.4.4. In this setup, the microcontroller receives plaintexts via an RS-232 interface, encrypts them, and returns the corresponding ciphertexts. As an attacker of the device, we therefore have access to the plaintext and the ciphertext. Consequently, we are quite flexible in step 1 of the DPA attack, where we need to choose an intermediate result of AES. We can choose any intermediate result that is a function of the plaintext or the ciphertext and a few key bits.

In the concrete attack, we have decided to choose the output byte of the first AES S-box in round one. This intermediate result is a function of the first byte of plaintext and the first byte of the secret key. After having chosen this intermediate result, we have recorded the power consumption of the microcontroller during the first round of AES while it has encrypted 1 000 different plaintexts. This second step of the DPA attack has led to a matrix $\mathbf{T}$ of power consumption values. The third step has then been to calculate hypothetical intermediate values based on the 1 000 known plaintexts. This means that we have calculated the values $v_{i,j} = S(d_i \oplus k_j)$, where $d_1, \ldots, d_{1\,000}$ are the first byte of each of the 1 000 plaintexts and $k_j = j - 1$ with $j = 1, \ldots, 256$. The matrix $\mathbf{V}$ has hence a size of $1\,000 \times 256$ values.

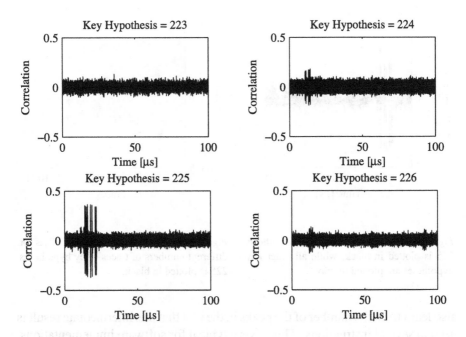

Figure 6.2. The rows of the matrix **R** that correspond to the key hypotheses 223, 224, 225, and 226.

The fourth step of the DPA attack has been to map **V** to a matrix **H** of hypothetical power consumption values. In the current attack, we have decided to use a very simple power model for this mapping. We have just considered the LSB of the values in **V**. Hence, we have used $h_{i,j} = LSB(v_{i,j})$ as our power model. Based on **H**, we have then performed the last step of the DPA attack. We have calculated the correlation coefficients between all columns of **H** and all columns of the recorded power consumption values **T**. The result of this calculation is a matrix **R** of correlation coefficients.

In practice, there exist several different ways to visualize this matrix **R**. One of these ways is to show each row of the matrix in a separate plot. In this case, each plot corresponds to one key hypothesis. Figure 6.2, for example, shows the plots for the key hypotheses 223 to 226 of the current attack. It can be observed that there are very high peaks in the plot for key hypothesis 225. In fact, these peaks are the highest ones of the entire matrix **R**. All other values of **R** are significantly smaller. The fact that these high peaks occur for key hypothesis 225 and the fact that the first peak occurs at $13.8\,\mu s$ provides a lot of information to an attacker.

First of all, the peaks indicate that the first byte of the secret key of the microcontroller is 225. Second, the microcontroller computes the output of the first AES S-box at position $13.8\,\mu s$ of the recorded traces. Furthermore, we can

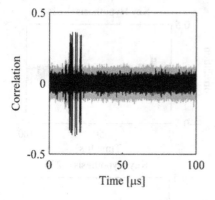

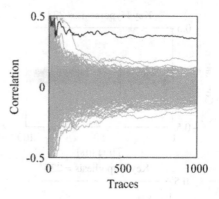

Figure 6.3. All rows of **R**. Key hypothesis 225 is plotted in black, while all other key hypotheses are plotted in gray.

Figure 6.4. The column of **R** at 13.8 μs for different numbers of traces. Key hypotheses 225 is plotted in black.

also learn from the number of the peaks in the plot that this intermediate result is used in several instructions. This is very typical for software implementations. After the attacked intermediate result has been calculated, it is usually moved from a register to memory and then later on loaded back into a register as operand for subsequent operations of the algorithm. Each time the microcontroller performs an operation that involves the attacked intermediate result, there occurs at least one peak in **R**. This is why there are so many peaks in the plot for key 225.

The peaks for key 225 are the most significant ones. However, when looking closely at the other plots of Figure 6.2, we can also observe some smaller peaks in these plots. These peaks occur because not all columns of **H** are independent. This means whenever a column of **H** leads to a high correlation coefficient, also some other columns lead to some correlation. However, this correlation is typically significantly smaller, and hence, the key of the device can be identified easily. In fact, the plots for all keys, except for key 225, look very similar in the current attack. This can also be observed in Figure 6.3. In this figure, the plots for all keys are shown. Key 225 is plotted in black, while all other keys are plotted in gray. There are no significant peaks in gray—only the plot for key 225 contains high peaks.

An important question when performing DPA attacks is how many traces are needed in order to obtain such high peaks in the matrix **R**. Figure 6.4 provides a preliminary answer to this question. This figure shows how the first peak in the plot for key 225 (*i.e.* the peak located at 13.8 μs in Figure 6.3) evolves as a function of the used number of traces. Key 225 is again plotted in black, while the correlation of all other key hypotheses at position 13.8 μs is plotted in gray.

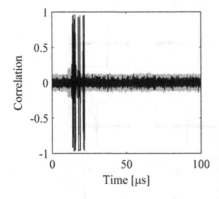
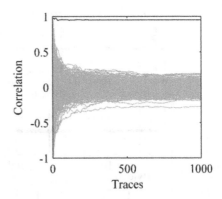

Figure 6.5. All rows of **R**. Key hypothesis 225 is plotted in black, while all other key hypotheses are plotted in gray.

Figure 6.6. The column of **R** at $13.8\,\mu s$ for different numbers of traces. Key hypothesis 225 is plotted in black.

Figure 6.4 nicely illustrates how the estimated correlation coefficients $r_{k,13.8}$ converge towards $\rho_{k,13.8}$ for $k = 1, \ldots, K$. The more traces are used, the better is the estimation of the correlation, see Section 4.6.5. The value $r_{225,13.8}$ converges to about 0.35, while all other correlation coefficients converge to values below 0.2. Key 225 leads to the highest correlation coefficient if about 160 or more traces are used. Hence, as a first estimate we can say that about 160 traces are required for a successful attack. A more precise number will be derived later in Section 6.4.

DPA Attack Using the Hamming-Weight Model

In the previous DPA attack on the microcontroller we have used the bit model $h_{i,j} = LSB(v_{i,j})$ to estimate the power consumption. However, from Chapter 4 we already know that the power consumption of the microcontroller is inversely proportional to the Hamming weight of the processed data. Hence, we can use this knowledge to improve the DPA attack on the S-box output. We now show results of a DPA attack that uses the same traces as before. However, in step 4 we have used the Hamming-weight model instead of the bit model, *i.e.* we have set $h_{i,j} = HW(v_{i,j})$.

Figure 6.5 shows the result of this attack. The plot for key 225 again contains the highest peaks. However, in comparison to the result before, the peaks are much higher. In the attack based on the Hamming-weight model, the value $r_{225,13.8}$ converges to about 0.95 instead of 0.35. Therefore, this peak can be detected with a much lower number of traces. Figure 6.6 shows that already after 20 traces, the value $r_{225,13.8}$ is bigger than the other correlation coefficients at $13.8\,\mu s$. Hence, DPA attacks on the microcontroller require significantly less

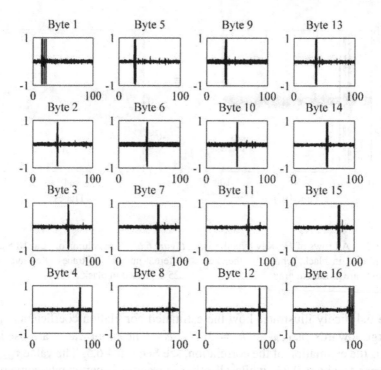

Figure 6.7. Plots for the correct key hypotheses in the DPA attacks on the 16 S-box outputs in the first round. The y-axes show the correlation and the x-axes show the time in μs.

traces, if the Hamming-weight model is used. This is because the Hamming-weight model describes the power consumption of the microcontroller much better than the bit model.

> The better the used power model describes the attacked device, the less traces are needed in DPA attacks.

The DPA attacks that we have presented so far have revealed the first byte of the secret key that is used by the microcontroller. We now also show results of DPA attacks on the remaining key bytes. In order to reveal all these key bytes, we have performed the same attack as described before on all 16 key bytes. This means that we have performed 16 DPA attacks in total. Each of these DPA attacks has targeted a different S-box output in the first round of AES based on the Hamming-weight model. For all attacks, the same power traces have been used.

Figure 6.7 shows the results of the attacks. Each of the 16 plots in this figure corresponds to the plot for the correct key hypothesis in the corresponding attack. Hence, the first plot is the same as in Figure 6.5. Based on the 16 plots in Figure 6.7, we can make some important observations. First, all 16 DPA attacks have been successful and lead to similar correlation values for the correct key hypothesis. This is due to the fact that the same instruction is used for all S-box look-ups. Therefore, also the same leakage occurs for all S-box outputs. The second important observation is that the S-box look-ups are performed sequentially. The software implementation performs one S-box look-up after the other, and hence, also one key byte after the other is used. Based on the traces shown in Figure 6.7 it is easy to see that the microcontroller performs the S-box look-ups in the following sequence: $1, 5, 9, 13, 2, 6, 10, 14, 3, 7, 11, 15, 4, 8, 12, 16$. The microcontroller hence processes one row of the AES state after the other, see Appendix B. Note that in the Appendix the bytes of the state are counted from 0 to 15.

A final interesting observation is that for some key bytes more peaks occur than for others. In particular, there are more peaks in the plots for the bytes 1 and 16. This is due to the fact that the software implementation performs some additional operations with the corresponding S-box outputs. Every peak in the plots has its source and an attacker can learn many things about the device by analyzing these peaks. In fact, it is often possible to perform a kind of reverse engineering of an implementation based on the observed peaks.

6.2.2 Examples for Hardware

The DPA attacks on the software implementation that we have presented in the previous section have been very effective. Using the Hamming-weight model, we have been able to reveal the entire key of the AES implementation with about 20 traces. We now investigate DPA attacks on a more challenging target. The target of the attacks in the current section is the AES ASIC that is described in Appendix B.3. We have performed several DPA attacks on this chip using the measurement setup described in Section 3.4.4.

In this setup, we have access to the plaintexts and the ciphertexts. Hence, we have again been quite flexible when choosing a suitable intermediate result for our DPA attacks. This time, we have decided to choose the inputs of the AES S-boxes in the last round. Each input byte of these S-boxes depends on one byte of the last round key of AES. Hence, 16 DPA attacks are necessary to reveal the entire key. Just like in the previous attacks, we have started with the first byte.

After having selected an intermediate result for our attack, we have recorded the power consumption of our AES ASIC while it has encrypted 100 000 random plaintexts. Based on the 100 000 resulting ciphertexts, we have then calculated the hypothetical intermediate values $v_{i,j} = S^{-1}(d_i \oplus k_j)$. The size of the resulting matrix $\mathbf{V}$ is $100\,000 \times 256$. The next step of the DPA attack has been to map $\mathbf{V}$ to a matrix $\mathbf{H}$ of hypothetical power consumption values. In case of the attack on the hardware implementation, this step is much more critical than

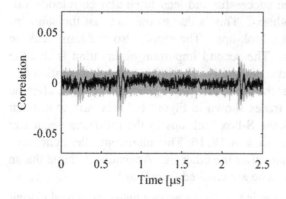

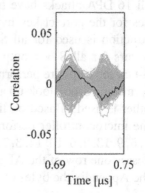

Figure 6.8. Result of the DPA attacks when using bit 3 as   *Figure 6.9.* Zoomed view of
power model.   the result.

in case of the previous attacks. This is why we now discuss the different power
models and the corresponding results in more detail.

Attacks Using the Bit Model or the Hamming-Weight Model

First, we have performed DPA attacks using the bit model. However, in
contrast to before, we have not only looked at the LSB of $v_{i,j}$, but we have
performed separate attacks for each of the eight bits of $v_{i,j}$. Figure 6.8 shows
the result of a DPA attack that has used bit 3 as power model. In this plot of the
matrix $\mathbf{R}$, there are significant peaks between $0.5\,\mu s$ and $1\,\mu s$. However, the
plot for the correct key hypothesis (plotted in black) does not lead to the highest
peak. Several key hypotheses lead to higher peaks as it can also be observed in
the zoomed view of the result in Figure 6.8. We have obtained similar results
for all eight bits we have attacked, except for the case when we have used bit
6 as power model. In case of bit 6, the correct key hypothesis has led to the
highest peak in $\mathbf{R}$. Figures 6.10 and 6.11 show the result of this DPA attack.

We have repeated the same attack for the other 15 bytes. This means that we
have attacked each input bit of each S-box in the last round. The results of these
attacks have all been quite similar. Most of the bits that we have attacked have
not revealed the key. There have only been very few bits, where the correct key
hypothesis has indeed lead to the highest peaks in $\mathbf{R}$. Hence, the results of the
DPA attacks based on the bit model have not been conclusive.

We have therefore switched to the Hamming-weight model. Using the
Hamming-weight model, we have again attacked each of the 16 input bytes
of the S-boxes in the last round. The results of these attacks have been better
than before. However, they have still not been conclusive. In eight of the 16
DPA attacks, the correct key hypothesis has lead to the highest peaks in $\mathbf{R}$,

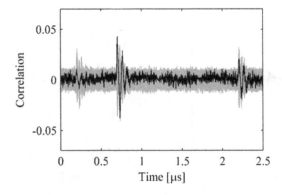

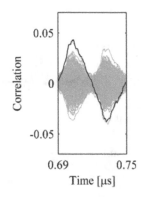

Figure 6.10. Result of the DPA attacks when using bit 6 as power model.

Figure 6.11. Zoomed view of the result.

while this was not the case in the other eight attacks. Hence, only half of the DPA attacks have been successful.

At first sight, the results of the DPA attacks using the bit model and the Hamming-weight model might look surprising. However, the fact that not all attacks have been successful can be explained as follows. As already discussed in Section 3.3, the bit model and the Hamming-weight model typically describe the power consumption of CMOS circuits only very badly. There is a correlation between the bits and the Hamming weights of the processed data and the power consumption. However, in general this correlation is not very big. The microcontroller that we have attacked in the previous section is an exception due to its precharged bus.

In case of the ASIC implementation, the bit model and the Hamming-weight model are not suitable. If these power models are used, the correlations between the columns of the hypotheses matrix **H** are often bigger than the correlations between the columns of **H** and **T**. In this case it can happen that all hypotheses lead to very similar peaks in **R** and that therefore a very high number of traces is necessary to determine the correct key hypothesis. We elaborate on this issue in more detail in Section 6.3. As for now, we switch to more suitable power models for the AES ASIC implementation and repeat our DPA attacks.

Attacks Using the Hamming-Distance Model

In Section 3.3, we have shown that the Hamming-distance model describes the power consumption of CMOS circuits much better than the Hamming-weight model. However, in order to use the HD model, the attacker requires some additional knowledge about the attacked device. Essentially, the attacker needs to know the state of a cell in the circuit before or after it processes the attacked intermediate result, see Section 3.3.1.

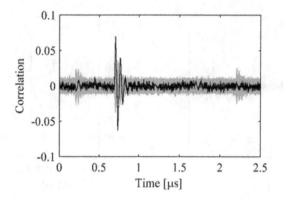 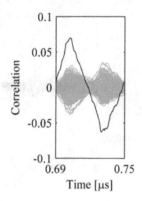

Figure 6.12. Result of the DPA attacks when using the HD model for byte one.

Figure 6.13. Zoomed view of the result.

In case of the AES ASIC implementation, we know from Appendix B.3 that the inputs of the S-boxes in the last round are stored in the same register as the ciphertext. At the beginning of the last round, this register stores the inputs of the S-boxes and at the end of the last round it stores the ciphertext. Based on this knowledge it is easy to build a suitable HD model for the attack. At the beginning of this section, we have already calculated the matrix **V** that contains hypothetical values for the input of the first S-box in the last round. The vector **d** stores the corresponding bytes of the ciphertexts. Hence, hypothetical power consumption values can be calculated as shown in (6.3). Each value $h_{i,j}$ is equal to the Hamming distance between the respective values that are stored in the register in the last round.

$$h_{i,j} = HD(v_{i,j}, d_i) = HW(v_{i,j} \oplus d_i) \qquad (6.3)$$

Figures 6.12 and 6.13 show the result of the DPA attack on the input of the first S-box. The correct key hypothesis leads to a higher correlation than before. Furthermore, this correlation can be easily distinguished from the correlations of the other key hypotheses. The attack therefore reveals the key byte that is used for the first S-box. We have also performed DPA attacks based on the HD model for all other key bytes. All these attacks have been successful. Hence, all 16 bytes of the last round key have been revealed. Figure 6.14 shows the plots for the correct key hypotheses. The height of the peaks for the different bytes differs significantly. Nevertheless, each of the shown peaks is higher than the peaks of the respective alternative hypotheses.

When comparing the results of this figure with the results of the DPA attacks on the software implementation (see Figure 6.7), we can make some important observations. In case of the software implementation, the S-box look-ups have been performed sequentially. This is different in case of the hardware

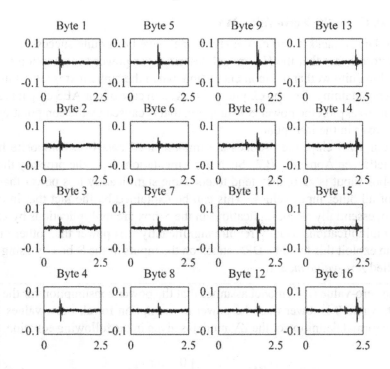

Figure 6.14. Plots for the correct key hypotheses in the DPA attacks on the 16 S-box inputs in the last round. The y-axes show the correlation and the x-axes show the time in μs.

implementation. It can be observed in Figure 6.14 that in each clock cycle four S-box look-ups are performed in parallel. In the first clock cycle, for example, the bytes 1, 2, 3, and 4 are processed. This parallelism strongly reduces the correlation coefficients that occur in the DPA attacks. In case of the microcontroller, we have observed correlations of almost 1. In case of the hardware implementation, the highest peaks are about 0.1.

Another important difference between the results is that in the attacks on the software implementation, the correlations for all 16 key bytes have been approximately the same. In case of the hardware implementation, there are significant differences. These differences occur due to the fact that there are four S-boxes in the AES chip. Each of these S-boxes has a different layout with different parasitics, and hence also the power consumption characteristics are different. Furthermore, in case of our AES ASIC implementation, different activities are happening in parallel during the four clock cycles that perform the S-box look-ups.

Attacks Using the Zero-Value Model

The DPA attacks based on the HD model have been quite successful. This is due to the fact that the HD model describes the power consumption of the AES chip quite well at the moment of time when the ciphertext is stored in the register containing the S-box inputs. However, in case of our AES chip it is also possible to exploit the power consumption that is caused by the computation of the S-boxes in the last round.

In our AES chip, the S-boxes are implemented based on composite field arithmetic, see Appendix B.3. Such implementations have the property that if the S-box input is zero, they tend to consume significantly less power than in case of all other input values. This can be explained by the fact that in case of zero, essentially all multiplications in the S-box are multiplications by zero. Such multiplications usually require significantly less power than other ones. We can exploit this fact in a DPA attack on the inputs of the S-box by using the so-called *zero-value model*.

The zero-value (ZV) model assumes that the power consumption for the data value 0 is lower than the power consumption for all other values. The formal definition of the ZV model is given in the following equation.

$$h_{i,j} = ZV(v_{i,j}) = \begin{cases} 0 & \text{for } v_{i,j} = 0 \\ 1 & \text{for } v_{i,j} \neq 0 \end{cases} \qquad (6.4)$$

Using the ZV model means that for each hypothetical S-box input that is zero, we set the hypothetical power consumption also to zero. In all other cases, we set it to one. We have used the ZV model to perform successful DPA attacks on all 16 input bytes of the S-boxes in the last round, *i.e.* the entire key has been revealed based on the zero-value model. The last two plots of Figure 6.15 show results of such DPA attacks. In order to compare all attacks we have presented in this section, we have also added other results to this figure. Figure 6.15 shows results of all attacks discussed in this section for the first two bytes of the S-boxes.

It can be observed that the bit model and the HW model only lead to small correlation coefficients while the HD and the ZV model lead to much higher values. Furthermore, it can be observed that in most attacks peaks occur in at least two consecutive clock cycles (1 clock cycle takes $0.5\,\mu s$). This can be explained in the following way. In the first clock cycle, the cells that process the attacked intermediate value switch from some prior state to the attacked intermediate value. In the subsequent clock cycle, the intermediate value is "overwritten" by some new value. Both operations, the switching from the prior value to the intermediate value and the switching from the intermediate value to the new value, lead to a power consumption that depends on the intermediate

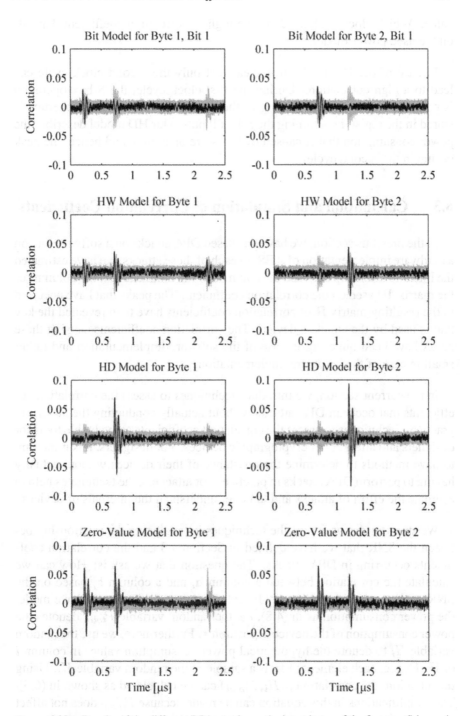

Figure 6.15. Results of the different DPA attacks on the input bytes of the first two S-boxes in the last round.

value. Which clock cycle leads to the highest correlation coefficient depends on the used power model.

In case of the HD model it is clear that only the second clock cycle can lead to a high correlation. During the first clock cycle, the S-box operation is performed and at the beginning of the second clock cycle the ciphertext is stored in the register containing the S-box inputs. Our HD model describes the power consumption that is caused by this store operation, and hence, the peak occurs in the second cycle.

6.3 Calculation and Simulation of Correlation Coefficients

In the previous section, we have discussed DPA attacks on a software and on a hardware implementation of AES. In each of these attacks, we have estimated the linear relationship between the columns of the matrix H and the columns of the matrix T based on the correlation coefficient. The peaks that have occurred in the resulting matrix R of correlation coefficients have then revealed the key that is used by the attacked device. The correlation coefficients causing these peaks have been quite big in case of the software implementation and rather small in case of the hardware implementation.

In the current section, we introduce techniques to assess the correlation co- efficients that occur in DPA attacks without actually conducting the attacks in practice. Techniques to assess the correlation coefficients are valuable tools for designers and attackers of cryptographic devices. For designers, it is important to have methods to determine the resistance of their devices without actually having to perform DPA attacks in practice. For attackers, the techniques help to estimate the effort of attacks and to better understand the internals of devices.

We start our discussion of the techniques by analyzing the relationship be- tween the SNR that we have defined in Section 4.3 and the correlation coef- ficients occurring in DPA attacks. The question that we ask is: How can we calculate the correlation between a column h_i and a column t_j based on the SNR of the traces at position j? In order to answer this question, we model the power consumption as in (4.8), *i.e.* the random variable P_{total} denotes the power consumption of the device at position j. Furthermore, we use the random variable H_i to denote the hypothetical power consumption values in column i of H. Hence, each element of h_i is a sample of the random variable H_i. Using this notation, the correlation $\rho(H_i, P_{total})$ can be calculated as shown in (6.5). The simplifications in this equation can be made because P_{const} does not affect the correlation coefficient and because the noise $P_{noise} = P_{sw.\,noise} + P_{el.\,noise}$ is statistically independent of P_{exp}.

$$\rho(H_i, P_{total}) =$$
$$= \rho(H_i, P_{exp} + P_{sw.\,noise} + P_{el.\,noise} + P_{const})$$
$$= \rho(H_i, P_{exp} + P_{sw.\,noise} + P_{el.\,noise})$$
$$= \rho(H_i, P_{exp} + P_{noise})$$
$$= \frac{E(H_i \cdot (P_{exp} + P_{noise})) - E(H_i) \cdot E(P_{exp} + P_{noise})}{\sqrt{Var(H_i) \cdot (Var(P_{exp}) + Var(P_{noise}))}}$$
$$= \frac{E(H_i \cdot P_{exp} + H_i \cdot P_{noise}) - E(H_i) \cdot (E(P_{exp}) + E(P_{noise}))}{\sqrt{Var(H_i) \cdot Var(P_{exp})}\sqrt{1 + \frac{Var(P_{noise})}{Var(P_{exp})}}}$$
$$= \frac{\rho(H_i, P_{exp})}{\sqrt{1 + \frac{1}{SNR}}} \tag{6.5}$$

The correlation $\rho(H_i, P_{total})$ can be calculated based on $\rho(H_i, P_{exp})$ and the SNR.
$$\rho(H_i, P_{total}) = \frac{\rho(H_i, P_{exp})}{\sqrt{1 + \frac{1}{SNR}}}$$

The result derived in (6.5) is quite nice. As discussed in Section 4.3, the SNR describes the leakage of a device in a given attack scenario. The SNR is independent of the power model that is used by the attacker to exploit the leakage. The fact how the leakage is exploited in a particular attack, is described by $\rho(H_i, P_{exp})$. This is the correlation between the hypothetical power consumption values of the attacker and the exploitable power consumption of the device. This correlation essentially depends on the fact how well the power model of the attacker describes the power consumption that is caused by the processing of the attacked intermediate result.

In many scenarios, the correlation $\rho(H_i, P_{exp})$ can be determined quite easily by performing simulated DPA attacks. The basic idea of this approach is the following. First, a vector **d** is generated that contains all data inputs for the attacked intermediate result that are possible. The exploitable power consumption of the device is then simulated based on the data inputs **d**, the key that is used by the device, and a suitable power model. The result of the simulation is a matrix **S**, where each row of **S** contains a simulated trace of the exploitable power consumption. The matrix **S** is then used to perform a DPA attack. This means that hypothetical power consumption values for all keys are created based on the power model of the attacker. Subsequently, the correlation between each column of **H** and each column of **S** is calculated to obtain a matrix **R** of correlation coefficients.

The nice thing about a matrix $\mathbf{R}$ that has been generated in this way is that each element $r_{i,j}$ of this matrix is equal to $\rho_{i,j}$. This is due to the fact that when a correlation coefficient r is calculated based on all data inputs, then $r = \rho$. The result of the simulated DPA attack hence corresponds to the correlations $\rho(H_i, P_{exp})$ that also occur in an actual attack.

It is important to realize that the person who performs a simulated DPA attack switches between the role of a designer and the role of an attacker. For the first step, *i.e.* the generation of the simulated power traces $\mathbf{S}$, the key is assumed to be known. This is the part of the designer. For the DPA attack that is performed subsequently, the key is assumed to be unknown. This is the part of the attacker. Simulated DPA attacks are very useful to quickly determine $\rho(H_i, P_{exp})$ for different scenarios. Since the whole attack is a simulation, the power consumption characteristics of the device as well as the power model of the attacker can be changed easily. The effects of the changes can then be analyzed immediately. We now use simulated DPA attacks and the relation shown in (6.5) to assess the correlation coefficients that have occurred in the attacks presented in the previous section.

6.3.1   Examples for Software

In Section 6.2.1 we have discussed DPA attacks on our microcontroller. In these attacks, we have targeted the outputs of the S-boxes in the first round. These attacks have been performed using the HW model and the bit model.

Hamming-Weight Model

We first analyze the DPA attack where we have used the HW model. In this DPA attack, the exploitable power consumption P_{exp} corresponds to the data-dependent power consumption P_{data} that is caused by the processing of the output of the S-box. In order to determine the correlation coefficients $\rho(H_i, P_{exp})$ that occur in this attack, we have performed a simulated DPA attack. This means we have generated a vector $\mathbf{d}$ that contains all input values for the attacked S-box, *i.e.* $\mathbf{d} = (0, \ldots, 255)'$. Subsequently, we have mapped these input values to simulated power consumption values according to (6.6).

$$s_{i,ct} = HW(S(d_i \oplus k_{ck})) \tag{6.6}$$

We can use the HW model to generate simulated traces because this model describes the data-dependent power consumption of the microcontroller quite accurately. In particular, it holds that $\rho(H_i, P_{exp}) = \rho(H_i, P_{data}) = \rho(H_i, \mathbf{s}_{ct})$. In the current example, we have only simulated the power consumption at the moment of time when the S-box output is processed, *i.e.* at position ct. The matrix $\mathbf{S}$ hence consists of the column $\mathbf{s}_{ct}$ only. The power consumption at position ct is the power consumption that leaks the largest amount of the

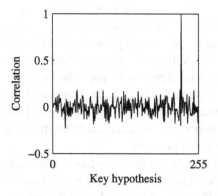

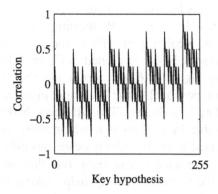

Figure 6.16. The correlation coefficients $\rho(H_i, P_{exp})$ occurring in a DPA attack using the HW model to attack an S-box output in the first round.

Figure 6.17. The correlation coefficients $\rho(H_i, P_{exp})$ occurring in a DPA attack using the HW model to attack an S-box input in the first round.

exploitable information. Therefore, this is the most relevant position of the power traces.

After having simulated s_{ct}, we have performed a DPA attack on this column. This means we have created hypothetical power consumption values for all keys based on the HW model. Subsequently, we have determined the correlation between each column of the matrix $\mathbf{H}$ and s_{ct}. This simulated DPA attack has lead to a matrix $\mathbf{R}$ of size 256×1. Figure 6.16 shows the result of this attack.

The 256 values that are plotted in this figure correspond to the correlations $\rho(\mathbf{h}_i, \mathbf{t}_{ct})$ for $i = 1, \ldots, 256$ that occur in DPA attacks using the HW model to attack the S-box output in the first round. In order to map these correlations to the correlations $\rho(\mathbf{h}_i, \mathbf{t}_{ct})$ that we have observed in Section 6.2.1, we additionally need to consider the noise that has occurred in the attacks in practice.

From Section 4.3.2, we know that the SNR in DPA attacks on the microcontroller is 22.89 if the HW model is used. Using this SNR in (6.5), we learn that the simulated correlations are scaled down by a factor of 0.98 in practice due to the electronic noise, *i.e.* $\rho(\mathbf{h}_i, \mathbf{t}_{ct}) = \rho(\mathbf{h}_i, \mathbf{t}_{ct}) \cdot 0.98$.

As it can be observed in Figure 6.16, the correlation for the correct key hypothesis ($k_{ck} = 225$) is 1 in the simulation and hence 0.98 in practice. All other correlation coefficients are smaller than 0.2. This confirms the observations we have already made for Figure 6.6. The estimated correlation coefficients in this figure converge to the correlation coefficients $\rho(\mathbf{h}_i, \mathbf{t}_{ct})$ shown in Figure 6.16.

We now take a brief look at a small variant of this attack in order to illustrate that the correlation coefficients $\rho(\mathbf{h}_i, \mathbf{t}_{ct})$ can also look completely different from the ones in Figure 6.16. In this variant, we perform a DPA attack on the S-box input instead of the S-box output. Such an attack can be simulated easily by setting $\mathbf{d}$ again to $(0, \ldots, 255)'$ and by using the power model $s_{i,ct} = HW(d_i \oplus k_{ck})$. Performing a DPA attack on these simulated traces based on calculating the HW of the S-box input for all key hypotheses leads to the matrix $\mathbf{R}$ that is plotted in Figure 6.17. While there is only one distinct peak in Figure 6.16, the correlation coefficients in Figure 6.17 are now inversely proportional to the Hamming distance between the key of the device and the key hypotheses. The difference between the correct key hypothesis and the other hypotheses is much smaller, and hence, an attacker needs to measure more traces in order to detect this difference in practice.

An attacker can learn from this small simulation that non-linear elements, like S-boxes, actually help making DPA attacks more effective. A one-bit difference at an S-box input leads to a difference of several bits at the output. Consequently, even if a key hypothesis is only wrong in one bit, the output of the S-box is different in several bits. When attacking the output of the S-box, the correlation for all wrong key hypotheses is therefore significantly smaller than the correlation for the correct one. The most effective attacks on AES can hence be mounted on intermediate results that occur after the S-boxes in the first round and before the S-boxes in the last round.

Bit Model

At the beginning of Section 6.2.1 we have presented a DPA attack that uses the bit model to attack the output of an S-box in the first round. We now also determine the correlation coefficients that occur in this attack by simulation and by using (6.5).

At first sight one might be tempted to assume that in this attack, P_{exp} corresponds to the power consumption that is caused by the LSB of the S-box output only. However, the eight bits of the S-box output are not independent from each other. Hence, the seven bits besides the LSB cannot be modeled as switching noise, but need to be modeled as part of P_{exp}. Only parts of the power consumption that are independent of the attacked intermediate result can be modeled as switching noise, see Section 4.3.1.

The correlation coefficients $\rho(H_i, P_{exp})$ that occur in the current attack can be determined in a very similar way as before. First, $\mathbf{d}$ is set to $(0, \ldots, 255)'$. Based on $\mathbf{d}$, the key k_{ck}, and the Hamming-weight model, the same simulated traces $\mathbf{S}$ are calculated as in case of the attack using the HW model. However, in contrast to this attack, we now use the HW model only to calculate $\mathbf{S}$. For the calculation of $\mathbf{H}$, we use the bit model. By calculating the correlation between the 256 columns of $\mathbf{H}$ and s_{ct}, we obtain the result shown in Figure 6.18.

The correlation coefficient for the correct key hypothesis ($k_{ck} = 225$) is again higher than all other correlation coefficients. However, it is now 0.35 instead of 1. Considering the electronic noise means again multiplying $\rho(H_i, P_{exp})$ with 0.98. When comparing Figure 6.18 with Figure 6.4, we can observe that the

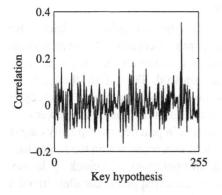
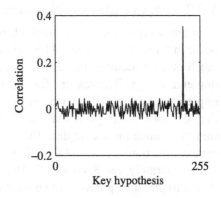

Figure 6.18. The correlation coefficients $\rho(H_i, P_{exp})$ occurring in a DPA attack using the bit model to attack the S-box output in the first round.

Figure 6.19. The correlation coefficients $\rho(H_i, P_{exp})$ when modeling the seven output bits of the S-box besides the LSB as switching noise.

simulated correlation coefficients match the correlation coefficients we have observed in practice again very well. The estimated correlation coefficients in Figure 6.4 converge to the simulated correlation coefficients $\rho(H_i, P_{exp}) \cdot 0.98$.

Simulation and reality match so nicely because we have accurately modeled the power consumption of the microcontroller and the S-box in our simulation. If we would have set P_{exp} to the LSB of the S-box output, we would have obtained the result that is shown in Figure 6.19. In this result, the correlation coefficients for the wrong hypotheses are significantly smaller than in Figure 6.18 and do not match reality.

We have generated Figure 6.19 by setting $\mathbf{d} = (0, \dots, 255)'$ and by then calculating the column $\mathbf{s}_{ct}$ using the bit model instead of the HW model. We have attacked $\mathbf{s}_{ct}$ using the same matrix $\mathbf{H}$ as before and have obtained the result $\mathbf{R}$. The seven output bits that switch besides the LSB have been modeled as switching noise, and hence, the SNR has been set to $1/7$. We have scaled down the matrix $\mathbf{R}$ by filling in this SNR into (6.5). Subsequently, we have scaled it further down by 0.98 in order to consider the electronic noise. This has lead to the result shown in Figure 6.19.

The difference between Figures 6.18 and 6.19 illustrates the importance of modeling P_{exp} correctly when simulating DPA attacks. It is important to always keep in mind that P_{exp} is not only caused by the parts of the circuit that store the attacked intermediate result directly, but also by the parts that store data that is correlated to the attacked intermediate result. This fact is even more important when attacking hardware instead of software implementations.

6.3.2   Examples for Hardware

In this section, we discuss methods to assess the correlation coefficients that occur in DPA attacks on our AES hardware implementation. These attacks are much more difficult to model and to analyze than the attacks on the software implementation. The reasons for this are manifold. First of all, the hardware implementation of AES performs many more operations in each clock cycle than the software implementation. This makes simulations of the power consumption much more complex. Remember that in practice we typically cannot measure the power consumption of operations that occur within the same clock cycle separately, see Section 3.5. Hence, all operations of a clock cycle contribute to the same points of the trace and appear to happen in parallel. In order to accurately simulate the power consumption of one clock cycle, all these operations need to be considered.

But it is not only the parallelism that makes the assessment of correlation coefficients more difficult in case of a hardware implementation. It is also important to consider the fact that the capacitive loads of the wires between the logic cells of a hardware implementation usually differ significantly. Therefore, each wire contributes differently to the overall power consumption. Furthermore, the switching activity of the wires is not independent. The data values that are processed within a clock cycle are partly correlated. In case of our AES hardware implementation, the AddRoundKey, the SubBytes, and the MixColumns operation are all executed within one clock cycle. Additionally, the result of these operations is stored in a register. The power consumption of the cells calculating the output of the SubBytes operation is strongly correlated to the one of the cells performing the MixColumns operation. Moreover, the power consumption caused by the output of the MixColumns operation is correlated to the power consumption of the cells storing this result.

On top of all these issues, it is also important to consider the effect of glitches. As pointed out in Section 3.1.3, the switching activity of combinational cells strongly depends on the timing properties of the input signals. Each cell of a CMOS circuit potentially switches its output several times in each clock cycle.

In summary, we face the following situation: We can only measure the total power consumption of the operations that occur within one clock cycle. The operations in this clock cycle process data that is partly correlated, and each data signal contributes differently to the power consumption due to the differences in the capacitive loads of the wires. Furthermore, the switching activity of the cell outputs depends on the timing properties of the corresponding input signals. The power simulation techniques for attackers that we have discussed in Section 3.3 are hence not suitable to simulate the power consumption of an entire hardware implementation of a cryptographic algorithm. Therefore, attackers can only make very rough estimates about the correlation coefficients that occur in DPA attacks on such implementations.

A precise calculation of the correlation coefficients is only possible for designers of cryptographic devices who have access to the netlist and the layout of the chip. We now discuss how attackers of hardware implementations can roughly assess the correlation coefficients that occur in DPA attacks. Subsequently, we use simulation techniques of designers to analyze the DPA attacks on our AES ASIC that we have presented in 6.2.2.

Estimation Technique for Attackers

According to (6.5), the highest correlation coefficient that occurs in a DPA attack is a function of $\rho(H_{ck}, P_{exp})$ and the SNR. Both factors are usually quite difficult to determine for an attacker. However, when making additional assumptions, (6.5) can be simplified to an equation that is more suitable for an attacker. The first assumption is that the correlation between the correct key hypothesis and P_{exp} is 1, *i.e.* $\rho(H_{ck}, P_{exp}) = 1$. The second assumption is that all wires of the circuit have the same capacitive load, and the third assumption is that the switching activity of all wires is independently and identically distributed.

These assumptions are of course very radical. However, an attacker usually cannot do much better due to the lack of knowledge about the device. Based on the three assumptions, (6.5) can be simplified as shown in (6.7). In this equation, a denotes the number of wires processing a bit of the attacked intermediate result, and n is the number of wires processing statistically independent bits. Notice that $SNR = a/n$ holds because all wires are independently and identically distributed. For example, if a device stores 32 independently and uniformly distributed bits in a register, the correlation coefficient for an attack using a perfect power model for the LSB is $\sqrt{1/32} = 0.18$. In this case $a = 1$ and $n = 31$.

$$\rho(H_{ck}, P_{total}) = \frac{\rho(H_{ck}, P_{exp})}{\sqrt{1 + \frac{1}{SNR}}} = \sqrt{\frac{\frac{a}{n}}{\frac{a}{n} + 1}} = \sqrt{\frac{a}{a + n}} \qquad (6.7)$$

Equation (6.7) is a nice tool to get a first estimation of an upper bound for the correlation coefficients occurring in an attack. For example, it is clear that we have to expect correlation coefficients lower than 0.18 when attacking a single bit of an intermediate result in a 32-bit AES architecture. In fact, it is likely that the correlation coefficients will actually be significantly lower than 0.18 because in a 32-bit architecture usually several operations are performed in the same clock cycle. This means that many more than 32 bits toggle. Equation (6.7) provides a very rough estimation of $\rho(H_{ck}, P_{total})$ for an attacker who has no detailed knowledge about the attacked device. If more information is available, also one of the following techniques can be applied.

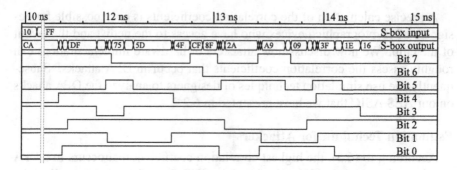

Figure 6.20. Simulation of the transitions occurring at the output of an AES S-box when the input changes from 10_{hex} to FF_{hex}.

Simulation Techniques for Designers

Designers of cryptographic devices can simulate the power consumption of the device much better than attackers, see Section 3.2. Hence, they can also determine the correlation coefficients that occur in DPA attacks in a much more precise way. However, it is important to point out that an accurate simulation of the power consumption of an entire chip is very time consuming. Hence, usually designers of cryptographic devices use certain assumptions to speed-up the simulations.

We now discuss a simulation based on a back-annotated netlist of an S-box that is a part of our AES ASIC implementation. We model the power consumption of the S-box by simply counting the number of transitions that occur in this circuit, see Section 3.2.2. The simulation of this small part of the circuit provides some insights about the attack results we have observed in Section 6.2.2.

In order to characterize the switching activities of our implementation of the S-box, we have performed a simulation for every input transition. The S-box takes an 8-bit input, and hence, there are 2^{16} different input transitions. Figure 6.20 shows the transitions of the S-box output when the input switches from 10_{hex} to FF_{hex}. In this figure, it can be observed that most of the output bits switch not only once, but several times: there occur many glitches on the output bits and also on the internal wires of the S-box.

After having counted the transitions that occur for the 2^{16} input transitions, we have calculated the average number of transitions that occur for the 256 input values of the S-box. The average number of transitions for the input 0 is the average of the number of transitions that are caused by the input transitions $0 \rightarrow 0, 1 \rightarrow 0, \ldots, 255 \rightarrow 0$. The average number of transitions for all 256 input values is plotted in Figure 6.21. It can be observed that if the S-box input is 0, the least transitions occur on average. This explains why the DPA attack based on the zero-value model has been working so well in Section 6.2.2. Obviously, the cells of the S-box switch significantly less for input zero.

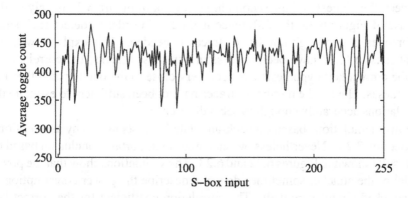

Figure 6.21. Average number of transitions occurring in the S-box for the 256 input values.

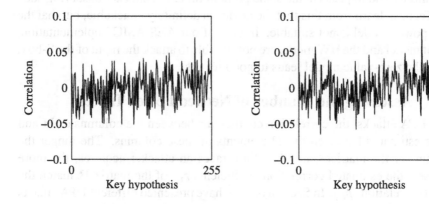

Figure 6.22. Correlation coefficients that occur when using the HW model to perform a DPA attack on the simulated toggle counts of the S-box.

Figure 6.23. Correlation coefficients that occur when using the bit model to perform a DPA attack on the simulated toggle counts of the S-box.

The simulated transition counts can also be used to determine correlation coefficients that occur in DPA attacks. We have seen in Section 6.2.2 that attacks on the input of an S-box based on the bit model or the HW model have not been successful. We now use the 2^{16} transition counts as matrix $\mathbf{S}$ for simulated DPA attacks in order to explain the results of Section 6.2.2. We have performed simulated DPA attacks based on the HW model and the bit model. Notice, that the 2^{16} input transitions are all transitions that can occur at an S-box input. Hence, for each element of the result $\mathbf{R}$ it holds that $r_{i,j} = \rho_{i,j}$. The results of the attacks are shown in Figures 6.22 and 6.23.

In both figures the correct key hypothesis leads to the highest peak. However, also other key hypotheses lead to peaks with similar heights. The difference between the highest and the second highest peak in Figure 6.22 is only 0.018. In case of Figure 6.23, the difference is 0.025. In order to detect such small differences, a very high number of traces is needed. This explains why not all attacks in 6.2.2 have been successful. In these attacks, the correlation ρ for the correct key hypothesis has been very close to the one for the other key hypotheses. Hence, the number of traces has not been sufficient to estimate the correlations accurately enough to see a difference.

Power simulations based on back-annotated netlists use many assumptions, see Section 3.2.2. Nevertheless, we can draw an important conclusion based on the results shown in Figures 6.22 and 6.23. The simulations show that the power model of the attacker sometimes does not describe the power consumption of the attacked circuit very well. The correlation coefficient for the correct key hypothesis does not lead to a significant peak in these cases. Instead, many key hypotheses lead to peaks at the same point in time. An attacker observing such an effect can learn from this that the device is definitely attackable, but that the used power model is not suitable. In case of our AES ASIC implementation, the bit model and the HW model are not suitable to attack the input of the S-box, while the zero-value model leads to good results.

6.4   Assessing the Number of Needed Power Traces

In DPA attacks, the correlation coefficients between the columns of $\mathbf{H}$ and $\mathbf{T}$ are estimated based on the D elements of these columns. The longer the columns of the matrices, *i.e.* the more traces an attacker acquires, the more precisely the estimated correlation coefficients $r_{i,j}$ of the matrix $\mathbf{R}$ match the actual correlations $\rho_{i,j}$. In Section 6.2, we have presented different DPA attacks and we have seen how the values $r_{i,j}$ approach $\rho_{i,j}$. In case of the software implementation, we have also been able to determine the correlations $\rho_{i,j}$ without performing the attack in practice, see Section 6.3.

However, the most important question in the context of DPA attacks is actually not how well an attacker can estimate $\rho_{i,j}$. The most important question is: how many traces does an attacker need to acquire in order to determine the key that is used by the device? In case of the correlation coefficient this means: how many traces are needed in order to find out which column of $\mathbf{H}$ has the strongest correlation to a column of $\mathbf{T}$? Although this question looks pretty simple at first sight, giving a precise answer is not straightforward. In fact, we can only provide a rule of thumb that is based on the following two assumptions.

The first assumption is that a DPA attack is successful if there is a significant peak in row ck of $\mathbf{R}$. The row $\mathbf{r}'_{ck} = (r_{ck,1}, \ldots, r_{ck,T})$ is the row that is calculated based on the correct key hypothesis. In practice, it holds for most attacks that k_{ck} is successfully revealed, if a peak occurs in the row $\mathbf{r}'_{ck}$. An exception to this are DPA attacks that are performed based on power models that do not fit the power consumption of the attacked device, see Section 6.2.2.

In this case, there occur peaks in many rows of $\mathbf{R}$ and the correct key usually cannot be identified. We do not consider this type of attack for our rule of thumb because for most devices suitable power models exist.

Our second assumption is that the number of traces that are needed to see a peak in the row $\mathbf{r}'_{ck}$ exclusively depends on $\rho_{ck,ct}$. Statistically speaking, we can model the calculation of $\mathbf{r}'_{ck}$ as follows. Each estimated correlation coefficient is distributed according to the sampling distribution discussed in Section 4.6.5. This sampling distribution is defined by the actual correlation ρ and the number of power traces n. In most attacks, the recorded traces are quite long compared to the interval during which the attacked intermediate result is processed. Usually many operations are executed during the recording that are completely independent of the attacked intermediate result. Therefore, most of the correlations $\rho_{ck,1}, \ldots, \rho_{ck,T}$ are usually zero in practice. The question of whether there is a peak in $\mathbf{r}'_{ck}$ or not can hence be viewed as the question of whether or not it is possible to determine that the correlation coefficient $\rho_{ck,ct}$ is different from zero or not.

6.4.1 Rule of Thumb

Our rule of thumb is that the number of needed traces for a DPA attack corresponds to the number of traces that is necessary to distinguish the estimated correlation coefficient for $\rho_{ck,ct}$ from an estimated correlation coefficient for $\rho = 0$ with a confidence of $\alpha = 0.0001$. In Section 4.6.6, we have discussed the number of traces that are needed to distinguish estimators for two correlation coefficients ρ_1 and ρ_0. Therefore, (4.44) can be directly used to assess the number of traces that are needed in a DPA attack. As shown in Table 4.3, the value of the quantile is $z_{0.9999} = 3.719$.

> As a rule of thumb, the number of traces n that are needed to mount a successful DPA attack can be calculated as follows:
>
> $$n = 3 + 8 \frac{z_{1-\alpha}^2}{\ln^2 \frac{1+\rho_{ck,ct}}{1-\rho_{ck,ct}}} \qquad (6.8)$$

In order to illustrate the relationship between $\rho_{ck,ct}$ and the calculated number of traces according to (6.8), Table 6.1 shows some calculated values. When looking at the numbers for $\rho_{ck,ct} \leq 0.2$ it can be observed that there is essentially a quadratic relationship between $\rho_{ck,ct}$ and n. Reducing $\rho_{ck,ct}$ by a factor of 10 means that 100 times more traces are needed. In fact, it is also possible to simplify (6.8) for the case that $\rho_{ck,ct} \leq 0.2$. For these low correlation coefficients, it holds that $n \approx 28/\rho_{ck,ct}^2$. In practice, many DPA attacks lead to correlation coefficients below 0.2. This is why we now summarize the relationships between the SNR, $\rho_{ck,ct}$, and n for these attacks. The relation-

Table 6.1. The calculated number of traces n for different values of $\rho_{ck,ct}$.

$\rho_{ck,ct}$	$n_{\alpha=0.0001}$	$\rho_{ck,ct}$	$n_{\alpha=0.0001}$	$\rho_{ck,ct}$	$n_{\alpha=0.0001}$
0.900	16	0.090	3 400	0.009	341 493
0.800	26	0.080	4 307	0.008	432 206
0.700	40	0.070	5 630	0.007	564 519
0.600	61	0.060	7 668	0.006	768 378
0.500	95	0.050	11 049	0.005	1 106 471
0.400	157	0.040	17 273	0.004	1 728 870
0.300	292	0.030	30 720	0.003	3 073 559
0.200	676	0.020	69 140	0.002	6 915 526
0.100	2 751	0.010	276 606	0.001	27 662 152

ships shown in (6.9) can be derived from (4.10), (6.5), and (6.8). Note that the approximation $\rho_{ck,ct}^2 \sim SNR$ is only valid for small SNRs.

For $|\rho_{ck,ct} \leq 0.2|$ and for small SNRs, the following relations hold:

$$n \approx \frac{28}{\rho_{ck,ct}^2} \sim \frac{1}{SNR} = \frac{Var(P_{sw.\,noise} + P_{el.\,noise})}{Var(P_{exp})} \tag{6.9}$$

- Halving $\rho_{ck,ct}$ means that four times more traces are needed.

- Doubling the amount of noise reduces $\rho_{ck,ct}$ by $\sqrt{2}$, and hence doubles the number of needed traces.

- Halving the capacitance of the wires processing the attacked intermediate results halves the amplitude of P_{exp}, quarters the SNR, and hence quadruples the number of traces.

6.4.2 Examples

We now assess the number of needed traces in two of the DPA attacks we have presented in Section 6.2. First, we look at the DPA attack on the AES software implementation that targets the output of an S-box in the first round based on the bit model. As it can be observed in Figure 6.3, and as we have calculated in Section 6.3.1, $\rho_{ck,ct}$ is $0.35 \cdot 0.98 = 0.34$ in this attack. Hence, according to (6.8), the number of needed traces is about 224.

The second attack we look at is the DPA attack using the HD model to attack an S-box input in the last round of our AES ASIC implementation. We have performed this attack in Section 6.2.2 based on 100 000 power traces. The maximum correlation coefficient that we have observed in the result shown in Figure 6.12 is 0.065. This value is only an estimator for $\rho_{ck,ct}$. However,

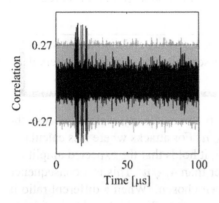 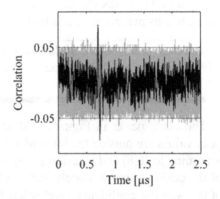

Figure 6.24. Result of the DPA attack on the microcontroller when using 224 traces.

Figure 6.25. Result of the DPA attack on the AES ASIC implementation when using 6 532 traces.

when using 100 000 traces to estimate a correlation coefficient of about 0.065, the estimation is already very close to the actual correlation, see Section 4.6.5. Therefore, we can use $\rho_{ck,ct} = 0.065$ to do the calculation of the needed traces. This calculation leads to 6 532 traces.

After having done these calculations, we have performed the two DPA attacks again. The first one has been based on 224 traces and the second one has been based on 6 532 traces. The results of these attacks are shown in Figures 6.24 and 6.25. In both results, the correct key hypothesis leads to clearly visible peaks.

An interesting observation when looking at these figures is that the points plotted in gray form a kind of "band". The gray points are the correlation coefficients for the 255 incorrect key hypotheses. At positions before and after the black peaks, these points essentially correspond to estimators for $\rho = 0$. It is therefore quite easy to explain why all these points are located within a well defined interval. From (4.36) and (4.37) we know that the estimators for $\rho = 0$ are normally distributed with $\mu = 0$ and $\sigma = 1/\sqrt{n-3} \approx 1/\sqrt{n}$. Furthermore, we know from Section 4.2.1 that samples from a normal distribution are located within $\pm 4 \cdot \sigma$ with a probability of 99.99%. This is the explanation why almost all gray points are located in the interval $\pm 4/\sqrt{n}$.

In case of Figure 6.24, this interval is $\pm 4/\sqrt{224} = 0.27$. In case of the second attack it is $\pm 4/\sqrt{6\,532} = 0.05$. As it can be observed in Figures 6.24 and 6.25, the calculated bounds describe the plotted result quite well. The calculated bounds are of course also valid for the black points before and after the peaks. At these locations ρ is also zero. It is important to mention that the

$\pm 4/\sqrt{n}$ bound holds for all DPA attacks. This bound can be observed in all attack results presented in this book.

> The estimators for all correlation coefficients before and after the attacked intermediate result is processed are essentially located in the interval $\pm 4/\sqrt{n}$.

The black peaks in Figures 6.24 and 6.25 are clearly visible because the peak values are outside the interval $\pm 4/\sqrt{n}$. For attacks where n is calculated according to (6.8) and where $\rho_{ck,ct} \leq 0.2$, it holds that the expected amplitude of the peaks is approximately 30% higher than $4/\sqrt{n}$. This is a consequence of the way the confidence level α has been chosen. When a different ratio is desired, a different value for α has to be chosen. The higher the confidence level, the more significant is the peak. Choosing $\alpha = 0.0001$ is a proposal we have found to be practical, if the goal is to calculate a number of needed traces that leads to a successful attack with high confidence.

6.5   Alternatives to the Correlation Coefficient

In the first article on DPA attacks [KJJ99], the so-called difference-of-means method instead of the correlation coefficient has been used to compare the columns of $\mathbf{H}$ and $\mathbf{T}$. This method is another way to determine the relationship between columns of two matrices. In this section, we briefly discuss this method and we also describe some other alternatives that have been proposed to determine the relationship between $\mathbf{H}$ and $\mathbf{T}$.

DPA attacks that are based on these alternative methods all work in a very similar way as the attacks we have already described in the previous sections. In fact, the steps 1, 2, and 3 of the attacks (see Section 6.1) are done exactly in the same way as before. The first difference occurs in step 4 when the matrix $\mathbf{V}$ is mapped to $\mathbf{H}$. In case of the correlation coefficient there are no special constraints for this mapping, while in case of the alternatives, only binary power models are possible. This means that the power model needs to be chosen in such a way that $\mathbf{h}_{i,j} \in \{0,1\} \,\forall i,j$. Hence, it is for example not possible to use the Hamming-weight model directly. This model needs to be reduced to a binary model before it can be used in DPA attacks based on the alternative methods. This can, for example, be done by setting $h_{i,j} = 1$, if $HW(v_{i,j}) \geq 4$ and by setting $h_{i,j} = 0$, if $HW(v_{i,j}) < 4$. However, it is clear that such a binary model describes the power consumption of the attacked device not as accurately as a non-binary model. This is also the main reason why DPA attacks based on the alternative methods are usually not as effective as DPA attacks based on the correlation coefficient.

Besides the different requirements for the power models, an attack based on alternative methods of course also differs in step 5. The calculation of the matrix $\mathbf{R}$ is not done based on the correlation coefficient. It is done based on some other statistical method. In case of all alternative methods, the matrix $\mathbf{R}$ has exactly the same size as in case of the correlation coefficient. However, the values of the matrix $\mathbf{R}$ differ and they also have a different meaning as in case of the correlation coefficient.

6.5.1   Difference of Means

The difference-of-means method has already been used in the example presented in Section 1.3. The basic idea of this method is to determine the relationship between the columns of $\mathbf{H}$ and $\mathbf{T}$ based on the following observation. An attacker creating a binary matrix $\mathbf{H}$ makes the assumption that the power consumption for certain intermediate values is different from the power consumption for all other values. The sequence of zeros and ones in each column of $\mathbf{H}$ is a function of the input data $\mathbf{d}$ and a key hypothesis k_i. In order to check whether a key hypothesis k_i is correct or not, an attacker can split the matrix $\mathbf{T}$ into two sets of rows, *i.e.* two sets of power traces, according to $\mathbf{h}_i$. The first set contains those rows of $\mathbf{T}$ whose indices correspond to the indices of the zeros in the vector $\mathbf{h}_i$. The second set contains all remaining rows of $\mathbf{T}$. Subsequently, the mean of the rows can be calculated. The vector $\mathbf{m}'_{0i}$ denotes the mean of the rows in the first set and $\mathbf{m}'_{1i}$ denotes the mean of the rows in the second set. The key hypothesis k_i is correct, if there occurs a significant difference between $\mathbf{m}'_{0i}$ and $\mathbf{m}'_{1i}$ at some point in time.

The difference between $\mathbf{m}'_{0i}$ and $\mathbf{m}'_{1i}$ indicates that there is a correlation between $\mathbf{h}_{ck}$ and some columns of $\mathbf{T}$. Just like in the previous attacks, this difference occurs exactly at the moments of time when the intermediate values that correspond to $\mathbf{h}_{ck}$ are processed. At all other moments of time the difference between the vectors is essentially zero. In case the key hypothesis is not correct, the difference between $\mathbf{m}'_{0i}$ and $\mathbf{m}'_{1i}$ is more or less zero at all moments of time. The result of a DPA attack based on the difference-of-means method is a matrix $\mathbf{R}$, where each row of $\mathbf{R}$ corresponds to the difference between the mean vectors $\mathbf{m}'_{0i}$ and $\mathbf{m}'_{1i}$ of one key hypothesis. Equations to calculate $\mathbf{R}$ according to the difference-of-means method are given in (6.10) to (6.14). In these equations, n denotes the number of rows of $\mathbf{H}$, *i.e.* the number of power traces that are used for the attack.

$$m_{1i,j} = \frac{1}{n_{1i}} \cdot \sum_{l=1}^{n} h_{l,i} \cdot t_{l,j} \tag{6.10}$$

$$m_{0i,j} = \frac{1}{n_{0i}} \cdot \sum_{l=1}^{n} (1 - h_{l,i}) \cdot t_{l,j} \tag{6.11}$$

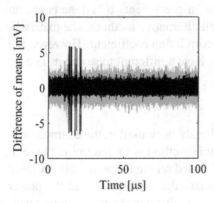
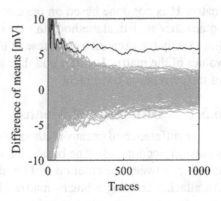

Figure 6.26. All rows of **R**. Key hypothesis 225 is plotted in black, while all other key hypotheses are plotted in gray.

Figure 6.27. The column of **R** at 13.8 μs for different numbers of traces. Key hypothesis 225 is plotted in black.

$$n_{1i} = \sum_{l=1}^{n} h_{l,i} \qquad (6.12)$$

$$n_{0i} = \sum_{l=1}^{n} (1 - h_{l,i}) \qquad (6.13)$$

$$\mathbf{R} = \mathbf{M_1} - \mathbf{M_0} \qquad (6.14)$$

In order to provide an example of an attack that is based on this method, we have again performed a DPA attack on the S-box output of the AES implementation running on our microcontroller. In fact, we have performed a similar attack as we have already presented in Chapter 1. However, instead of the MSB, we have chosen to attack the LSB this time, *i.e.* $h_{i,j} = LSB(v_{i,j})$. This model is binary. Hence, it can be used for the difference-of-means method without adaptation. The result of the attack has been calculated based on 1 000 traces, see Figures 6.26 and 6.27. It can be observed that the correct key hypothesis $k_{ck} = 225$ can again be easily identified.

However, note that the difference-of-means method only considers differences of values and not the corresponding variances. As it can be seen in (6.2), the correlation coefficient considers differences as well as variances. Therefore, DPA attacks based on the difference-of-means method require more traces than DPA attacks based on the correlation coefficient. The difference between the two methods can be observed by comparing Figure 6.27 and Figure 6.4.

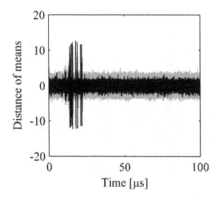 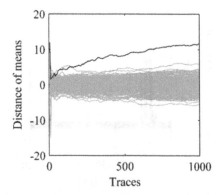

Figure 6.28. All rows of **R**. Key hypothesis 225 is plotted in black, while all other key hypotheses are plotted in gray.

Figure 6.29. The column of **R** at 13.8 μs for different numbers of traces. Key hypothesis 225 is plotted in black.

6.5.2 Distance of Means

The distance-of-means method is an improvement of the difference-of-means method because this method also takes the variances into account. The distance-of-means method is based a commonly used hypothesis test to determine whether two distributions have an equal mean or not. We have already discussed this test in Section 4.6.4.

In DPA attacks based on the distance-of-means method, the matrix **T** is split into two sets of rows for each key hypothesis exactly as in the previous section. The difference between the previous method and this method is that now the means of the two sets are compared according to the distance-of-means test and not just based on subtracting the means. The elements of the matrix **R** are hence calculated as shown in (6.15), where $s_{i,j}$ is the standard deviation of the difference distribution of the two sets according to (4.32).

$$r_{i,j} = \frac{m_{1i,j} - m_{0i,j}}{s_{i,j}} \qquad (6.15)$$

Based on this formula, we have again attacked the LSB of the first S-box output of our AES implementation on the microcontroller. The result of this attack is shown in Figures 6.28 and 6.29. The attack requires about the same number of traces as the attack based on the correlation coefficient. The two methods essentially lead to the same results if a binary power model is used. However, as already pointed out before, the drawback of the distance-of-means method is that it cannot be used for non-binary power models.

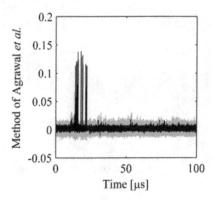

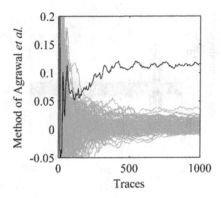

Figure 6.30. All rows of **R**. Key hypothesis 225 is plotted in black, while all other key hypotheses are plotted in gray.

Figure 6.31. The column of **R** at 13.8 μs for different numbers of traces. Key hypothesis 225 is plotted in black.

6.5.3   Generalized Maximum-Likelihood Testing

Another method to perform DPA attacks has been proposed by Agrawal *et al.* in [ARR03]. This approach is based on the statistical concept of generalized maximum-likelihood testing, see for example [Kay98]. In the approach of Agrawal *et al.*, the matrix **H** is extended by an additional column $\mathbf{h}_{K+1}$. This column contains a random sequence of zeros and ones and is referred to as the null hypothesis. In step 5 of the DPA attack, the initial K key hypotheses are compared to the null hypothesis $\mathbf{h}_{K+1}$. Equation (6.16) shows the method that is used for the comparison. In order to calculate $r_{i,j}$, the formulas given in (6.10) to (6.13) need to be used. Furthermore, $s_{0i,j}$ and $s_{1i,j}$ denote the standard deviations of the points of the rows that have been used to calculate the corresponding means $m_{0i,j}$ and $m_{1i,j}$. For a detailed description of (6.16), refer to [ARR03].

$$
\begin{aligned}
r_{i,j} \;=\; & \frac{\left((m_{1i,j} - m_{0i,j}) - (m_{1K+1,j} - m_{0K+1,j})\right)^2}{\dfrac{s_{0K+1,j}}{n_{0K+1}} + \dfrac{s_{1K+1,j}}{n_{1K+1}}} \\[2mm]
& -ln\left(\frac{\dfrac{s_{0i,j}}{n_{0i}} + \dfrac{s_{1i,j}}{n_{1i}}}{\dfrac{s_{0K+1,j}}{n_{0K+1}} + \dfrac{s_{1K+1,j}}{n_{1K+1}}}\right)
\end{aligned}
\tag{6.16}
$$

In order to provide an example of an attack based on this metric, we have again attacked the LSB of the first S-box output of our AES implementation. Figures 6.30 and 6.31 show the corresponding result. Obviously, the method of Agrawal *et al.* reveals the key k_{ck} easily. Figure 6.31 also indicates that a similar number of traces is needed as in case of the distance-of-means method and as in case of the correlation coefficient. However, the method of Agrawal *et al.* is not deterministic due to the randomly generated null hypothesis h_{K+1}. Furthermore, it only allows binary power models.

6.6 Template-Based DPA Attacks

In the previous sections, we have assumed that the attacker has only limited knowledge about the power consumption characteristics of the attacked device. We have used very simple power models, like the bit model, the HW model, the HD model, or the ZV model to map **V** to **H**. However, it is clear that the better the power model fits the actual power consumption, the more effective an attack becomes.

In the current section, we assume that the attacker has the chance to characterize the power consumption of the device based on templates. This leads to so-called template-based DPA attacks which constitute the strongest kind of DPA attacks. Template-based DPA attacks have first been introduced by Agrawal *et al.* in [ARR03]. They are essentially an extension of the template-based SPA attacks that have already been discussed in Section 5.3.

Templates are the optimal way to describe the power consumption characteristics of a device. Hence, template-based DPA attacks are also optimal DPA attacks in the sense that they minimize the probability of error when determining the key of the device. It is important to point out though, that this optimality only holds for the multivariate-Gaussian model (see Section 4.4.2) and only for a given selection of points of interest. An optimal strategy for selecting the points of interest in general is not available.

6.6.1 General Description

We start the description of template-based DPA attacks by first only considering one trace, *i.e.* one row t_i' of **T**. In this case, the attacker is essentially interested in answering the following question: Given the trace t_i', what is the probability that the key of the device equals k_j where $j = 1, \ldots, K$? This conditional probability $p(k_j | t_i')$ can be calculated using *Bayes' theorem*, see for example [Kay98]. Bayes' theorem allows us to calculate the probability $p(k_j | t_i')$ based on the prior probability $p(k_l)$ and the probability $p(t_i' | k_l)$ for $l = 1, \ldots, K$. This is shown in (6.17).

$$p(k_j | t_i') = \frac{p(t_i' | k_j) \cdot p(k_j)}{\sum_{l=1}^{K} (p(t_i' | k_l) \cdot p(k_l))} \tag{6.17}$$

The prior probabilities are the probabilities for the different keys without considering the trace t_i'. Bayes' theorem can hence essentially be viewed as an update function for probabilities. The input of the function are the prior probabilities $p(k_l)$ that do not consider t_i'. The output are the posteriori probabilities $p(k_j|\mathbf{t}'_i)$ that consider $\mathbf{t}'_i$. Note that sums of the priori probabilities and the posteriori probabilities are always 1, *i.e.* $\sum_{l=1}^{K} p(k_l) = \sum_{l=1}^{K} p(k_l|\mathbf{t}'_i) = 1$.

Given just one trace $\mathbf{t}'_i$, the best guess for the key that is used by the device is the key k_j that leads to the highest probability $p(k_j|\mathbf{t}'_i)$. Guessing the key of the device based on this strategy is called maximum-likelihood approach. This approach is optimal in the sense that it minimizes the probability of error, see Section 5.3.1. However, although the approach is optimal, attacks based on a single trace do not always succeed. Often there is just not enough information available in a single trace to reveal the key of the device. This is why we now extend this approach to multiple traces. We are hence interested in determining the probability $p(k_j|\mathbf{T})$. This is the probability that, given a matrix $\mathbf{T}$ of power consumption values, the device uses the key k_j. We denote the number of rows of $\mathbf{T}$ by D.

The extension from $p(k_j|\mathbf{t}'_i)$ to $p(k_j|\mathbf{T})$ is not very difficult. Since the power traces are statistically independent, we can multiply the probabilities that correspond to different traces and fill the product into (6.17), see (6.18). An alternative way to derive this formula is to apply Bayes' rule iteratively. This means $p(k_j|\mathbf{T}) = p(k_j|\mathbf{t}'_D)$, if the prior probability $p(k_j)$ is set to $p(k_j|\mathbf{t}'_{i-1})$ when calculating $p(k_j|\mathbf{t}'_i)$.

$$p(k_j|\mathbf{T}) = \frac{\left(\prod_{i=1}^{D} p(\mathbf{t}'_i|k_j)\right) \cdot p(k_j)}{\sum_{l=1}^{K} \left(\left(\prod_{i=1}^{D} p(\mathbf{t}'_i|k_l)\right) \cdot p(k_l)\right)} \qquad (6.18)$$

Equation (6.18) is the basis for template-based DPA attacks. It leads to the probabilities $p(k_j|\mathbf{T})$ for $j = 1, \ldots, K$ that can be used to guess the key of the device based on the maximum-likelihood approach. The important question we have not answered so far is how an attacker can actually determine all the probabilities $p(k_j)$ and $p(\mathbf{t}'_i|k_j)$ that are needed to calculate $p(k_j|\mathbf{T})$. Finding the prior probabilities $p(k_j)$ is usually straightforward. Typically, all keys are equally likely and hence $p(k_j) = 1/K$. The probabilities $p(\mathbf{t}'_i|k_j)$ are more difficult to determine. However, this is exactly the place where templates come into play. In the following, we discuss different methods to determine $p(\mathbf{t}'_i|k_j)$ based on templates.

Using Templates to Determine $p(\mathbf{t}'_i|k_j)$

In order to determine the probabilities $p(\mathbf{t}'_i|k_j)$, we can essentially use the same methods we have already used for template-based SPA attacks, see Section 5.3.2. This means, we can use templates for pairs of data and key, or we can use templates for intermediate values. We focus on the latter approach first.

When using templates for intermediate values, the attacker first needs to choose an intermediate result using the same criteria as in the previously described DPA attacks, see Section 6.1. Subsequently, the power consumption of this intermediate result needs to be characterized. This means that it is necessary to build a template h_v for all values v of the intermediate result.

After having built the templates for the attacked intermediate result, the steps 2 and 3 of a standard DPA attack are performed, *i.e.* the attacker measures the traces $\mathbf{T}$ and calculates $\mathbf{V}$. Based on $\mathbf{T}$ and $\mathbf{V}$, the probabilities $p(\mathbf{t}'_i|k_j)$ are then calculated according to (6.19) for $i = 1, \ldots, D$ and $j = 1, \ldots, K$. Subsequently, these probabilities are used to calculate $p(k_j|\mathbf{T})$ according to (6.18) for all $j = 1, \ldots, K$. Based on the maximum-likelihood principle, the key k_j that leads to the highest probability $p(k_j|\mathbf{T})$, is then used as guess for the key that is used by the device.

$$p(\mathbf{t}'_i|k_j) = p(\mathbf{t}'_i; h_{v_{i,j}}) \tag{6.19}$$

In case templates are built for pairs of data and key, template-based DPA attacks work almost exactly the same way. First, a template $h_{d,k}$ needs to be built for each combination of the data inputs d_i and the keys k_j, where $i = 1, \ldots, D$ and $j = 1, \ldots, K$. During the attack, the probability $p(\mathbf{t}'_i|k_j)$ is then set to $p(\mathbf{t}'_i; h_{d_i,k_j})$. Notice that in this case it is not necessary to calculate a matrix $\mathbf{V}$ of hypothetical intermediate values. The data inputs $\mathbf{d}$ and the key hypotheses $\mathbf{k}$ are directly used as parameters of the template.

6.6.2 Examples for Software

In order to illustrate the effectiveness of template-based DPA attacks we have again attacked the AES software implementation on our microcontroller. We show two examples of template-based DPA attacks. In both attacks, we use templates for the output of an S-box in the first round in order to determine $p(\mathbf{t}'_i|k_j)$.

The first attack considers all eight output bits of the S-box. Since we already know that the microcontroller leaks the Hamming weight of the data it processes, we have built templates for the nine Hamming weights $0, \ldots, 8$ of the S-box output, see Section 5.3.2. We refer to these templates as $h_{0,\ldots,8}$ and we set $p(\mathbf{t}'_i|k_j) = p(\mathbf{t}'_i; h_{HW(v_{i,j})})$. After having built templates based on these considerations, we have performed a DPA attack according to (6.18). The result of this attack is shown in Figure 6.32.

The figure shows how the probabilities $p(k_j|\mathbf{T})$, where $j = 1, \ldots, 256$, evolve as a function of the number of traces. The correct key hypothesis leads to a probability of almost 1 already after seven traces. This is a very low number compared to the previous attacks. However, the plot shown in Figure 6.32 strongly depends on the input values that have been used for the power traces. There are input values for which the correct key is even determined with less traces and there are input values that need up to ten traces. Nevertheless, on average the attack is still stronger than the previously discussed attacks.

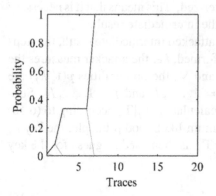
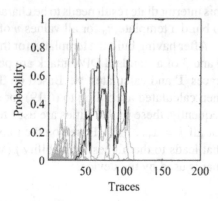

Figure 6.32. The probabilities $p(k_j|\mathbf{T})$ as a function of the number of traces when using templates based on the HW. $p(k_{ck}|\mathbf{T})$ is plotted in black.

Figure 6.33. The probabilities $p(k_j|\mathbf{T})$ as a function of the number of traces when using templates based on the LSB. $p(k_{ck}|\mathbf{T})$ is plotted in black.

For the second example, we have only used the LSB of the S-box output to build templates. Clearly, this is something an attacker would not do in practice, if there is also the chance to build templates based on byte values. However, this example illustrates how results of template-based DPA attacks look like if there is more noise in the traces. For this attack, we have built the two templates h_0 and h_1 and we have hence set $p(\mathbf{t}_i'|k_j) = p(\mathbf{t}_i'; h_{LSB(v_{i,j})})$. The result of this attack is shown in Figure 6.33. About 110 traces are necessary until the correct key hypothesis leads to the highest probability. On average less than 150 power traces have been required.

6.7   Notes and Further Reading

Considerations for DPA Attacks.   DPA attacks consist of several steps. In each of these steps several considerations have to be taken into account. In Section 6.1 we have listed the most important considerations. We now go through the steps again and discuss some more aspects.

Step 1: In step 1 of a DPA attack, the attacker chooses an intermediate result of the algorithm. This intermediate result must be a function of some known, non-constant data and a small part of the key.

In practice, attackers typically know either the plaintext or the ciphertext. Sometimes they even know both. However, in most symmetric algorithms knowing the plaintext (or the ciphertext) only allows targeting intermediate results that occur at the beginning of the first or the last encryption round. If the attacker is able to choose the plaintext or the ciphertext, then more intermediate results or even more rounds can be targeted. This depends on the structure of the encryption algorithm. An example for a chosen plaintext attack that allows targeting intermediate results after the MixColumns operation in AES was discussed in [Jaf06b].

The situation is similar in asymmetric algorithms. Given a plaintext or a ciphertext, the attacker can target intermediate results that occur either in the first or the last step of the cryptographic operation.

Step 2: In step 2 of a DPA attack, the attacker measures the power consumption of the cryptographic device. The cryptographic device executes the cryptographic algorithm using the known plaintexts (or ciphertexts) with the unknown, constant key.

In practice, attackers typically try to measure only the power consumption that is related to the targeted intermediate result. This means, if the plaintext is known, the attacker sets the trigger of the oscilloscope to the sending of the plaintext from the PC to the cryptographic device and records the power consumption for a short period of time. If the ciphertext is known, the attacker sets the trigger to the sending of the ciphertext from the cryptographic device to the PC and records the power consumption for a short period of time. Hence, the decisions in step 1 and step 2 depend on each other. The latter method for triggering is sometimes more convenient, because triggering on the communication that is initiated by the device typically leads to power traces that are better aligned, see Section 8.2.

Step 3: In step 3 of a DPA attack, the attacker calculates hypothetical intermediate values. This means, for all keys and for all plaintexts (or ciphertexts) the corresponding intermediate value is computed.

At a first glance, this step seems to be computationally infeasible, because we have to run through all values of the key. However, due to the way how cryptographic algorithms are designed, the attacker only has to guess a small part of the key in order to calculate the targeted intermediate value. For example, in AES, the first operation that mixes the key with the plaintext is the AddRoundKey operation. This operation works bit-wise. Hence, one bit of an intermediate value that occurs after AddRoundKey can be calculated based on one bit of the plaintext and one bit of the key. Even after the next operation, which is SubBytes, an intermediate value only depends on 8 bits of the key. Also in asymmetric algorithms, only small parts of the key are mixed with the plaintext. For instance, in typical implementations of RSA, the key is either used bit-wise (square-and-multiply algorithm) or block-wise (windowing algorithms), see [MvOV97].

Summarizing, in practice it is possible to choose intermediate values that only depend on a small part of the key for most kinds of cryptographic algorithms.

Step 4: In step 4 of a DPA attack, the attacker maps the hypothetical intermediate values to hypothetical power consumption values. For this mapping, an appropriate power consumption model needs to be chosen.

In practice, attackers typically can decide between the HW, the HD, and the ZV model. If the attacker has no idea about which model is appropriate, then simply all models should be tried. Attacking combinational logic is typically more difficult. In this case, the power model typically needs to be determined by simulations, see Section 3.3. If the power model is determined by characterization, a template-based DPA attack is conducted.

Step 5: In step 5 of a DPA attack, the attacker compares the hypothetical power consumption values with the power traces. This comparison is done with the help of statistical tests. The last step reveals the key.

Sometimes, DPA attacks produce high correlation coefficients for many or even all key hypotheses at the time when the targeted intermediate result is processed. There are two reasons why such a behavior can occur. First, one might simply have not used enough traces. Second, the power model might be wrong. If the power model is wrong, the correlation between the hypothetical power consumption values dominates the correlation between the hypothetical power consumption values and the traces. However, even if the power model is correct, it can happen that several key hypotheses lead to high correlation coefficients. These high correlation peaks for wrong keys are sometimes referred to as "ghost peaks". However, these peaks are not spooky. These peaks occur because the hypothetical intermediate values are correlated. The height of these correlations depends on the intermediate result that is attacked. For instance, in AES there is only a low correlation between the output values of the S-boxes but there is a high correlation between the outputs of the AddRoundKey operation for different key hypotheses. Hence, it is typically easier to determine the key in attacks on S-boxes than in attacks on AddRoundKey. In DES, other S-boxes are used than in AES. The outputs of the DES S-boxes are much more correlated than the outputs of the AES S-boxes. Hence, in attacks on the DES S-boxes typically more key hypotheses have high correlation coefficients.

Basic DPA Techniques. The concept of DPA was introduced by Kocher *et al.* in [KJJ99]. In this article, they explained a DPA attack on DES with the difference-of-means method. They also pointed out that standard statistical methods can be used. Furthermore, they provided hints that they had already developed "higher-order DPA" and "template DPA". Coron *et al.* [CKN01] discussed the application of several standard statistical tests in power analysis attacks.

Messerges *et al.* [MDS99a] investigated DPA attacks on DES implementations as well. They pointed out that DPA attacks can also be mounted on address information. Chari *et al.* [CJRR99a] discussed DPA attacks on the AES candidate algorithms. They were the first to use the term "power model" explicitly. Furthermore they used the covariance as statistical method. Akkar *et al.* [ABDM00] analyzed some commonly used power models in the context of smart cards. Brier *et al.* [BCO04] observed that also the correlation coefficient can be used as statistical method. Bevan and Knudsen [BK03] suggested using a least-square test on the result of the DPA attack. Their idea can be roughly described as follows. In addition to the DPA attack, the attacker performs simulated DPA attacks on the intermediate result that was targeted in the DPA attack. The attacker simulates DPA attacks for all keys and then compares, which of the simulated DPA attacks matches best to the DPA attack on the real traces. The one that matches best gives the correct key. They use a least-square test for the matching.

Naming Conventions for DPA Attacks. It is important to understand that the concept of DPA attacks is independent of the statistical tests or the type of attacked data. In other words, no matter which statistical test is used or which kind of data is targeted, all attacks that correspond to the definition in Section 6.1 are DPA attacks. It has become popular in the scientific literature, to introduce new names for attacks that are minor variations of standard DPA attacks, e.g. attacks that just use another statistical test. In this book, we refrain from introducing these names.

Advanced DPA Techniques. Agrawal *et al.* [ARR03] introduced the concept of generalized maximum-likelihood testing for power analysis attacks, see Section 6.5. In addition, they applied the general concept of template attacks, which was introduced by Chari *et al.* [CRR03], to DPA.

Schindler *et al.* [SLP05] introduced the concept of stochastic models. In this concept, the attacker models the power consumption based on some suitable functions. For example, a stochastic model for our microcontroller could consist of a linear function with eight unknowns $f(\mathbf{x}) = \sum_{i=1}^{8} c_i x_i$. Each unknown x_i represents a bit in the microcontroller. The weights c_i are derived by characterizing the device. This concept equals template attacks in many aspects. However, templates estimate the parameters of the probability distribution that describes the power consumption. In contrast to this, stochastic models estimate the parameters of a certain predefined function that describes the power consumption. Templates provide the "best" description of the power consumption of a device (assuming a multivariate normal distribution and given a set of interesting points). Stochastic models require the attacker to choose a certain function that describes the power consumption. Hence, stochastic models are in the best case, *i.e.* when the function describes the power consumption perfectly, as strong as templates.

Higher-order DPA attacks are DPA attacks that combine several intermediate values. We discuss them in Chapter 10.

DPA Attacks on Symmetric Cryptosystems.   Most of the articles that are related to basic DPA techniques target implementations of DES, see first paragraph of this section. Few articles discuss how to target unprotected implementations of AES. Örs *et al.* [OGOP04] discussed DPA attacks on an AES hardware implementation that has no countermeasures. Jaffe [Jaf06b] discussed DPA attacks after the MixColumns operation.

Lemke *et al.* [LSP04] discussed DPA attacks on implementations of IDEA, RC6, and HMAC constructions. Ha *et al.* [HKM+05] discussed DPA attacks on ARIA. Yoo *et al.* [YKH+04] discussed DPA attacks on SEED. Lano *et al.* [LMPV04] discussed DPA attacks on synchronous stream ciphers.

Prouff [Pro05] provided a mathematical proof that S-boxes indeed help DPA attackers. In other words, DPA attacks work better if intermediate results after the S-box are targeted.

DPA Attacks on Asymmetric Cryptosystems.   Few articles discuss how to attack unprotected implementations of asymmetric cryptosystems. Messerges *et al.* [MDS99b] discussed DPA attacks on implementations of RSA. They listed different variants. Some can be seen as extended SPA attacks (trace-pair analysis), one is a classical DPA attack as described in Section 6.1. Jaffe [Jaf06b] gave the probably most comprehensive overview of DPA attacks on RSA.

Coron [Cor99] discussed DPA attacks for implementations of ECC. Oswald [Osw05] surveyed state-of-the-art power analysis attacks for implementations of ECC.

Page and Vercauteren [PV04] reported on DPA attacks on implementations of pairing-based cryptosystems. Furthermore, Page and Stam [PS04] discussed DPA attacks on implementations of XTR.

DPA Attacks Using EM Leakage.   The same information that leaks through the power consumption also leaks through the electromagnetic field of a device. However, the EM field of a device usually also contains some additional information. As pointed out in Section 3.5.2, the parasitics that are attached to the power supply grid act as a filter for the power consumption that can be measured at the power supply pins. The EM field of a device therefore typically contains more information, in particular at high frequencies. This information can be exploited using a wideband receiver and an oscilloscope. Essentially, the receiver performs some demodulation and filtering before the oscilloscope measures the power traces for the DPA attack, see Agrawal *et al.* [AARR03].

The first to report on successful attacks that exploit this leakage were Gandolfi *et al.* [GMO01] and Quisquater and Samyde [QS01]. Mangard [Man03b] also reported on practical implementations of DPA attacks using EM leakage.

Comparing DPA Attacks. Whenever DPA attacks are published in the literature they are compared to previously published attacks. These comparisons are important because they are used to qualify some attacks as "better" than other attacks. Hence, it is important to have some objective criteria that allow fair and meaningful comparisons.

First, when comparing DPA attacks it is important to look at which types of attacks are compared. For example, a DPA attack and a template-based DPA attack cannot be compared straightforwardly because the assumptions that are made are entirely different. Second, also the used power models need to be considered. There is a big difference between the bit model, the HW model, and all the other models that exist. Third, it is important to take the type of the compared devices into account. For example, an AES implementation in software on an 8-bit microcontroller has much less components that work in parallel than an AES implementation in hardware with a 32-bit architecture.

> When comparing DPA attacks, the types of DPA attacks, the power models, the devices, and the implementations have to be taken into account.

Still, even if the same type of DPA attacks is performed on comparable devices with comparable implementations, the remaining question is which parameter should be used for the comparison. In most articles, the number of traces (that are required for a successful attack) is used. However, there are many ways to derive this number. In the context of this book, we have introduced two methods for this. The first method was an intuitive approach. We have repeatedly performed the DPA attack with an increasing number of traces. We have plotted how the highest correlation coefficient evolves as a function of the number of traces, see for example Figure 6.31. The number of traces where the highest correlation coefficient stands out "significantly" can be considered as the number of traces that are required for a successful attack. It should be clear that deriving the number of traces in this way is quite subjective. The second method that we introduced was the rule of thumb that maps $\rho_{ck,ct}$ to the number of samples, see Section 6.4. This rule of thumb is based on some simplifications, which make sense from our point of view. However, it is subjective as well. The only parameters that are not subjective are the correlation coefficients $\mathbf{R}$ and in particular $\rho_{ck,ct}$.

Table 6.2. Distinction of power analysis attacks and types of exploited leaks.

	Without characterization	With characterization
Few traces	SPA (SPA leak)	Template-based SPA (DPA leak)
Many traces	DPA (DPA leak)	Template-based DPA (DPA leak)

We suggest to compare DPA attacks based on correlation coefficients.

Quantifying Leakage. In the context of this book, we have used the SNR as measure for leakage. Recall that the correlation $\rho(H_i, P_{total})$ can be calculated based on $\rho(H_i, P_{exp})$ and the SNR. Hence, the SNR is well suited for DPA attacks.

However, for template-based DPA attacks the SNR is not well suited. Recall that in template-based DPA attacks, the template matching leads to the conditional probabilities $p(t'_i|k_j)$, which are used to derive the probabilities for the keys $p(k_j|t'_i)$. Hence, in template-based DPA attacks quantifying the leakage must be done with the conditional probabilities. Standaert *et al.* [SPAQ06] provided the first article that tackles this issue.

Philosophy: SPA vs. DPA vs. Template Attacks. Different researchers use different approaches to distinguish between SPA, DPA, and template attacks. The two approaches that are most useful are the following. The first approach distinguishes power analysis attacks based on the assumptions that are made about the attacker. This means, the criterion is whether an attacker needs only few power traces or many power traces, and whether the attacker is able to characterize the device or not. We use this approach in the context of this book.

The second approach distinguishes the type of leakage that is exploited. The two types of leaks are SPA leaks and DPA leaks. According to Jaffe [Jaf06a], an SPA leak is "a leak that is easily observable from a small number of traces using visual inspection", and a DPA leak is a leak that "requires statistical analysis" and "visual inspection is not sufficient to break the device". In other words, an SPA leak is a very strong leak that can be directly seen in a trace, whereas a DPA leak is small and cannot be directly seen in a trace.

Table 6.2 shows how the different types of power analysis attacks, which we have discussed so far, are related to the first approach. The table also shows which type of leakage they exploit. According to Table 6.2, we call all attacks that are based on few traces SPA attacks. If the attacker cannot characterize the device, the attacks indeed only exploit SPA leaks. If the attacker can characterize the device, then we call the attacks template-based SPA attacks. These template-based SPA attacks actually exploit DPA leaks. We call attacks that use many traces DPA attacks. All DPA attacks exploit DPA leaks. We call DPA attacks in which the attacker characterizes the device template-based DPA attacks.

Chapter 7

HIDING

Power analysis attacks work because the power consumption of cryptographic devices depends on intermediate values of the executed cryptographic algorithms. Therefore, the goal of countermeasures is to avoid or at least to reduce these dependencies. In case of hiding, this is done by breaking the link between the power consumption of the devices and the processed data values. Hence, cryptographic devices that are protected by hiding execute cryptographic algorithms in the same way as unprotected devices. In particular, they calculate the same intermediate values. Yet, the hiding countermeasures make it difficult for an attacker to find exploitable information in power traces.

In this chapter, we first provide a general description of hiding. Subsequently, we analyze how hiding can be implemented at the architecture level of cryptographic devices. In particular, we discuss examples of countermeasures for software and for hardware implementations of cryptographic algorithms. Finally, we analyze hiding at the cell level. We discuss different logic styles that have been proposed to counteract power analysis attacks.

7.1 General Description

The goal of hiding countermeasures is to make the power consumption of cryptographic devices independent of the intermediate values and independent of the operations that are performed. There are essentially two approaches to achieve this independence. The first approach is to build devices in such a way that the power consumption is random. This means that in each clock cycle a random amount of power is consumed. The second approach is build devices that consume an equal amount of power for all operations and for all data values. Hence, equal amounts of power are consumed in each clock cycle.

> The power consumption of a cryptographic device is independent of the performed operations and the processed data values, if it has one of the following two properties:
>
> - The device consumes random amounts of power in each clock cycle.
> - The device consumes equal amounts of power in each clock cycle.

The ideal goal of making the power consumption perfectly random or equal cannot be reached in practice. However, there are several proposals on how to get close to this goal. These proposals can be divided into two groups. The first group of proposals randomizes the power consumption by performing the operations of the executed cryptographic algorithms at different moments of time during each execution. These proposals hence only affect the time dimension of the power consumption. This is different in case of the second group. The goal of these proposals is to make the power consumption random or equal by directly changing the power consumption characteristics of the performed operations. The second group of proposals hence affects the amplitude dimension of the power consumption.

7.1.1   Time Dimension

In step 2 of the description of DPA attacks (see Section 6.1), we have pointed out that the recorded power traces should be correctly aligned for these attacks. This means that the power consumption of each operation should be located at the same position in each power trace. If this condition is not fulfilled, DPA attacks require significantly more power traces. This observation is the motivation for designers of cryptographic devices to randomize the execution of the cryptographic algorithms, *i.e.* the devices perform the operations of the algorithms at different moments of time during each execution. This makes the power consumption appear to be more or less random for an attacker. The more random the execution of an algorithm is, the more difficult it becomes to attack the device. The most commonly used techniques to randomize the execution of cryptographic algorithms are the random insertion of dummy operations and the shuffling of operations.

> The power consumption a cryptographic device can be randomized by performing the operations of the executed algorithm at different moments of time during each execution. This can be done by randomly inserting dummy operations or by shuffling.

Random Insertion of Dummy Operations

The basic idea of this technique is to randomly insert dummy operations before, during, and after the execution of the cryptographic algorithm. Each time the algorithm is executed, randomly generated numbers are used to decide how many dummy operations are inserted at these different positions. It is important that the total number of inserted operations is equal for all executions of the algorithm. In this way, attackers cannot get any information about the number of inserted operations by measuring the execution time of the algorithm.

In an implementation that is protected by this approach, the position of each operation depends on the number of dummy operations that have been inserted before this operation. This number randomly varies from execution to execution. The more this position varies, the more random the power consumption appears. However, it is also clear that the more dummy operations are inserted, the lower is the throughput of the implementation. This is why in practice, a suitable compromise needs be found for every implementation.

Shuffling

An alternative to the insertion of dummy operations is the shuffling of operations of the cryptographic algorithm. The basic idea of this approach is to randomly change the sequence of those operations of a cryptographic algorithm that can be performed in arbitrary order. In case of AES, 16 S-box look-ups need to be performed in every round. These look-ups are independent of each other. Hence, they can be performed in arbitrary order. Shuffling these operations means that during each execution of AES, randomly generated numbers are used to determine the sequence of the 16 S-box look-ups.

Shuffling randomizes the power consumption in a similar way as the random insertion of dummy operations. However, shuffling does not affect the throughput as much as the random insertion of dummy operations. The disadvantage of shuffling is that it can only be applied to a certain extent. The number of operations that can be shuffled in a cryptographic algorithm are limited. This number depends on the algorithm and on the architecture of the implementation. In practice, shuffling and the random insertion of dummy operations are often combined.

7.1.2 Amplitude Dimension

Cryptographic devices can be protected against power analysis attacks by making the power consumption either equal or random during all clock cycles. In the previous section, we have discussed methods to randomize the execution of cryptographic algorithms. Now, we describe techniques to directly change the power consumption characteristics of the performed operations. These techniques lower the leakage of a cryptographic device by lowering the signal-to-noise ratio (see Section 4.3.2) of the performed operations. Ideally, the goal is to set the SNR to 0. This can be achieved by setting $Var(P_{exp}) = 0$ or by increasing $Var(P_{sw.\,noise} + P_{el.\,noise})$ to infinity. In order to reduce $Var(P_{exp})$ to 0, the power consumption needs to be exactly equal for all operations and data values. Increasing $Var(P_{sw.\,noise} + P_{el.\,noise})$ to infinity means that the amplitude of the noise needs to be infinitely increased.

In practice, both approaches can only be implemented to a certain degree. This means that $Var(P_{exp})$ can be reduced to small values and $Var(P_{sw.\,noise} + P_{el.\,noise})$ can be increased to large values. However, there is no countermeasure that reaches the ideal goal of $SNR = 0$. We now provide a brief overview of the ideas that have been proposed to increase the noise and to lower the signal of the performed operations.

The signal-to-noise ratio of an operation, *i.e.* the leakage of an operation, can be lowered by either increasing the noise or by lowering the signal.

- **Increasing the Noise:** The goal is to build a device where random switching activity dominates the power consumption.

- **Reducing the Signal:** The goal is to build a device where all operations require an equal amount of power for all data inputs.

Increasing the Noise

The simplest way to increase the noise of operations is to perform several independent operations in parallel. Hence, hardware architectures of cryptographic algorithms with a wide datapath are better than architectures with a narrow datapath. For example, it is more difficult to attack the power consumption of a single bit in a 128-bit architecture of AES than in a 32-bit architecture.

Another way to increase the noise is to use dedicated noise engines. Noise engines perform random switching activities in parallel to the actual operations. This increases $Var(P_{sw.\,noise})$, and hence, the SNR is lowered.

Reducing the Signal

At first sight, it might not seem to be a challenging task to build a device that consumes an equal amount of power for all operations and for all data values. However, when looking at this task in more detail, it immediately turns out that this task is not trivial. DPA attacks are based on large numbers of power traces. Hence, these attacks can exploit very small differences in the power consumption. Only a countermeasure that makes the power consumption of a device exactly the same for all operations and all data values provides perfect protection against DPA attacks. There exist two strategies to approach this goal in practice.

Table 7.1. Hiding countermeasures to make the power consumption of cryptographic devices random or equal during all clock cycles.

	Equal power consumption	Random power consumption
Time dimension	-	Dummy operations, shuffling
Amplitude dimension	Reduction of signal	Increase of noise

The most commonly used strategy is to employ dedicated logic styles for the cells of cryptographic devices. As discussed in Section 3.1, the overall power consumption of a cryptographic device is the sum of the power that is consumed by its cells. If each cell is built in such a way that its power consumption is constant, the overall power consumption is constant. The second strategy to approach the goal $Var(P_{exp}) = 0$, is to filter the power consumption of the cryptographic device. The basic idea of this approach is to remove all data-dependent and operation-dependent components of the power consumption by a filter.

7.1.3 Methods to Implement Hiding

The goal of hiding is to make the power consumption of cryptographic devices independent of the performed operations and independent of the processed data. In practice, there are three ways to approach this goal. The first way is to make the power consumption appear to be random by randomly changing the sequence of the performed operations. In this case, the power consumption characteristics of the operations do not need to be changed. This approach only affects the time dimension of the power consumption. The second way to protect a device is to randomize the power consumption by performing random switching activities in parallel to the actual operation. The goal here is to increase $Var(P_{sw.noise})$. The third way is to try to make the power consumption equal during all clock cycles. This means that the signal of the performed operations is reduced, *i.e.* $Var(P_{exp})$ is reduced. The different ways to protect cryptographic devices by hiding are summarized in Table 7.1. Except for shuffling, all hiding countermeasures are independent of the cryptographic algorithms and the protocols that are implemented on the cryptographic device.

In practice, there exist many ways to implement hiding countermeasures. It is surprising though, that there is only a small number of scientific publications on concrete implementations of hiding countermeasures. It seems that hiding countermeasures are far more popular in industry than in the scientific community. An exception to this are the recent articles that have been published on different logic styles to counteract power analysis attacks.

7.2 Architecture Level

At the architecture level, we distinguish between countermeasures for software and for hardware implementations of cryptographic algorithms. We first describe hiding countermeasures for software and then we discuss proposals for countermeasures that can only be implemented in hardware.

7.2.1 Software

In software, the options to alter the power traces of a cryptographic device are very limited. The power consumption characteristics of the instructions that are executed on a device are defined by the underlying hardware. Nevertheless, there are some hiding countermeasures that can also be implemented in software.

Time Dimension

The most commonly used hiding countermeasure for software is to randomize the execution of the algorithms. The random insertion of dummy operations as well as shuffling can be implemented easily in software. However, it is important to point out that these countermeasures require random numbers. These random numbers need to be generated on the cryptographic device. In general, the insertion of dummy operations and shuffling do not provide a high level of protection against power analysis attacks.

Amplitude Dimension

The power consumption characteristics of the instructions of a cryptographic device are defined by the hardware of the device. However, it is the software that determines which instructions are used to implement cryptographic algorithms. In practice, there are usually several ways to implement cryptographic algorithms in software. By selecting the instructions that are used for the implementation, it is possible to alter the power traces of the device. This can help to prevent SPA attacks, like the visual inspection of power traces. However, it is usually not sufficient to provide protection against DPA attacks. We now provide a short list of methods to reduce the leakage of devices when implementing cryptographic algorithms in software.

- **Choice of Instructions:** In general, not every instruction of a cryptographic device leaks the same amount of information about its operands and its results. Therefore, the instructions to implement cryptographic algorithms should be chosen carefully. Only those instructions should be used that leak small amounts of information.

- **Program Flow:** Changes in the program flow can usually be detected easily when performing a visual inspection of power traces, see Section 5.2.

In particular, attackers can easily detect conditional jumps and patterns of repeated instruction sequences. Programmers should therefore avoid conditional jumps that depend on the key or on data that is related to the key. The sequence of the executed instructions should always be independent of the key.

- **Memory Addresses:** Cryptographic devices leak information about memory addresses, just like they leak information about all other data values. Therefore, the used memory addresses should not depend on the key. If it is necessary to use key-dependent addresses, only addresses should be used that leak similar information. For example, memory addresses with the same Hamming weight should be used, if the device leaks the Hamming weight of the addresses.

- **Parallel Activity:** The leakage of a cryptographic device can not only be lowered by lowering $Var(P_{exp})$. It can also be lowered by increasing the noise. Programmers can increase the noise by performing activities in parallel to the execution of the cryptographic algorithm. For this purpose, components such as coprocessors or communication interfaces can be used.

7.2.2 Hardware

Hiding countermeasures that are implemented in software protect cryptographic devices only to a limited degree. In hardware, there are significantly more options to protect cryptographic devices by hiding. We now provide an overview of hardware countermeasures that can be implemented at the architecture level. We again distinguish countermeasures that affect the time dimension of the power consumption and countermeasures that affect the amplitude dimension.

Time Dimension

In hardware, there are essentially two groups of proposals to randomly change the moments of time at which the operations of the cryptographic algorithm are executed. The first group of proposals is similar to the countermeasures we have already discussed in the context of software implementations. The goal of these countermeasures is to randomly insert dummy operations or cycles and to shuffle the performed operations.

- **Dummy Operations and Shuffling:** The random insertion of dummy operations and shuffling can essentially be done in the same way as in software. This means that random numbers are used to determine the positions where dummy operations are inserted. Furthermore, random numbers are used to determine the sequence of operations of the executed cryptographic algorithms.

■ **Random Insertion of Dummy Cycles:** This method is a variant of the insertion of dummy operations. While dummy operations can take several clock cycles, the goal of this method is to only insert individual clock cycles. For this purpose, usually the registers of the protected cryptographic device are duplicated. The original registers are used to store the intermediate values of the performed cryptographic algorithm. The other registers are used to store random values. Dummy cycles can hence be inserted as follows. During the execution of the cryptographic algorithm, a random number is generated in each clock cycle. This random number is used to determine whether the clock cycle is a dummy cycle or not. If it is a dummy cycle, the device performs computations with the random data stored in the duplicated registers. Otherwise, the device continues with the execution of the cryptographic algorithm.

The second group of proposals affects the clock signal. These proposals randomly change the clock signal in order to make the alignment of the power traces more difficult. The most commonly used techniques to manipulate the clock signal of cryptographic devices are listed below.

■ **Skipping Clock Pulses:** The basic idea of this approach is to insert a kind of filter into the path of the clock signal. This filter randomly skips pulses of the clock signal that is provided to the cryptographic device. Random numbers are used to determine which clock pulses are skipped and which are not.

■ **Randomly Changing the Clock Frequency:** An alternative to the skipping of clock pulses is to generate a clock signal with a randomly changing frequency directly on the cryptographic device. This can for example be done by controlling the frequency of an internal oscillator based on random numbers.

■ **Multiple Clock Domains:** In this case, several clock signals are generated on the cryptographic device. The alignment of the operations of the cryptographic algorithm is destroyed by randomly switching between the clock signals.

For all countermeasures that affect the time dimension of the power traces, it is crucial that attackers cannot identify the countermeasures. This means that attackers must not be able to detect the inserted dummy operations or cycles, the shuffling, or the manipulations of the clock signal.

Amplitude Dimension

In hardware, the power consumption characteristics of the operations of a cryptographic device can be changed more easily than in software. At the architecture level, the power consumption of cryptographic devices can be made equal for all operations and all data values by filtering the power consumption. Alternatively, noise can be added to the power traces in order to counteract power analysis attacks.

- **Filtering:** The goal of this approach is to remove exploitable components of the power consumption by filtering. This means that a filter is inserted between the power supply pins of the cryptographic device and the circuit that is computing the cryptographic algorithm. In practice, the power consumption can be filtered by using switched capacitors, constant current sources, and all other kinds of circuits that regulate the power consumption.

- **Noise Engines:** An alternative method to filtering is to generate noise in parallel to the computation of the cryptographic algorithm. Noise engines are typically built based on random number generators. In order to have a big impact on the power consumption, the random number generators need to be connected to a network of large capacitors. The random charging and discharging of this network leads to noise in the power consumption that makes power analysis attacks more difficult.

When implementing countermeasures at the architecture level, it is important to understand that the SNR not only depends on the cryptographic device, but also on the measurement setup that is used for the attack. If a countermeasure reduces the SNR for one measurement setup, it does not necessarily reduce the SNR for all setups. For example, filtering the power consumption makes attacks more difficult, if the power consumption is measured via a resistor. However, an attacker who measures the electromagnetic emanation of the device is not significantly affected by this countermeasure. For the same reason, noise engines should be spread over the entire device instead of being placed only in one corner of the device.

7.3 Cell Level

Counteracting power analysis attacks at the cell level has been one of the first reactions of the semiconductor industry after the publication of these attacks. In the scientific community, it took much longer until the first proposals for cell-level countermeasures appeared. During the last years, several proposals for logic styles to counteract power analysis attacks have been made. Many of these logic styles are based on the concept of hiding.

Applying the concept of hiding to the cell level means that the logic cells of a circuit are implemented in a such way that their power consumption is independent of the processed data and the performed operations. This independence is typically achieved by making the power consumption of the logic cells constant in each clock cycle for all processed logic values. Constant in each clock cycle means that the instantaneous power consumption of a cell is the same in each clock cycle. A consequence of this behavior is that the logic cells always consume the maximum amount of power in each clock cycle. The total power consumption of a cryptographic device is constant in this case. Thus, it is independent of the processed data and the performed operations.

Logic styles with a constant power consumption counteract SPA attacks and DPA attacks. However, such logic styles are often just called *DPA-resistant logic styles*, because counteracting DPA attacks is usually their main field of application. Logic styles with a constant power consumption are typically implemented as *dual-rail precharge (DRP)* logic styles. In this section, we give a general description of DRP logic styles and we discuss how they are used to make the power consumption of circuits constant. Furthermore, we discuss how DRP logic styles can be used for semi-custom circuit design.

7.3.1   General Description of DRP Logic Styles

DRP logic styles combine the concepts of dual-rail (DR) logic and precharge logic. The resulting functionality builds the basis for logic cells that have a constant power consumption in each clock cycle. Besides this functionality, the DRP cells and the wires between these cells must be built in a special balanced way to achieve a constant power consumption.

> DRP logic styles are used for cryptographic devices to make the power consumption of the logic cells in the device constant in each clock cycle. This means that the logic cells always consume the maximum amount of power.

Dual-Rail Logic

In contrast to single-rail (SR) logic, where a logic signal a is carried on a single wire, DR logic uses a pair of wires for this purpose. In DR logic, one wire usually carries the non-inverted signal a while the other wire carries the inverted (complementary) signal $\overline{a}$. This type of encoding is also known as differential encoding. In this case, a valid logic signal is only present when the two wires carry complementary values, *i.e.* one wire is set to 1 while the other wire is set to 0. The two wires of a DR wire pair are often referred to as complementary wires. When referring to the logic values that are present on a DR wire pair, we put them into parentheses, e.g. (0, 1). This example means that the wire carrying the non-inverted signal is set to 0 while the wire carrying the inverted signal is set to 1.

Figure 7.1 shows a 2-input SR cell and a corresponding 2-input DR cell. The input signals a, b and the output signal q of the SR cell are carried on single

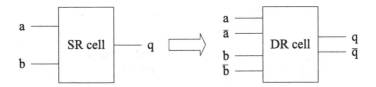

Figure 7.1. A 2-input SR cell and a corresponding 2-input DR cell.

wires. In case of the DR cell, the input signals and the output signal are carried on complementary wires.

Note that no inverters are required in DR circuits and thus also not in DRP circuits. A logic signal is inverted simply by swapping the two complementary wires that carry this logic signal.

Precharge Logic

In a precharge circuit, all logic signals alternate between a so-called precharge value and the logic values that are processed. The precharge value is either 0 or 1. The phase in which all signals in the circuit are set to the precharge value is called the precharge phase. The phase in which all signals of the circuit are set to their current logic values is called the evaluation phase.

The sequence of the precharge phase and the evaluation phase is typically controlled by the clock signal. In most cases, the logic value of the clock signal defines the phase in which a circuit currently is. This means that during one clock cycle, the signals in the circuit are precharged and evaluated according to the given mapping between the value of the clock signal (0/1) and the state of the circuit (precharge phase/evaluation phase).

Dual-Rail Precharge Logic

The combination of DR logic and precharge logic leads to DRP circuits where all logic signals are encoded on complementary wires. During the evaluation phase, the values on the complementary wires are set according to the processed data to (0, 1) or to (1, 0). When the circuit is switched to the precharge phase, the values on the complementary wires are set to the precharge value, which is either 0 or 1. Note that during the precharge phase, no complementary values are present on the DR wire pairs in the circuit.

If we assume that the precharge value is 0 and that the first half of the clock cycle corresponds to the evaluation phase, then always one complementary output of a DRP cell performs the transitions $0 \rightarrow 1 \rightarrow 0$ during a clock cycle. The other complementary output stays at 0. This means that a DRP cell always performs the same transitions at its outputs during each clock cycle. These transitions either occur at output q or at output $\bar{q}$ depending on the input values

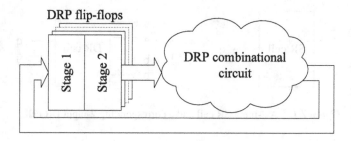

Figure 7.2. DRP flip-flops consist of two stages, which are precharged alternately.

and the functionality of the DRP cell. This behavior allows making the power consumption of DRP cells constant.

DRP Flip-Flops

The flip-flops in DRP circuits consist of two stages, as shown in Figure 7.2. Stage 1 is in the precharge phase when stage 2 and the combinational DRP cells are in the evaluation phase. In this phase, stage 2 of the DRP flip-flops provides the stored logic values to the combinational DRP cells. The combinational DRP cells calculate their output values according to the respective input values. Just before stage 2 of the DRP flip-flops and the combinational DRP cells are precharged, stage 1 of the DRP flip-flops stores the logic values at its inputs. This behavior ensures that all cells of the DRP circuit are precharged during a clock cycle and that the processed logic values are not lost during the precharge phase.

7.3.2   Constant Power Consumption of DRP Logic Styles

As discussed in the previous section, the transitions that occur at the complementary outputs of a DRP cell are the same in each clock cycle. They occur either at output q or at output $\bar{q}$. According to Section 3.1, the power consumption of a logic cell is proportional to the capacitance at the output where a transition occurs. Thus, the first step towards a constant power consumption is to balance the capacitances at the complementary outputs of a DRP cell. This ensures that a DRP cell charges the same capacitance in each clock cycle. Additionally, it is necessary to ensure a constant power consumption in each clock cycle for charging and discharging internal nodes of a DRP cell.

Glitches in DRP circuits would most likely interfere with the requirement of a constant power consumption of the DRP cells in each clock cycle. Thus, glitches are usually avoided, which can be done quite easily in DRP circuits. In the following, we discuss in detail how the complementary outputs and the internal power consumption of DRP cells can be balanced.

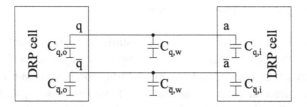

Figure 7.3. The capacitances at the complementary output of a DRP cell.

Balancing the Complementary Outputs

Let us denote the capacitances at the complementary outputs of a DRP cell by C_q and by $C_{\bar{q}}$. C_q and $C_{\bar{q}}$ both consist of three parts: The output capacitance C_o of the DRP cell, the capacitance C_w of the wire to the subsequent cells, and the sum of the input capacitances C_i of these cells. Figure 7.3 shows the situation for both complementary outputs q and $\bar{q}$ of a DRP cell that drives one subsequent cell with the complementary inputs a and $\bar{a}$. In order to balance C_q and $C_{\bar{q}}$, the three parts of these capacitances need to be pairwise balanced.

For modern process technologies of digital circuits, the biggest contribution to the capacitance at the output of a logic cell is typically C_w. Therefore, balancing $C_{q,w}$ and $C_{\bar{q},w}$ is the most essential task. This is done during placing and routing of the DRP cells. There are different methods to pairwise balance the capacitances of complementary wires. Two of them are differential routing and backend duplication. They are briefly introduced in Section 7.3.3.

> The degree to which the power consumption of DRP cells is constant mainly depends on the pairwise balancing of the capacitances of the complementary wires.

The capacitances $C_{q,o}$ and $C_{\bar{q},o}$ can be balanced by carefully designing the DRP cells. This means that both cell outputs need to be driven by an equal number of transistors with the same parameters (width, *etc.*). Furthermore, the wires within the cells connecting to the complementary outputs need to have the same capacitance. The capacitances $C_{q,i}$ and $C_{\bar{q},i}$ can also be balanced during the design of the DRP cells. The complementary wires of an input need to be connected to an equal number of gate terminals of transistors with the same parameters, and the capacitances of the input wires within a cell need to be equal.

Note that for a constant instantaneous power consumption of a DRP cell, the resistances of the paths over which C_q and $C_{\bar{q}}$ are charged and discharged must be balanced in the same way as the different capacitances. Fortunately, the balancing of the capacitances normally has the side effect that the corresponding resistances are also balanced.

Balancing the Internal Power Consumption

In each clock cycle, DRP cells also consume a small amount of power to charge and discharge internal nodes. During the design of the DRP cells, it must be ensured that this internal power consumption is constant in each clock cycle. This requirement can for example be achieved by ensuring that all internal nodes of the DRP cells are charged and discharged during each clock cycle.

7.3.3   Semi-Custom Design and DRP Logic Styles

A very common implementation approach for cryptographic devices is *semi-custom circuit design based on standard cells*, see Section 2.2.2. This approach can also be used for DRP logic styles, because such a countermeasure can be applied independently of the used cryptographic algorithms and the chosen circuit architectures. Therefore, the high-level design of cryptographic devices can be done independently of the used DRP logic styles, which are automatically employed later in the design process.

There is one very important issue when designing cryptographic devices that are implemented in DRP logic styles. It must be ensured that no sensitive intermediate results of the used cryptographic algorithms are leaving the secured cryptographic devices. It must be assumed that all circuits connected to the outputs of cryptographic devices are not sufficiently protected against power analysis attacks. Usually, these circuits are SR circuits. Furthermore, because of significantly increased area requirements and power consumption, often not all components of cryptographic devices are implemented in DRP logic styles. Therefore, any sensitive intermediate results that leave the secured components of cryptographic devices would cause a vulnerability to power analysis attacks.

When using DRP logic styles, some extensions to a standard semi-custom design flow as depicted in Figure 2.4 are necessary. There are essentially three reasons for these extensions. First, logic synthesizers typically can not use DRP cells directly. Most logic synthesizers are specialized in using SR cells (like CMOS cells) and do not support DRP cells. The usual solution to this problem is that logic synthesis of the high-level design is performed using an SR cell library. The resulting SR cell netlist is then converted to a DRP cell netlist. This process is called *logic style conversion*. The second reason for extensions of the design flow is that in most cases, an SR interface needs to be added to the DRP circuits. Third and last, the requirement of balanced complementary wires must be enforced during the design process. The necessary extensions to a standard semi-custom design flow are shown in Figure 7.4 in gray. In the following, a semi-custom design flow for DRP circuits is discussed in more detail.

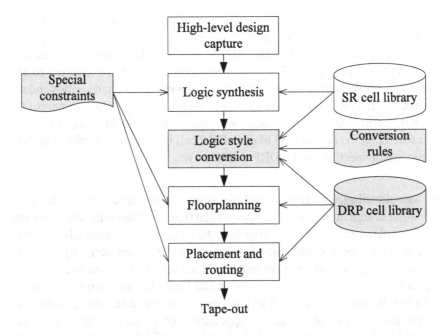

Figure 7.4. Semi-custom design flow using DRP logic styles.

- **High-Level Design Capture:** The main issue during high-level design is that sensitive intermediate results must not leave the components of crypto-graphic devices that are implemented in DRP logic styles.

- **Logic Synthesis:** Logic synthesizers typically cannot use DRP cells di-rectly. Therefore, logic synthesis is done using an appropriate SR cell li-brary. Special constraints are applied to ensure that only such SR cells are used for which corresponding DRP cells exist.

- **Logic Style Conversion:** The main steps of the logic style conversion process are cell substitution, adaptation of the signal nets, and addition of an SR interface to the DRP circuit. During cell substitution, all SR cells are replaced by the corresponding DRP cells. The logic function, the timing behavior, and the layout of the DRP cells is provided by a DRP cell library. The mapping between SR cells and DRP cells is defined by conversion rules. Note that SR inverters are removed because their functionality in DRP circuits is achieved by swapping the respective two complementary wires. In the second step, the signal nets of the SR circuit are duplicated to get complementary wires. The additional wires are connected to the appropriate inputs and outputs of the DRP cells according to the conversion rules. The duplication is done for all signal nets in the SR circuit except for dedicated SR signals like the clock signal or the asynchronous reset signal.

Furthermore, if necessary, all combinational DRP cells are connected to the clock tree. In the third step of the logic style conversion process, typically an SR interface is added to the DRP circuit. This interface ensures that all SR signals entering the DRP circuit are precharged and converted to DR signals. Furthermore, the SR interface removes the inverted signals and the effects of the precharge phase from the complementary output signals of the DRP circuit. At the end of the logic style conversion process, it must be ensured that the output load of each DRP cell in the circuit is still below the maximum specified in the DRP cell library.

- **Floorplanning, Placement, and Routing:** During these steps of the semi-custom design flow, it must be ensured that the capacitance and the resistance of complementary wires are pairwise the same. This is basically achieved by putting special constraints on these steps. However, perfectly balanced complementary wires are impossible to achieve in practice because their exact capacitances and resistances on the final chip are simply not known during the design process. The used models of the wires always introduce inaccuracies. Furthermore, the capacitances of a wire are not only located between the wire and the supply lines (V_{DD} and GND) but also between the wire and its neighboring wires. These capacitances are called cross-coupling capacitances. Thus, the effective capacitance of a wire depends on the states of the neighboring wires, which means that it is data dependent. Unfortunately, the cross-coupling capacitances become more and more dominant over the capacitances to the supply lines in modern process technologies.

  Two of the proposed methodologies to place and route DRP circuits in a balanced way are *differential routing* [TV04b] and *backend duplication* [GHMP05]. In the differential routing approach, the complementary wires are routed in parallel. A prototype chip that has been implemented using this approach is presented in [THH+05]. In the backend duplication method, two circuits are produced with the same layout, and one circuit always processes the complementary signals of the other circuit. In general, the routing process is more complex in DRP circuits than in SR circuits because the number of wires is essentially doubled.

7.4 Examples of DRP Logic Styles

In the following, two DRP logic styles that are used for DPA-resistant circuits are presented in detail. The first logic style is *Sense Amplifier Based Logic (SABL)* and the second one is *Wave Dynamic Differential Logic (WDDL)*. For each logic style, the functionality and the properties of cells and circuits are discussed. Furthermore, one combinational cell (NAND) and one sequential cell (D-flip-flop) are presented in detail.

DPA-resistant logic styles can in general be categorized into two groups: DPA-resistant logic styles which require the implementation of new logic cells from scratch and DPA-resistant logic styles whose logic cells are based on SR cells that are available in existing standard-cell libraries. These existing standard-cell libraries are typically provided and maintained by semiconductor foundries. The advantage of the first approach is that the functionality and the layout of the logic cells is fully customizable. The advantage of the second approach is a significantly reduced design effort. The second approach makes changes of the used process technology easier but it usually leads to logic cells with a lower DPA resistance.

7.4.1 Sense Amplifier Based Logic

Sense Amplifier Based Logic has been introduced by Tiri *et al.* in [TAV02]. SABL cells are specifically designed to have a constant internal power consumption independent of the processed logic values. Furthermore, the combinational SABL cells are designed in a way that their time-of-evaluation (TOE) is data independent, *i.e.* the cells only evaluate after all input signals have been set to complementary values. Due to their special design, SABL cells must be implemented from scratch.

A characteristic of SABL circuits is that all SABL cells are connected to the clock signal and that all of them are precharged simultaneously. Thus, very high current peaks occur during the precharge phase. The current peaks during the evaluation phase are smaller because the combinational SABL cells do not evaluate simultaneously. Since combinational SABL cells only evaluate after all input signals have been set to complementary values, the evaluation events of the different cells are spread out over the evaluation phase. In balanced SABL circuits, the propagation delays of the complementary wires are pairwise identical. Since the TOE of the combinational SABL cells is independent of the data, the combinational SABL cells always evaluate at fixed moments of time in each clock cycle.

The area requirements of SABL circuits are at least doubled compared to corresponding CMOS circuits, while the maximum clock rates are typically halved. The power consumption of SABL circuits is increased significantly, but a general factor for the increase cannot be given. The reason is that the actual increase of the power consumption depends on many different aspects: e.g. the size of the circuit; the ratio between the number of combinational and sequential cells, which defines the increase of the clock tree compared to a CMOS circuit; the circuit architecture and the input data statistics, which define the switching activities of the nodes in the circuit including glitches. The DPA resistance of SABL circuits is typically very high (if all complementary wires are sufficiently balanced), because the internal power consumption of the SABL cells is constant and the SABL cells always evaluate at fixed moments of time in

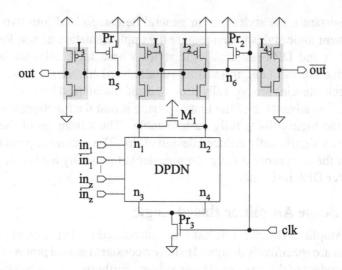

Figure 7.5. Transistor schematic of a generic n-type SABL cell.

each clock cycle. Examples of SABL circuits and their properties are presented in [TAV02] and [TV03].

General Description of SABL Cells

Figure 7.5 shows the transistor schematic of a generic n-type SABL cell. In order to achieve the same transitions at the cell outputs *out* and $\overline{out}$ in each clock cycle, the n-type SABL cell consists of the *differential pull-down network* (DPDN) and the *cross-coupled inverters* I_1 and I_2. The DPDN is made of NMOS transistors. Thus, the SABL cell is called an n-type cell. If the input signals $in_1, \overline{in_1}, \ldots, in_z, \overline{in_z}$ of the DPDN are set to complementary values, one of the nodes n_1 and n_2 is connected to one of the nodes n_3 and n_4. Which nodes are connected is determined by the structure of the DPDN. This structure defines the logic function of the n-type SABL cell. The inverters I_1 and I_2 are called cross coupled because the output of one inverter is connected to the input of the other inverter and vice versa. Therefore, if one inverter switches its output signal to a specific logic value, the output signal of the other inverter is switched to the opposite value. Such a structure is also known as *sense amplifier* [RCN03].

An n-type SABL cell is in the evaluation phase when the clock signal is 1. During the evaluation phase, the input signals $in_1, \overline{in_1}, \ldots, in_z, \overline{in_z}$ of the DPDN are set to complementary values and the MOS transistor Pr_3 is turned on. As a result, the outputs n_5 and n_6 of the cross-coupled inverters I_1 and I_2 and furthermore the cell outputs *out* and $\overline{out}$ are switched to complementary values. An n-type SABL cell is in the precharge phase when the clock signal is 0. During the precharge phase, the MOS transistors Pr_1 and Pr_2 ensure that all internal nodes of an n-type SABL cell are set to 1. As a result, the inverters I_3 and I_4 produce a precharge value of $(0, 0)$ at the complementary outputs of an n-type SABL cell. A precharge value of $(0, 0)$ is necessary, if n-type SABL cells are cascaded. It ensures that the NMOS transistors in the DPDNs of the subsequent n-type SABL cells are not accidentally providing a conducting path to *GND* at the onset of the next evaluation phase. This way of cascading precharged cells is known as *domino style* [RCN03].

There exist also p-type SABL cells which use a differential pull-up network (DPUN) instead of a DPDN. The DPUN consists of PMOS transistors. A SABL circuit consists either completely of n-type cells or of p-type cells. All considerations we make for the n-type SABL cells are equivalently valid for the p-type SABL cells. In the following, the behavior of an n-type SABL cell during a clock cycle, *i.e.* during an evaluation phase and a subsequent precharge phase, is discussed in more detail.

Evaluation Phase. At the end of the precharge phase, all internal nodes of an n-type SABL cell have been set to 1. Additionally, the input signals in_1, $\overline{in_1}$, ..., in_z, $\overline{in_z}$ of the DPDN have all been precharged to 0. At the onset of the evaluation phase (*i.e.* the clock signal switches from 0 to 1), the PMOS transistors Pr_1 and Pr_2 are turned off (the connection between the drain and source terminal is insulating) and the NMOS transistor Pr_3 is turned on (the connection between the drain and source terminal is conducting). Thus, the nodes n_3 and n_4 of the DPDN are set to 0. Since the input signals in_1, $\overline{in_1}$, ..., in_z, $\overline{in_z}$ of the DPDN are still all 0, the NMOS transistors of the DPDN are still turned off. Neither the node n_1 nor the node n_2 is connected to 0 via the DPDN and all other internal nodes of an n-type SABL cell beside n_3 and n_4 stay at 1.

The situation changes as soon as the input signals in_1, $\overline{in_1}$, ..., in_z, $\overline{in_z}$ are set to complementary values. This happens when the preceding n-type SABL cells have evaluated their output values. Now, either the node n_1 or the node n_2 is connected to 0 via the DPDN according to the input signals and the actual structure of the DPDN. If n_1 is set to 0, the inverter I_1 is operational. Its output signal n_5 is switched to 0 since the input signal n_6 of inverter I_1 is still 1. Node n_5 also works as input to inverter I_2. Thus, the output signal n_6 of inverter I_2 stays at 1 and the cross-coupled inverters have been set to a specific state: $n_1 = 0 \Rightarrow n_5 = 0, n_6 = 1$. If n_2 is connected to 0 via the DPDN, the opposite state occurs: $n_2 = 0 \Rightarrow n_5 = 1, n_6 = 0$. The NMOS pass-transistor M_1 acts as a resistor that ensures that both nodes n_1 and n_2 are eventually set to 0 when one of these nodes is connected to 0 via the DPDN. The resistance between the drain and source terminal of M_1 is chosen high enough so that the cross-coupled inverters switch to the correct state. The reason for using the pass-transistor M_1

is explained in more detail later. After the cross-coupled inverters have been set to a specific state, the SABL cell has evaluated and the outputs of the cell have been set to complementary values. Note that the two inverters I_3 and I_4 at the outputs of the SABL cell only invert the complementary values of n_5 and n_6, so the output signals *out* and $\overline{out}$ are still complementary values.

Precharge Phase. At the onset of the precharge phase, the clock signal switches from 1 to 0. The NMOS transistor Pr_3 is turned off, which disconnects the nodes n_3 and n_4 from *GND*. Simultaneously, the PMOS transistors Pr_1 and Pr_2 are turned on. Thus, the nodes n_5 and n_6 are set to 1. As a result, the output signals *out* and $\overline{out}$ are set to the precharge value 0. Via the NMOS transistors of I_1 and I_2, which are also turned on, the nodes n_1 and n_2 are set to 1. Furthermore, one path of the DPDN to node n_3 or node n_4 is still conducting because the input signals in_1, $\overline{in_1}$, ..., in_z, $\overline{in_z}$ are still set to complementary values. The reason for this is the delay between the outputs and the inputs of connected SABL cells and the fact that all SABL cells are precharged at the same time. Therefore, when an SABL cell is switched to the precharge phase, the input signals in_1, $\overline{in_1}$, ..., in_z, $\overline{in_z}$ stay in a complementary state for a short period of time. Within this time, the nodes n_3 and n_4 are set to 1 via the conducting path of the DPDN. Afterwards, the input signals in_1, $\overline{in_1}$, ..., in_z, $\overline{in_z}$ are set to the precharge value 0 and the conducting path in the DPDN is removed. This event finishes the precharge phase.

Constant Power Consumption of SABL Cells

For a constant power consumption of n-type SABL cells, the DPDN in the cells must fulfill four requirements, and the internal structure of the cells must be balanced. In the following, we discuss these issues in more detail.

DPDN Requirement 1. A constant internal power consumption of n-type SABL cells is achieved by charging all internal nodes during the precharge phase and discharging all internal nodes (except one of n_5 and n_6) during the evaluation phase. To achieve this, the DPDN of n-type SABL cells must be built in such a way that every internal node of the DPDN is connected to one of the four output nodes n_1, n_2, n_3, or n_4 of the DPDN for complementary input signals in_1, $\overline{in_1}$, ..., in_z, $\overline{in_z}$. Together with the NMOS pass-transistor M_1, this structure ensures that all internal nodes of the DPDN are discharged to 0 during the evaluation phase and charged to 1 during the precharge phase.

DPDN Requirement 2. The DPDN must be implemented in such a way that every possible conducting path in the DPDN has the same resistance. This is ensured by the employment of the same number of transistors with identical parameters (width, *etc.*) in every conducting path of the DPDN. The same resistance for each such path ensures that the instantaneous power consumption for charging and discharging events through the DPDN is the same.

DPDN Requirement 3. Both wires of every complementary input wire pair must be connected to the same number of gate terminals of transistors with identical parameters. This ensures that the capacitances of complementary inputs of SABL cells are pairwise balanced. As already discussed, this is a requirement for a constant power consumption of DRP cells in general.

DPDN Requirement 4. A conducting path through the DPDN may only occur after all input signals in_1, $\overline{in_1}$, ..., in_z, $\overline{in_z}$ have been set to complementary values. This ensures that the n-type SABL cells only evaluate after all input signals have been set to complementary values. As a result, the TOE of the n-type SABL cells is data independent.

Balanced Internal Structure. As discussed before, all internal nodes of an n-type SABL cell are precharged to 1 during the precharge phase. In the subsequent evaluation phase, all internal nodes are discharged to 0 except for exactly one of the two nodes n_5 and n_6. Therefore, the capacitances of these two nodes must be balanced to achieve a constant internal power consumption of n-type SABL cells in each clock cycle. The capacitances of these two nodes can be balanced by designing the layouts of the inverters I_1 and I_2, the inverters I_3 and I_4, and the PMOS transistors Pr_1 and Pr_2 in a pairwise identical manner. An identical layout of the inverters I_3 and I_4 also ensures that the capacitances and resistances at the complementary outputs out and $\overline{out}$ of n-type SABL cells are balanced.

Examples of SABL Cells

In the following, an n-type SABL NAND cell and an n-type SABL D-flip-flop are presented in detail.

N-Type SABL NAND Cell. Figure 7.6 shows the transistor schematic of an n-type SABL NAND cell. The specific implementation of the DPDN realizes the NAND function. Whenever at least one of the two input signals a and b is 0, *i.e.* $\overline{a}$ or/and $\overline{b}$ is 1, the left branch of the DPDN is conducting. Thus, the output q of the SABL cell is set to 1 during the evaluation phase. The output $\overline{q}$ stays at 0. Only when both input signals a and b are 1, the right branch of the DPDN is conducting. In this case, the output $\overline{q}$ of the SABL cell is set to 1 and the output q stays at 0. Note that an n-type SABL NOR cell can easily be derived from the n-type SABL NAND cell by inverting the input signals a and b and the output signal q of the SABL NAND cell. This conforms to swapping of the respective complementary wires.

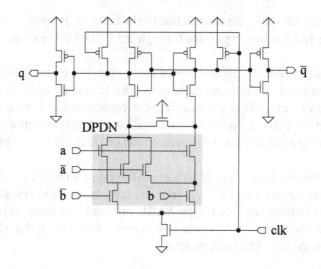

Figure 7.6. Transistor schematic of an n-type SABL NAND cell.

The presented DPDN fulfills all four requirements formulated in the last section. First, for complementary input signals, all internal nodes of the DPDN are connected to one of the four output nodes of the DPDN. Second, every conducting path goes through two NMOS transistors and thus has the same resistance if all NMOS transistors in the DPDN are equally sized. Third, all complementary input wires are connected to the same number of transistors. The inputs a and $\overline{a}$ are connected to two transistors each. The inputs b and $\overline{b}$ are connected to one transistor each. This results in pairwise balanced input capacitances. Fourth and last, a conducting path occurs in the DPDN only after all input signals are set to complementary values. Note that the original implementation of the DPDN of the SABL NAND cell as presented in [TAV02] did not fulfill the last requirement. The improved version as shown in Figure 7.6 has been presented in [TV05b].

N-Type SABL D-Flip-Flop. An n-type SABL D-flip-flop is shown in Figure 7.7. It consists of two stages, as described in Section 7.3.1. Stage 1, a p-type SABL latch, is precharged when the clock signal is 1. Stage 2, an n-type SABL latch, is precharged when the clock signal is 0. Note that the input signals d and $\overline{d}$ are provided by some n-type SABL cell, which is also precharged when the clock signal is 0. Since the n-type SABL latch at the output defines how signals are provided, the whole cell is an n-type SABL D-flip-flop.

At the negative clock edge, the p-type SABL latch stores the current complementary data values at the inputs d and $\overline{d}$, while the n-type SABL latch and other n-type SABL cells are precharged. At the subsequent positive clock

Stage 1: p-type SABL latch | Stage 2: n-type SABL latch

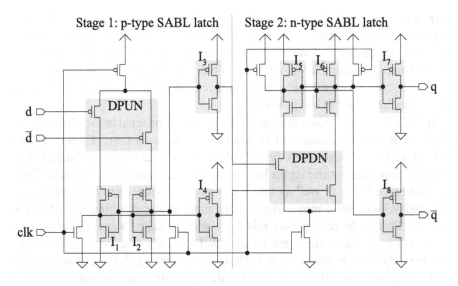

Figure 7.7. Transistor schematic of an n-type SABL D-flip-flop.

edge, the n-type SABL latch stores the data values from the p-type SABL latch (which gets precharged) and provides the data values at the outputs q and $\bar{q}$. The delay between the n-type SABL cell providing the inputs d and $\bar{d}$ and the p-type SABL latch must be high enough to ensure that the complementary data values are stored before the n-type SABL cell gets precharged. The same is true for the delay between the p-type SABL latch and the n-type SABL latch. If the delay between the p-type SABL latch and the n-type SABL latch is too small, an even number of inverters can be added after I_3 and I_4, respectively.

For a constant power consumption of the n-type SABL D-flip-flop, the capacitances at the outputs of the inverters I_1 and I_2, ..., I_7 and I_8 must be pairwise balanced. This can be achieved by a careful design of the layout of the n-type SABL D-flip-flop. Note that in the n-type SABL D-flip-flop, no pass-transistors (like M_1 in Figure 7.5) are required to charge all nodes of the DPUN and to discharge all nodes of the DPDN during the evaluation phase of the respective latch. The reason is that when the p-type SABL latch or the n-type SABL latch are in the evaluation phase, the preceding cells are in the precharge phase. Thus, all transistors of the DPUN or the DPDN of the respective latch are turned on anyway.

7.4.2 Wave Dynamic Differential Logic

Wave Dynamic Differential Logic has been introduced by Tiri and Verbauwhede in [TV04a]. WDDL cells are built based on SR cells that are available in existing standard-cell libraries. The structure of WDDL cells is much simpler than that of SABL cells (see Section 7.4.1). This leads in general to less complex and significantly smaller circuits. However, this comes at the expense of the DPA resistance of WDDL circuits. The lower DPA resistance is mainly due to the combinational WDDL cells. Their internal power consumption and their TOE are data dependent. An advantage of WDDL circuits is that they can also be realized on FPGAs [TV04c].

In WDDL circuits, just the sequential cells are connected to the clock signal. Only these cells precharge and evaluate all at the same time. Combinational WDDL cells precharge when their inputs have been set to the precharge value. They evaluate, when their inputs have been set to complementary values. Thus, the precharge value as well as the complementary values provided by the sequential WDDL cells move like a wave through the combinational WDDL circuit. This is the reason why this logic style is called *Wave* Dynamic Differential Logic. Since the combinational WDDL cells precharge and evaluate successively, the current peaks in WDDL circuits are much lower than those in SABL circuits. Even if the propagation delays of the complementary wires in balanced WDDL circuits are pairwise identical, the TOE of the combinational WDDL cells is still varying a little bit depending on the processed data. This effect is also known as early propagation. The reason is the already mentioned data-dependent TOE of the combinational WDDL cells.

Similar to SABL, the area requirements of WDDL circuits are at least doubled compared to CMOS circuits with equivalent functionality. Also the power consumption of WDDL circuits increases significantly. The maximum clock rates in WDDL circuits are approximately the same as in corresponding CMOS circuits. However, due to the fact that the WDDL D-flip-flop consists of two stages of SR D-flip-flops (for details see below), it is necessary to double the clock frequency to achieve the same throughput as in CMOS or SABL circuits. The drawback of doubling the clock frequency is an increased power consumption. Examples of WDDL circuits and their properties are presented in [TV04a] and [THH$^+$05]. The use of WDDL in a semi-custom design flow is presented in [TV05a] and [TV06].

General Description of WDDL Cells

In the following, we discuss combinational WDDL cells. The structure of sequential WDDL cells is quite different from the one of combinational WDDL cells. It is discussed later by means of an example.

Figure 7.8 shows the generic structure of a combinational WDDL cell. A combinational WDDL cell basically consists of two circuits that realize the Boolean functions F_1 and F_2. These functions must be defined in the following way: If the input signals in_1, $\overline{in_1}$, ..., in_z, $\overline{in_z}$ are set to complementary values, complementary output values are calculated according to the intended logic function of the cell. This means, for complementary input values, F_1 and

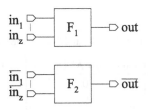

Figure 7.8.   Generic structure of a combinational WDDL cell.

F_2 must satisfy the following equation:

$$F_1(in_1, ..., in_z) = \overline{F_2(\overline{in_1}, ..., \overline{in_z})} \qquad (7.1)$$

In order to achieve the same transitions at the cell outputs *out* and $\overline{out}$ in each clock cycle, F_1 and F_2 must be *positive monotonic* Boolean functions. The second important effect of using positive monotonic Boolean functions is that combinational WDDL cells set their complementary outputs automatically to the precharge value when their complementary inputs have been set to the precharge value. In the following, we first discuss positive monotonic Boolean functions in general. Afterwards, we look in detail at the behavior of a combinational WDDL cell during a clock cycle, *i.e.* during an evaluation phase and a subsequent precharge phase. In all considerations, we assume that the precharge value is 0.

Monotonic Boolean Functions.   A *monotonic* Boolean function is a function that changes its output value in a monotonic way whenever its input values also change in a monotonic way. Monotonic changes mean that the logic values change only in one direction, *i.e.* either only $0 \rightarrow 1$ or only $1 \rightarrow 0$ transitions occur. As a result, an output value only performs either a single $0 \rightarrow 1$ or a single $1 \rightarrow 0$ transition as long as the input values of the function are changed in a monotonic way. An example of a monotonic Boolean function is the AND function, see Table 7.2. An example of a Boolean function that is not monotonic is the XOR function.

A *positive monotonic* Boolean function is a monotonic Boolean function where the direction of the monotonic input changes and the monotonic output change is the same. Accordingly, a *negative monotonic* Boolean function is a monotonic Boolean function where the direction of the monotonic input changes and the monotonic output change is opposite. An example of a positive monotonic Boolean function is the OR function, see Table 7.2. An example of a negative monotonic Boolean function is the NAND function.

Positive monotonic Boolean functions have the property that for all input values set to 0, the output value of the function is 0. Otherwise, $0 \rightarrow 1$ changes of the input values could not cause a $0 \rightarrow 1$ change of the output value of the

Table 7.2. Truth tables of 2-input AND cells and 2-input OR cells.

Input a	Input b	Output $q = \text{AND}(a, b)$	Output $q = \text{OR}(a, b)$
0	0	0	0
0	1	0	1
1	0	0	1
1	1	1	1

function, as the output value would already be 1. With the same argumentation, it can be shown that for all input values set to 1, the output of the positive monotonic Boolean function is 1.

Evaluation Phase. At the onset of the evaluation phase, all complementary input signals of a combinational WDDL cell have been set to the precharge value 0 because of the preceding precharge phase. Since only positive monotonic Boolean functions are allowed, the complementary output signals of the cell have also been set to 0. When the input signals are then set to complementary values, only $0 \rightarrow 1$ transitions occur at the inputs of the WDDL cell and thus at the inputs of the positive monotonic Boolean functions F_1 and F_2. As a result, only one transition can occur at each of the two complementary outputs *out* and $\overline{out}$ of a WDDL cell. But since F_1 and F_2 always calculate complementary output values for complementary input values as indicated in (7.1), only one complementary output of the WDDL cell performs a $0 \rightarrow 1$ transition while the other output stays at 0.

Precharge Phase. In the subsequent precharge phase, all complementary input signals of the combinational WDDL cell that have been set to 1 are now set back to 0. Thus, only $1 \rightarrow 0$ transitions occur at the inputs of the positive monotonic Boolean functions F_1 and F_2. As a result, only one $1 \rightarrow 0$ transition can occur at the output of each function. As already mentioned, when all complementary input signals of a combinational WDDL cell are set to 0, also the complementary output signals are set to 0. Therefore, only one $1 \rightarrow 0$ transition occurs at the complementary outputs of the WDDL cell. This happens when the output that has been set to 1 in the preceding evaluation phase is set back to 0.

Constant Power Consumption of WDDL Cells

The instantaneous power consumption of combinational WDDL cells is not entirely constant mainly due to three reasons: First, the resistances of the charging and discharging paths of internal nodes and of complementary outputs is not the same for all input values. For different input values, usually different con-

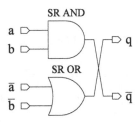

Figure 7.9. Cell schematic of a WDDL NAND cell.

ducting paths with a varying number and arrangement of MOS transistors are established. Second, the internal power consumption of combinational WDDL cells is typically not constant. In general, not all internal nodes of a WDDL cell are charged and discharged in the same way in each clock cycle for all data values. This behavior leads to the so-called memory effect, *i.e.* the charges stored at internal nodes of a cell depend on the processed data values. Third, the TOE of combinational WDDL cells depends on the input values.

The reason for all these problems is that the combinational WDDL cells are built in a simple way from basic SR cells like AND cells and OR cells. Combinational WDDL cells that avoid all these problems would have a significantly increased size and complexity.

Examples of WDDL Cells

In the following, a combinational WDDL cell implementing the NAND function is presented. Furthermore, as an example of a sequential WDDL cell, a WDDL D-flip-flop is shown.

WDDL NAND Cell. Figure 7.9 shows the cell schematic of a WDDL NAND cell. It consists of a 2-input SR AND cell and a 2-input SR OR cell. The truth table of a WDDL NAND cell for complementary input values is shown in Table 7.3. The AND function and the OR function fulfill (7.1) for complementary input values. Furthermore, both are positive monotonic Boolean functions. Thus, this cell has the expected behavior of a combinational WDDL cell that we have discussed above.

WDDL D-Flip-Flop. Figure 7.10 shows the cell schematic of a WDDL D-flip-flop. It consists of four SR D-flip-flops arranged in two stages as described in Section 7.3.1. One of the stages always stores the precharge value while the other stage stores complementary logic values. The WDDL D-flip-flop alternates between the precharge phase and the evaluation phase always at the positive edge of the clock signal. In the clock cycle where the precharge value is stored in stage 2, the WDDL D-flip-flop is in the precharge phase. In the

Table 7.3.   Truth table of a WDDL NAND cell for complementary input values.

a	b	$\bar{a}$	$\bar{b}$	$q = OR(\bar{a},\bar{b})$	$\bar{q} = AND(a,b)$
0	0	1	1	1	0
0	1	1	0	1	0
1	0	0	1	1	0
1	1	0	0	0	1

clock cycle where the complementary values are stored in stage 2, the WDDL D-flip-flop is in the evaluation phase.

During the evaluation phase of a WDDL circuit, stage 1 stores the pre-charge value and stage 2 provides complementary logic values to the subsequent WDDL cells. At the end of the evaluation phase, complementary logic values are provided to the inputs of stage 1. With the next positive clock edge, the WDDL circuit enters the precharge phase. At this clock edge, stage 1 stores the complementary logic values provided to its inputs and stage 2 stores the pre-charge value from stage 1. As a result, stage 2 provides the precharge value to the WDDL cells connected to the outputs of the WDDL D-flip-flop. At the end of the precharge phase, also the WDDL cell providing the inputs to stage 1 has been precharged. At the next positive clock edge, stage 1 stores the precharge value provided at its inputs. At the same time, stage 2 stores the complementary logic values from stage 1.

Using four SR D-flip-flops instead of one entails a significant increase of the area requirements of WDDL D-flip-flops. However, it is necessary to also shift the precharge value through the two stages of the WDDL D-flip-flop to precharge it correctly. When the operation of a WDDL circuit is started, it must be ensured that the two stages of the WDDL D-flip-flops in the circuit are reset correctly. One stage needs to be set to the precharge value while the other stage needs to be set to the actual complementary reset value. Note that a WDDL D-flip-flop avoids most of the problems of combinational WDDL cells concerning a constant power consumption (e.g. data-dependent TOE).

7.5   Notes and Further Reading

Hardware Countermeasures at the Architecture Level.   The basic idea of hiding countermeasures was already mentioned in the first article on power analysis attacks [KJJ99] by Kocher *et al.* During the last years, several concrete proposals for hiding countermeasures have been published. We now provide a brief survey of hardware countermeasures that can be implemented at the architecture level.

In [CKN01], Coron *et al.* analyzed the effect of an RLC filter that was inserted into the power supply line of the cryptographic device. However, the authors

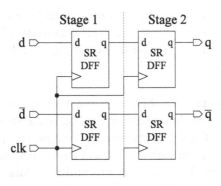

Figure 7.10. Cell schematic of a WDDL D-flip-flop.

conclude that this is not an effective way to prevent leakage. Shamir presented the idea of decoupling the power supply of cryptographic devices using two capacitors [Sha00]. The basic idea of this approach is that one capacitor is charged by the power supply while the other capacitor powers the device. The capacitors are periodically switched. Hence, the device is never connected to the power supply directly. A similar idea that uses a three-phase charge pump to power the device was presented in [CPM05].

Another approach to reduce the leakage of cryptographic devices is to use active circuits on the devices to flatten the power consumption. Rakers *et al.* [RCCR01] discussed this approach in the context of RFIDs. A circuit to suppress exploitable signals on contact-based systems was presented in [RWB04] by Ratanpal *et al.* Similar approaches were also presented by Muresan *et al.* [MVZG05] and by Mesquita *et al.* [MTT$^+$05].

Benini *et al.* studied how energy-aware design techniques can be used to counteract power analysis attacks in [BMM$^+$03b] and [BMM$^+$03a]. The basic idea of this approach is to have different components on the chip that have a similar functionality. Random numbers are then used to decide which component is used to perform the operations of the executed cryptographic algorithm. As the number of components is usually low, also the degree of randomness that is induced by this countermeasure is typically quite low.

Saputra *et al.* studied how to integrate secure instructions into a non-secure processor in [SVK$^+$03] and [SOV$^+$05]. The secure instructions are implemented using dual-rail precharge circuits.

Besides the countermeasures that affect the amplitude dimension of the power consumption, there are also proposals to randomize the execution of cryptographic algorithms in hardware. In [MMS01a], May *et al.* proposed to use non-deterministic processors to counteract power analysis attacks. These processors randomly change the sequence of the executed program during each execution. Of course, only the sequence of independent instructions can be changed. The non-determinism of processors can be increased by also randomly inserting additional instructions. This was studied in [IPS02]. Yang *et al.* proposed to randomly change the supply voltage and the clock frequency of circuits in [YWV$^+$05].

Asymmetric Cryptography. Many asymmetric cryptographic schemes require the calculation of the binary algorithm, e.g. square-and-multiply, double-and-add, or a variant of it, e.g. k-ary, sliding window, *etc.* In the case of RSA, the binary algorithm is done with elements of a finite ring and is called square-and-multiply algorithm. It is used to perform the modular exponentiation. In the case of ECC, the binary algorithm is done in the additive group of elliptic curve points and called double-and-add algorithm. It is used to perform the point multiplication. As noted in Section 5.5, the conditional execution of operations that might occur in the binary algorithm (or a variant of it) can give away the entire exponent (scalar) if the operations (square and multiply, double and add) can be distinguished in the power trace. Hiding countermeasures can be applied in three different ways to the binary algorithm (or a variant of it). In the subsequent paragraphs we use the term exponentiation algorithm to refer to the binary algorithm and variants of it.

First, one can randomize the sequence of operations (square and multiply, double and add) in the exponentiation algorithm. This is typically done by randomizing the representation of the exponent in RSA and by randomizing the representation of the scalar in ECC. Walter [Wal02a] gave an example of this idea for RSA, Oswald and Aigner [OA01] and Ha and Moon [HM02] gave examples for ECC.

Second, one can fix the sequence of operations in the exponentiation algorithm. The simplest way to do this is to ensure that the multiplication (addition) operation is always executed [Cor99]. This is easy to implement but imposes a serious penalty in terms of performance. A better way to ensure that the sequence of operations is fixed during exponentiation is to use the Montgomery ladder. Montgomery [Mon87] introduced this technique for ECC. Lopez and Dahab [LD99] gave an optimized version of it for binary elliptic curves. Brier and Joye [BJ02] published versions for Weierstrass elliptic curves over finite fields with characteristic $\neq 2$ or 3. Chevallier-Mames *et al.* [CMCJ04] published another idea that they called *side-channel atomicity* which works as follows. Each operation is implemented as the repetition of blocks of instructions that look alike in the power trace. Based on these blocks, the code of the exponentiation algorithm is unrolled such that it appears as a repetition of the same atomic block. The sequence of blocks does then not depend on the exponent that is used in the exponentiation algorithm. Möller [Möl01] discussed how exponentiation algorithms (in ECC) that require precomputing certain elliptic curve points (and the recoding of the exponent) can be used in a natural way to get an efficient double-and-add-always algorithm. Thériault [Thé06] improved on Möller's work such that the exponentiation can also be performed by scanning the bits from the MSB down to the LSB. This implies that the recoding of the exponent can be done on the fly (and not in a precomputation step).

Third, one can try to ensure that the operations (square and multiply, double and add) lead to indistinguishable power traces. This seems to be easy in case of square and multiply. One simply uses the same piece of hardware (or piece of code) to calculate the square and the multiply operation. Realizing the same idea in case of double and add is more interesting. The simplest approach, of which an example is given in [CMCJ04], is to insert dummy instructions to make double and add look alike. Brier and Joye showed in [BJ02] how to unify the formulas for double and add in case of Weierstrass elliptic curves over finite fields with characteristic $\neq$ 2 or 3. Liardet and Smart [LS01] pointed out that for elliptic curves in the Jacobi form, the addition and double operation work in the same way. Hence, curves in this form inherently have indistinguishable double and add operations. Joye and Quisquater [JQ01] presented a similar study for Hessian elliptic curves. Hasan [Has00] discussed countermeasures for Koblitz curves.

Bajard *et al.* [BILT04] pointed out that implementations using residue number systems allow randomizing the representation of finite field elements. They observed that it is either possible to randomly choose the initial base elements of the residue number system or to randomly change the bases before or during an exponentiation. Ciet *et al.* [CNPQ03] investigated how using residue number systems can help to design more power analysis resistant architectures.

DRP Logic Styles. In addition to SABL (see Section 7.4.1) and WDDL (see Section 7.4.2), there exist various other proposals for DRP logic styles. Bystrov *et al.* [BSYK03] and Sokolov *et al.* [SMBY04, SMBY05] presented the so-called *Dual-Spacer Dual-Rail (DSDR)* logic style. In this DRP logic style, both possible precharge values are used in an alternating way. This means that if a dual-rail wire pair is precharged to (0, 0) in one clock cycle, it is precharged to (1, 1) in the next clock cycle and vice versa. In doing so, it is ensured that both complementary outputs of a cell are switched in each clock cycle. As a result, the energy consumption of a logic cell during a complete clock cycle is not influenced by any imbalance of the complementary outputs. However, especially for low clock rates, also the variations of the energy consumed in each half of a clock cycle can be observed in power measurements.

A DRP logic style that works similarly to DSDR logic was proposed by Bucci *et al.* [BGLT06]. The logic style is called *Three-Phase Dual-Rail Precharge Logic (TDPL)*. During a clock cycle, TDPL cells run through three operation phases. In the first phase, the complementary outputs of a TDPL cell are charged to (1, 1). Next, in the evaluation phase, one output is discharged to

0 according to the input values and the function of the TDPL cell. In the third phase, also the second output is discharged to 0. As discussed for DSDR logic, this leads to a constant energy consumption in each clock cycle independent of the balancing of the complementary outputs.

Another DRP logic style is *3-state Dynamic Logic (3sDL)* presented by Trifiletti *et al.* [AMM$^+$05]. This logic style uses $V_{DD}/2$ as precharge value. At the end of the evaluation phase, always one complementary output wire of a cell is charged to V_{DD} while the other output wire is discharged to GND. In the subsequent precharge phase, the two complementary wires are connected together. If both wires have the same capacitance, the voltage level at both wires settles to $V_{DD}/2$. This approach saves power. Another specialty of 3sDL is that the inverted output of a cell is not routed through the circuit. Instead, a dummy capacitor is added to the inverted output whose capacitance matches the capacitance of the non-inverted output. A major drawback of this approach is that this matching must be done individually for every cell in the circuit.

The DPA resistance of WDDL cells, DSDR cells, and similar DRP cells is degraded due to mainly two effects: early propagation and the memory effect. Guilley *et al.* [GHM$^+$04] proposed a special structure for DRP cells that avoids these effects. Drawbacks are a reduced cell speed and a significantly increased cell area.

So far, almost all articles on DRP logic styles have been published by people from academia. One of the very few scientific articles explicitly stating that also industry uses this technique is [FS03].

Asynchronous Logic Styles.   Asynchronous (or self-timed) circuits have also been proposed as a countermeasure against power analysis attacks. Using such circuits for this purpose has for instance been discussed by Moore *et al.* [MAC$^+$02] and by Yu *et al.* [YFP03]. Asynchronous circuits that counteract power analysis attacks are usually implemented as DRP circuits. Therefore, the power consumption of these asynchronous circuits can be balanced in the same way as discussed for synchronous DRP circuits in Section 7.3. Further considerations for balancing the power consumption in asynchronous circuits have for example been made by Kulikowski *et al.* [KSS$^+$05]. Unfortunately, the DPA resistance of asynchronous circuits still relies on the balancing of complementary wires. Additionally, asynchronous circuits are hard to verify and there is still a lack of mature EDA tools that support the design of such circuits.

A design flow for DPA-resistant asynchronous circuits was proposed by Kulikowski *et al.* [KST06]. An asynchronous implementation of AES to resist power and timing attacks was presented by Yu and Brée [YB04]. Gürkaynak *et al.* [GOK$^+$05] proposed the use of a globally-asynchronous locally-synchronous (GALS) design to increase the DPA security of circuits.

Current-Mode Logic Styles. In CML circuits, the output value of a logic cell is defined by currents that are passing through the cell. The sum of these currents is rather constant and more or less independent of the actual output value. This makes CML logic styles interesting for DPA-resistant circuits. In [TL05], Toprak and Leblebici proposed the use of *MOS Current-Mode Logic (MCML)* [YY92] as a DPA-resistant logic style. A drawback of MCML is its increased static power consumption. A CML logic style that avoids this drawback is *Dynamic Current-Mode Logic (DyCML)* [AE01]. The use of DyCML in DPA-resistant circuits has been proposed by Mace *et al.* in [MSH+04].

Chapter 8

ATTACKS ON HIDING

The goal of hiding countermeasures is to make the power consumption of cryptographic devices independent of the performed operations and the processed values. However, in practice this goal can only be achieved to a certain degree, see Chapter 7. Attacks on protected devices are therefore still possible. In most cases though, these attacks require significantly more effort than attacks on unprotected devices.

In this chapter, we first discuss the effectiveness of hiding countermeasures in general. In particular, we analyze how the different types of hiding countermeasures increase the number of power traces that are needed for DPA attacks. Subsequently, we look at two specific countermeasures in detail. We discuss DPA attacks on countermeasures that destroy the alignment of power traces, and we look at the effectiveness of the DRP logic styles presented in Section 7.4.

8.1 General Description

In Chapter 7, we have introduced two types of hiding countermeasures. On the one hand, there are countermeasures that randomize the execution of cryptographic algorithms. On the other hand, there are countermeasures that reduce the SNR of the executed operations.

We now analyze the effectiveness of these two types of countermeasures against DPA attacks by determining their effect on $\rho_{ck,ct}$. Recall that this is the correlation between the hypothetical power consumption for the correct key hypothesis H_{ck} and the power consumption at the moment of time ct. This is the moment of time when the device processes the attacked intermediate result. As pointed out in Section 6.4, the correlation $\rho_{ck,ct}$ determines the number of power traces that are needed to perform DPA attacks. For our analysis, we model the power consumption of the cryptographic device as in (4.8). This means that we use P_{total} to denote the power consumption at the moment ct. The correlation $\rho_{ck,ct}$ therefore corresponds to $\rho(H_{ck}, P_{total})$.

8.1.1 Time Dimension

The random insertion of dummy operations and shuffling change the execution of the cryptographic algorithm randomly. The attacked intermediate result is therefore processed at a different moment of time in each power trace, *i.e.* ct is randomly distributed. The statistical distribution of ct depends on the way the random insertion of dummy operations and the shuffling are implemented. If only shuffling is used, ct is typically uniformly distributed. The random insertion of dummy operations often leads to binomial or uniform distributions of ct. If the random insertion of dummy operations and shuffling are combined, the resulting distribution is the superposition of the corresponding distributions.

Independent of the shape of the distribution of ct, we denote the maximum of this distribution by $\hat{p}$. Furthermore, we denote the power consumption that is located at this position by $\hat{P}_{total}$. Hence, $\hat{P}_{total}$ has the following properties. With probability $\hat{p}$, $\hat{P}_{total}$ corresponds to the power consumption of the attacked intermediate result, *i.e.* with probability $\hat{p}$ it holds that $\hat{P}_{total} = P_{total}$. With probability $(1 - \hat{p})$, $\hat{P}_{total}$ corresponds to the power consumption of some other operations. We refer to the power consumption of these operations as P_{other}. The covariance $Cov(H_{ck}, \hat{P}_{total})$ can therefore be calculated as follows:

$$Cov(H_{ck}, \hat{P}_{total}) = \hat{p} \cdot Cov(H_{ck}, P_{total}) + (1 - \hat{p}) \cdot Cov(H_{ck}, P_{other})$$

Since $\hat{p}$ is the maximum probability, the correlation between H_{ck} and $\hat{P}_{total}$ leads to the highest correlation coefficient that occurs in a DPA attack on a protected device. The correlation $\rho(H_{ck}, \hat{P}_{total})$ hence determines the number of power traces that are needed for the attack. This correlation can be calculated based on $\rho(H_{ck}, P_{total})$ as shown in (8.1). The simplifications in this equation are possible because we assume that P_{other} and P_{total} are independent.

$$
\begin{aligned}
\rho(H_{ck}, \hat{P}_{total}) &= \frac{\hat{p} \cdot Cov(H_{ck}, P_{total}) + (1 - \hat{p}) \cdot Cov(H_{ck}, P_{other})}{\sqrt{Var(H_{ck}) \cdot Var(\hat{P}_{total})}} \\
&= \frac{\hat{p} \cdot Cov(H_{ck}, P_{total})}{\sqrt{Var(H_{ck}) \cdot Var(\hat{P}_{total})}} \\
&= \rho(H_{ck}, P_{total}) \cdot \hat{p} \cdot \sqrt{\frac{Var(P_{total})}{Var(\hat{P}_{total})}} \quad (8.1)
\end{aligned}
$$

The effect of shuffling and the random insertion of dummy operations mainly depends on $\hat{p}$. The probability $\hat{p}$ linearly reduces the correlation $\rho(H_{ck}, \hat{P}_{total})$. Halving $\hat{p}$, halves this correlation and hence quadruples the number of needed power traces, see Section 6.4. For example, if the 16 S-box look-ups of an AES implementation are shuffled, $\hat{p} = 1/16$ and $16^2 = 256$ times more traces are needed.

Besides $\hat{p}$, $\rho(H_{ck}, \hat{P}_{total})$ also depends on the way the variance of the power traces is changed by the random displacement of ct. The bigger $Var(\hat{P}_{total})$ is compared to $Var(P_{total})$, the lower $\rho(H_{ck}, \hat{P}_{total})$ becomes. However, in practice the variance of the power consumption does not change significantly due to the misalignment. Usually, the goal of designers is to build devices that have a similar distribution of the power consumption in each clock cycle in order to make the alignment of the traces difficult, see Section 8.2. A side effect of this is that $Var(P_{total})/Var(\hat{P}_{total})$ is close to one.

For DPA attacks on misaligned power traces, the correlation for the correct key hypothesis can be calculated as follows.

$$\rho(H_{ck}, \hat{P}_{total}) = \rho(H_{ck}, P_{total}) \cdot \hat{p} \cdot \sqrt{\frac{Var(P_{total})}{Var(\hat{P}_{total})}} \qquad (8.2)$$

8.1.2 Amplitude Dimension

Countermeasures that change the power consumption characteristics of the operations which process the attacked intermediate result, change the SNR of these operations. In Section 6.3, we have already discussed how the SNR can be mapped to a correlation coefficient. The effectiveness of hiding countermeasures that affect the amplitude dimension of the power consumption can hence be calculated according to (6.5).

Countermeasures that lower the SNR of the operations processing the attacked intermediate result reduce the correlation for the correct key hypothesis as follows.

$$\rho(H_{ck}, P_{total}) = \frac{\rho(H_{ck}, P_{exp})}{\sqrt{1 + \frac{1}{SNR}}} \qquad (8.3)$$

For low SNRs it holds that the correlation is proportional to $\sqrt{SNR}$.

The Effectiveness of DPA-Resistant Logic Styles

DPA-resistant logic styles are a popular hiding countermeasure against power analysis attacks. Therefore, we now discuss the effectiveness of DPA-resistant logic styles in more detail.

Consider a digital circuit that consists of l unprotected CMOS cells. We denote the power consumption of these cells by $P_1, \ldots, P_l$. Furthermore, we assume that the power consumption of the cells is independent and identically distributed and that P_1 is the target of a DPA attack. The SNR of the power consumption of cell 1 can hence be calculated as shown in (8.4).

$$
\begin{aligned}
SNR &= \frac{Var(P_{exp})}{Var(P_{sw.\,noise}) + Var(P_{el.\,noise})} \\
&= \frac{Var(P_1)}{Var(P_2 + \ldots + P_l) + Var(P_{el.\,noise})} \\
&= \frac{Var(P_1)}{Var(P_2) + \ldots + Var(P_l) + Var(P_{el.\,noise})}
\end{aligned}
\tag{8.4}
$$

The l unprotected cells are now replaced by DPA-resistant cells. The variances of the power consumptions of the cells are therefore reduced by a factor $a > 0$, where we assume a to be equal for all cells. Note that this reduction affects the variance of the exploitable power consumption P_1 and the variances of $P_2, \ldots, P_l$. In this scenario, the SNR for the power consumption of cell 1 can be calculated as shown in (8.5). It can be observed in this equation that the employment of a DPA-resistant logic style is equivalent to increasing the electronic noise by a.

$$
\begin{aligned}
SNR_a &= \frac{\frac{1}{a} \cdot Var(P_1)}{\frac{1}{a} \cdot Var(P_1) + \ldots + \frac{1}{a} \cdot Var(P_l) + Var(P_{el.\,noise})} \\
&= \frac{\frac{1}{a} \cdot Var(P_1)}{\frac{1}{a} \cdot Var(P_1 + \ldots + P_l) + Var(P_{el.\,noise})} \\
&= \frac{\frac{1}{a} \cdot Var(P_{exp})}{\frac{1}{a} \cdot Var(P_{sw.\,noise}) + Var(P_{el.\,noise})} \\
&= \frac{Var(P_{exp})}{Var(P_{sw.\,noise}) + a \cdot Var(P_{el.\,noise})}
\end{aligned}
\tag{8.5}
$$

$$
\frac{SNR_a}{SNR} = \frac{Var(P_{sw.\,noise}) + Var(P_{el.\,noise})}{Var(P_{sw.\,noise}) + a \cdot Var(P_{el.\,noise})}
\tag{8.6}
$$

The ratio of (8.5) and (8.4) is shown in (8.6). Based on this ratio, two observations can be made. First, if no switching noise is present, *i.e.* $Var(P_{sw.\,noise}) = 0$, a DPA-resistant logic style reduces the SNR by a. This is the maximum reduction that can be achieved. Second, if no electronic noise is present, *i.e.* $Var(P_{el.\,noise}) = 0$, DPA-resistant logic styles do not reduce the SNR.

In practice, the reduction of the SNR that is achieved by using a DPA-resistant logic style, is always between these two extremes. However, as this reduction depends on the amount of switching noise and electronic noise, the SNR is not suitable to compare the effectiveness of different logic styles. In this book, we compare logic styles based on the reduction of the variance of the power consumption they achieve. The bigger the factor a of a logic style is, the higher is its DPA resistance. Notice that this method of comparing logic styles does not make any assumptions about the power model that is used to exploit the leakage of the logic styles.

> We measure the effectiveness of logic styles based on the factor by which the logic styles reduce the variance of the power consumption in relation to CMOS.

8.2 DPA Attacks on Misaligned Power Traces

In practice, there are essentially three different ways to perform DPA attacks on misaligned power traces. The first one is to perform DPA attacks on the power traces as they are. In this case, the effectiveness of the DPA attacks can be calculated according to (8.1). The second way to perform DPA attacks is to try to align the power traces before performing the attacks. In case this alignment is successful, the effectiveness of the DPA attacks is significantly increased. The third way of performing DPA attacks is to preprocess the power traces in such a way that the effect of the misalignment is reduced.

In this section, we first discuss reasons for the misalignment of power traces in practice. Subsequently, we provide an overview of techniques to align and to preprocess power traces. Finally, we provide examples of DPA attacks on misaligned power traces. In these examples, we attack an implementation of AES that is protected by the random insertion of dummy operations and by shuffling, respectively.

8.2.1 Reasons for Misalignment

In practice, there are several reasons why power traces can be misaligned. Countermeasures, like the random insertion of dummy operations and shuffling, are just some of the possible reasons. In practice, the fact whether the recorded power traces are aligned or not, essentially depends on the trigger signal that is used by the oscilloscope to record the power traces. In order to obtain aligned power traces, it is necessary that the recording of the power consumption always starts at the same position in relation to the processing of the attacked intermediate result. Finding a suitable trigger signal for this purpose is not always easy. In practice, there are essentially two signals the attacker can use to trigger the oscilloscope. These are the communication and the power consumption of the cryptographic device.

Triggering the oscilloscope directly based on the power consumption is difficult. The power consumption of most devices does not contain signal patterns that can be used directly as trigger signals for oscilloscopes. Therefore, the most commonly used trigger signal is the communication. Usually, the oscilloscope is configured to start recording the power consumption after the last bit of plaintext has been sent to the cryptographic device. Alternatively, the oscilloscope can be configured to record the power consumption during the time interval immediately before the device returns the ciphertext. In both cases, the attacker measures the power consumption that is caused by the execution of the cryptographic algorithm or at least of a part of it. Power traces that are recorded in such a way can be misaligned due to the following two reasons.

The first reason is that the communication is not always performed synchronously to the clock signal. Therefore, the time interval between the trigger signal and the processing of the attacked intermediate result is not constant. Each recording of the power consumption potentially starts at a different position in relation to the processing of the attacked intermediate result. The second reason for misaligned power traces is that often conditional operations are performed between the processing of the attacked intermediate result and the communication. With conditional operations we refer to operations that are not performed in the same way during each execution, *i.e.* these operations do not always require the same amount of clock cycles. Countermeasures, like the random insertion of dummy operations or shuffling, are such operations. However, also interrupts of timers or conditional statements in the executed software can have similar effects.

Using the communication to trigger the oscilloscope is not optimal for DPA attacks. However, in practice it is usually the only option. Typically, there is no trigger signal available that directly indicates the processing of the attacked intermediate result. It is therefore quite common that attackers have to work with misaligned power traces.

8.2.2   Alignment of Power Traces

The best way to cope with misaligned power traces is to align the traces before performing DPA attacks. In case this alignment is successful, the correct key hypothesis leads to the correlation $\rho(H_{ck}, P_{total})$ instead of $\rho(H_{ck}, \hat{P}_{total})$. As there is usually a big difference between these two correlation coefficients, it is worth for an attacker to spend time and effort on the alignment of power traces. If it is possible to correctly align power traces, the effects of countermeasures, like the insertion of dummy operations and shuffling, are completely removed.

In practice, the alignment of power traces is usually done in two steps. First, a pattern that occurs in the first power trace is selected. Subsequently, the attacker tries to find this pattern in all other power traces. This means that for each power trace, the attacker determines the position of the trace that matches the selected pattern best. Having determined the most probable location of the pattern in all power traces, the attacker shifts all power traces in such a way that the pattern occurs at the same position in all power traces. In case that the pattern in the power traces is caused by the same operations of the cryptographic algorithm, the power traces have been aligned successfully.

> The alignment of power traces is usually done based on pattern matching. This means that a part of the first power trace is selected as pattern. Subsequently, the attacker tries to find this pattern in all other power traces.

The basic idea of the alignment of power traces is very simple. However, when applying alignment techniques in practice it turns out that this task can be very challenging. We now discuss the most important issues that need to be considered for the selection of the pattern and for the matching of the pattern.

Selecting the Pattern

Selecting a suitable pattern is the most critical issue when aligning power traces. It is essentially the selection of the pattern that determines whether the alignment is successful or not. However, there is no general rule on how to select an optimal pattern. In practice, patterns need to be selected specifically for each device by analyzing the power traces. The most important methods to find suitable patterns are the visual inspection of power traces and the plotting of histograms at different positions of the traces. When selecting a pattern, the following properties need to be considered.

- **Uniqueness:** It is crucial that the selected pattern has some unique properties, like characteristic peaks or minima. Clearly, the better the pattern can be distinguished from the rest of the power trace, the better the alignment works. Consider for example a software implementation of AES. If the attacker selects the power consumption that is caused by an S-box look-up as pattern, this a bad pattern. There are 16 S-box look-ups in each round, and hence, the pattern will match at 16 positions. In contrast to this, the attacker can select the power consumption that is caused by the loading of the key at the beginning of the algorithm. Instructions that load data from internal or external memory typically lead to characteristic shapes in the power consumption. Furthermore, the loading of the key is only done at the beginning of the execution of the algorithm. Therefore, this pattern is much more suitable for the alignment.

- **Data Dependency:** The power consumption of the attacked device depends on the processed data. Therefore, most operations that are executed during the cryptographic algorithm lead to a slightly different power consumption during each execution. For the matching phase it is best, if the pattern does not depend on intermediate results of the cryptographic algorithm. Ideally, the pattern corresponds to the power consumption of operations like jumps or operations that only involve the key. Operations that depend on intermediate results typically degrade the matching results.

- **Length:** At first sight one might think, the longer the pattern is, the better the alignment will work. However, this is usually not the case. When using the power consumption of many operations as pattern, this pattern typically contains also many operations that depend on intermediate results of the algorithm. The power consumption of these operations degrades the matching results.

- **Distance to the Attacked Intermediate Result:** In case a countermeasure, like the random insertion of dummy operations, is used by the attacked device, it is important that the selected pattern is located close to the processing of the attacked intermediate result. Otherwise, it can happen that although the power traces are correctly aligned for the pattern, they are not correctly aligned for the processing of the attacked intermediate result.

All in all, many things have to be considered when selecting a pattern. However, the attacker typically has only a very limited amount of knowledge about the cryptographic device. This is why it can take a lot of time and trials until a suitable pattern is found. It can also happen that there is no suitable pattern at all. Clearly, the more uniform the power traces are, the more difficult it is to find a pattern. This fact can be used by designers of cryptographic devices to prevent attackers from aligning power traces. Designers should try to build their devices in such a way that there is no unique pattern that can be used for the alignment.

Matching the Pattern

After the attacker has selected the pattern, it is necessary to search for the pattern in each power trace. Usually, attackers limit the search to an interval around the position of the pattern in the first trace. This helps to increase the matching accuracy and it also saves time. For the actual matching of the pattern at each position of the search interval, different techniques can be used. The most common ones are the following two.

- **Least Square:** In case of this approach, the difference between the pattern and the power trace is calculated. The elements of the resulting vector are then squared and summed up. The position of the search interval that leads to the lowest sum is assumed to be the position of the pattern. This approach essentially corresponds to template matching with reduced templates, see

Section 5.3.3. The pattern can be viewed as template that is matched at each position of the search interval.

- **Correlation Coefficient:** Another approach is to calculate the correlation coefficient between the pattern and the power trace at each position of the search interval. In this case the position that leads to the highest correlation is assumed to be the correct one.

In practice, all these techniques work reasonably well. As already pointed out, the matching results depend more on the selection of the pattern than on the pattern-matching technique.

Alternative Alignment Techniques

Pattern matching is the most commonly used technique to align power traces. However, there are also other techniques. In particular, there exists an effective technique to reduce the misalignment of power traces of cryptographic devices that are protected by the random insertion of dummy operations or shuffling. As pointed out in Section 7.1.1, these two countermeasures are based on random numbers. The misalignment of the traces can therefore be reduced by biasing the random numbers.

At first sight, this looks like a very difficult task for an attacker. However, the random numbers that are used by the countermeasures need to be generated by the device and they need to be processed by the device. Hence, the power consumption of the device depends on the random numbers at some positions of the traces. If an attacker is able to find out these positions, the random numbers can be biased by only using the traces for the attack that have similar properties at these positions. For example, an attacker can decide to only use traces that are below or above a certain threshold at these positions. By doing this selection, the attacker biases the distribution of the random numbers in the used traces. Hence, also the distribution of ct is biased and the misalignment is reduced, *i.e.* $\hat{p}$ is increased.

8.2.3 Preprocessing of Power Traces

DPA attacks on misaligned power traces are significantly less effective than attacks on aligned traces, see (8.1). Therefore, the goal of attackers is always to align the traces before performing DPA attacks. However, this is not always possible. For example, it can happen that the power traces are too noisy to be aligned successfully. In this case, attackers usually resort to other preprocessing techniques. We now analyze how the integration of power traces can help to increase the effectiveness of DPA attacks on misaligned power traces. Subsequently, we give an overview of preprocessing techniques that have a similar effect.

Integration

We start our discussion about the integration of power traces by first looking at DPA attacks that are based on aligned traces. This means we analyze how the correlation coefficient for the correct key hypothesis changes, if the attacker integrates the power traces before performing DPA attacks. By integration we mean that the attacker sums up the power consumption of l clock cycles. We denote the power consumption of these clock cycles by the random variables $P_1, \ldots, P_l$, and we assume that the attacked intermediate result is processed in the first clock cycle. This assumption can be made without loss of generality. Furthermore, we assume that the power consumptions $P_1, \ldots, P_l$ are independent of each other and that $P_2, \ldots, P_l$ are also independent of H_{ck}. Therefore, it holds for $i = 2, \ldots, l$ that $E(H_{ck} \cdot P_i) - E(H_{ck}) \cdot E(P_i) = 0$. The correlation coefficient for the correct key hypothesis can hence be calculated as follows:

$$
\rho\left(H_{ck}, \sum_{i=1}^{l} P_i\right) = \frac{E(H_{ck} \cdot \sum_{i=1}^{l} P_i) - E(H_{ck}) \cdot E(\sum_{i=1}^{l} P_i)}{\sqrt{Var(H_{ck}) \cdot (Var(\sum_{i=1}^{l} P_i))}}
$$

$$
= \frac{E(H_{ck} \cdot P_1 + H_{ck} \cdot \sum_{i=2}^{l} P_i) - E(H_{ck}) \cdot (E(P_1) + E(\sum_{i=2}^{l} P_i))}{\sqrt{Var(H_{ck}) \cdot Var(P_1)} \sqrt{\frac{\sum_{i=1}^{l} Var(P_i)}{Var(P_1)}}}
$$

$$
= \frac{E(H_{ck} \cdot P_1) - E(H_{ck}) \cdot E(P_1)}{\sqrt{Var(H_{ck}) \cdot Var(P_1)} \sqrt{\frac{\sum_{i=1}^{l} Var(P_i)}{Var(P_1)}}}
$$

$$
= \frac{\rho(H_{ck}, P_1)}{\sqrt{\frac{\sum_{i=1}^{l} Var(P_i)}{Var(P_1)}}} \tag{8.7}
$$

A DPA attack on integrated power traces is obviously less effective than a DPA attack on aligned, non-integrated traces. The fact how much the effectiveness is reduced, depends on the variances of the power consumption values that are summed up. In case the variances of the power consumption values of all integrated clock cycles are equal, the integration lowers the correlation coefficient by $\sqrt{l}$.

If the power consumption of l clock cycles is independently distributed and if the variances of the power consumption in all cycles are equal, the sum of the power consumptions leads to the following correlation coefficient for the correct key hypothesis.

$$
\rho\left(H_{ck}, \sum_{i=1}^{l} P_i\right) = \frac{\rho(H_{ck}, P_1)}{\sqrt{l}} \tag{8.8}
$$

While the effectiveness of DPA attacks is reduced in case of aligned traces, the integration of power traces can increase the effectiveness in case of misaligned traces. Consider for example a cryptographic device that randomly inserts up to $(a - 1)$ clock cycles, where the number of inserted clock cycles is uniformly distributed. When recording a power trace of this device, there are a positions where the power consumption of the attacked intermediate result can be located. All a positions are equally likely. A DPA attack on this device leads to a correlation that can be calculated according to (8.2) with $\hat{p} = 1/a$. Assuming that $Var(\hat{P}_{total}) = Var(P_{total})$, the correlation $\rho(H_{ck}, \hat{P}_{total})$ is equal to $\rho(H_{ck}, P_{total})/a$.

Now consider an attacker who sums up the power consumption of the a positions in each power trace before performing the DPA attack. This means that in each trace, the attacker sums up all positions where the attacked intermediate result can be located. Obviously, there is no longer a misalignment when performing a DPA attack like this. Assuming that the variances of the summed power consumption values are equal, the correlation for the correct key hypothesis can be calculated according to 8.8. This means that the correlation is $\rho(H_{ck}, P_{total})/\sqrt{a}$. It is hence significantly higher than in case of the DPA attack without integration. The misalignment reduces the correlation linearly, while the integration reduces the correlation with the square root of a.

This is an important observation that can be used to improve DPA attacks on misaligned power traces in general. In practice, misaligned power traces are usually integrated and attacked as follows. The attacker first chooses a window size for the integration, *i.e.* the attacker decides how many clock cycles are summed up. Subsequently, the attacker slides the integration window along the power traces. At each position of the traces, the attacker sums up the power consumption of all clock cycles that are in the integration window. Finally, a DPA attack on the integrated power traces is performed. This kind of DPA attack has been introduced in [CCD00].

Such a DPA attack works best, if the size of the integration window corresponds to the width of the distribution of *ct*. As the attacker usually does not know the distribution of *ct*, several window sizes need to be tried out until the best one is found. In the best case for the attacker, the linear reduction of the correlation coefficient according to (8.2) is changed to a reduction according to (8.8).

The integration of misaligned power traces increases the effectiveness of DPA attacks. The misalignment of the traces reduces the correlation linearly, while the integration reduces the correlation only with the square root of the number of integrated clock cycles.

Other Techniques

The integration of power traces is the most commonly used preprocessing technique to reduce the effects of misalignment. However, there are also other techniques. Most of these techniques are linear and can be modeled as a convolution of the power traces and a window function. There are also proposals to transform the power traces into the frequency domain in order to avoid the need to align them.

- **Convolution:** A powerful method to preprocess power traces is to calculate the convolution of the traces and some window function. In the time-discrete domain, a window function is a vector of length l. We denote this vector by $\mathbf{w} = (w_1, \ldots, w_l)$. The convolution of a vector $\mathbf{x}$ and the window function $\mathbf{w}$ is defined in (8.9). The result of the convolution is the vector $\mathbf{y}$.

$$y_i = \sum_{j=1}^{l} w_j \cdot x_{i-j+1} \qquad (8.9)$$

  In practice, there are many suitable window functions to preprocess power traces. If a rectangular window function, e.g. $w = (1, 1, 1, \ldots)$, is used, the preprocessing corresponds to the integration of the power traces. By using a function that looks like a comb, e.g. $w = (1, 0, 1, 0, \ldots)$, it is possible to sum up points that have a certain distance between them. For example, such a function can be used to sum up points of power traces that are separated by one or more clock cycles. Of course, it is also possible to use different weights in the window function.

- **Fast Fourier Transformation (FFT):** Another way to preprocess power traces is to map the power traces into the frequency domain. In this case, there is no need to consider the alignment of the traces any more. However, this technique does not always lead to good results. Whether this preprocessing technique works well or not essentially depends on the spectral characteristics of the leakage and the noise.

8.2.4 Examples

In this section, we provide two examples of DPA attacks on our microcontroller. First, we illustrate the effect of misaligned power traces on DPA attacks by looking at an implementation of AES that randomly inserts dummy operations. In the second example, we attack an implementation of AES that is protected by shuffling.

Random Insertion of Dummy Operations

For this example, we have implemented AES in such a way that NOP instructions are randomly inserted before the start of AES. A NOP instruction

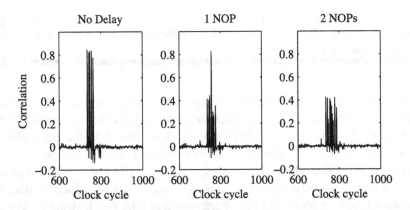

Figure 8.1. Results of DPA attacks on the output of an S-box. For the first attack, the power traces have been aligned correctly. The traces for the second and the third attack have been misaligned by randomly inserting one and two NOP instructions, respectively.

is an instruction that does not perform any operation. On our microcontroller, it takes one instruction cycle to execute it. We have performed three different DPA attacks on this implementation of AES on the microcontroller. Each of these attacks has been performed based on 10 000 compressed power traces.

For the first DPA attack, we have configured the microcontroller to not insert any NOP instructions. Hence, the recorded power traces have not been misaligned. Based on the power traces, we have performed a DPA attack on the output of the S-box look-up for byte one in the first round. This attack has been performed using the Hamming-weight model. The left plot of Figure 8.1 shows the correlation between the power traces and the hypothetical power consumption for the correct key hypothesis.

For the second DPA attack, we have configured the microcontroller to randomly insert one NOP instruction. Consequently, the power consumption of the attacked S-box look-up has occurred at two positions in the recorded power traces. Both positions have been equally likely. The plot in the middle of Figure 8.1 shows the result of the DPA attack based on these misaligned power traces. As it can be observed, the shape of the peak has changed compared to the previous attack. However, the maximum of the correlation is still the same. This can be explained by the fact that the S-box look-up takes more than one instruction cycle. In fact, this operation leaks the Hamming weight of the S-box output in two consecutive instruction cycles. Therefore, the misalignment of one instruction cycle does not lower the correlation for the correct key hypothesis. In case of our AES implementation, operations that take at least two instruction cycles are needed to lower the correlation.

> When using the random insertion of dummy operations as countermea-
> sure, it is crucial that the inserted operations are sufficiently long to
> indeed lower the correlation.

For the third DPA attack, we have configured our microcontroller to randomly insert two NOP instructions, *i.e.* either no instruction or two instructions have been inserted. The result of this DPA attack is shown in the right plot of Figure 8.1. As it can be observed, the correlation is halved compared to the correlation in the first attack. This corresponds to the effect of misaligned power traces that we have derived in (8.1). The correlation can be reduced further by randomly inserting more dummy operations that take two instruction cycles. Of course, also other instructions than NOP can be used for this purpose. In general, instructions should be used that are difficult to detect for an attacker. If the dummy operations can be found in the power traces, attackers can easily align the traces, see Section 8.2.2.

Shuffling

We now discuss DPA attacks on an implementation of AES that is protected by shuffling. In this implementation, the sequence of the S-box look-ups changes randomly for each execution of the algorithm. In order to illustrate the effect of shuffling, we have implemented the countermeasure in a scalable way. This means that it is possible to define the number of S-box look-ups that are shuffled. The target of all DPA attacks we have performed on this implementation has been the output of S-box one in the first round. For all attacks, 10 000 traces have been recorded and the Hamming-weight model has been used.

Figure 8.2 shows results of DPA attacks for different configurations of the microcontroller and for different preprocessing techniques. The plots on the left side of the figure show results of DPA attacks that have been performed based on raw power traces, *i.e.* no preprocessing has been performed. The plots on the right side are the result of DPA attacks that are based on convoluted power traces. For the convolution, we have used a window function that looks like a comb. The distance between the teeth in the window function has been chosen in such a way that the leakage of consecutive S-box look-ups is summed up.

The two plots in the first row of Figure 8.2 show results of DPA attacks on power traces that have been recorded while the shuffling has been turned off. The two plots are identical because for this configuration no preprocessing has been performed. The second row contains plots showing the result of DPA attacks based on power traces that have been recorded while shuffling has been activated for the first two S-boxes. The correlation shown in the left figure is therefore halved compared to the first row. The preprocessing for the result on the right side has been done in such a way that the power consumption of two

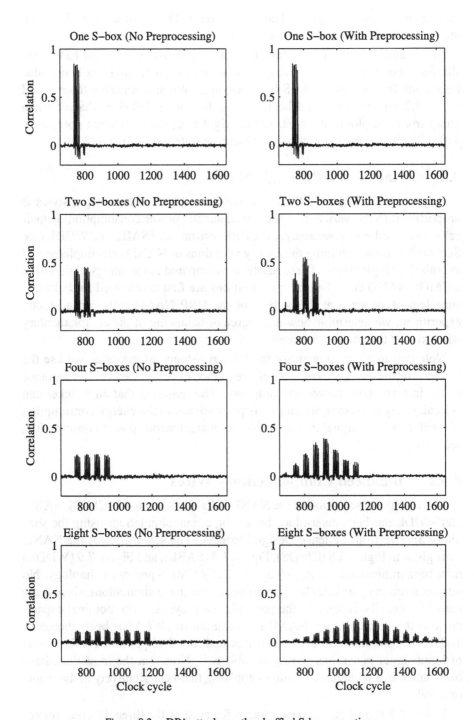

Figure 8.2. DPA attacks on the shuffled S-box operations.

subsequent S-box look-ups has been summed up. Therefore, the correlation on the right side is reduced by $\sqrt{2}$, see (8.8).

In the third and the fourth row, four and eight S-box look-ups have been shuffled. For the results on the right side, the preprocessing has hence also been done for four and eight S-box look-ups. When comparing the plots of Figure 8.2, it can be observed that the correlation on the left side is halved when going from one plot to the next. On the right side, the correlation goes down with $\sqrt{2}$. Hence, the correlation is reduced as stated in (8.2) and (8.8).

8.3   Attacks on DRP Logic Styles

As discussed in Section 8.1.2, the DPA resistance of DRP logic styles is proportional to the variance of the instantaneous power consumption of their cells. In the following, we analyze the DPA resistance of SABL and WDDL (see Section 7.4) by determining the energy variations of NAND cells implemented in both DRP logic styles. These results are compared to the energy variation of a CMOS NAND cell. The energy variations are first determined for balanced complementary wires at the outputs of the DRP NAND cells. In a second experiment, we determine how the degree of balancing of the complementary wires affects the energy variations.

Note that in order to quantify the DPA resistance of the cells, we use the variance of the energy consumption over a clock cycle instead of the variance of the instantaneous power consumption. The reason is that an attacker can typically only measure a signal that is proportional to the energy consumption of a cell and not a signal that resembles its instantaneous power consumption, see Section 3.5.2.

8.3.1   Balanced Complementary Wires

The energy consumption of the NAND cells implemented in CMOS, SABL, and WDDL has been determined by analog circuit simulations using the simulator *Spectre* from Cadence Design Systems. The schematics of the NAND cells given in Figure 2.8 (CMOS), Figure 7.6 (SABL), and Figure 7.9 (WDDL) have been implemented using a 0.35 μm, 3.3 V CMOS process technology. No intra-cell routing parasitics have been considered in the simulations, since these parasitics heavily depend on the particular cell layouts. The nominal capacitance at the outputs of the NAND cells (output to *GND*) has been chosen as 100 fF. The simulations have been performed for all 16 possible combinations of input signal transitions shown in Table 8.1. Note that the transitions have been sorted in a sequence that allows applying them consecutively to the inputs of a cell.

Figure 8.3 shows the 16 power traces for each of the three different implementations of the NAND cell. The power traces of the DRP NAND cells show

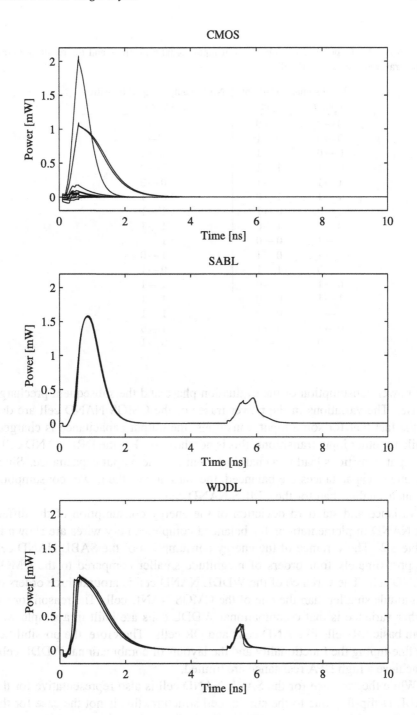

Figure 8.3. Simulated power traces of a CMOS NAND cell, an SABL NAND cell, and a WDDL NAND cell for different input transitions.

Table 8.1. All 16 possible combinations of input signal transitions and the resulting output signal transitions of a NAND cell.

Input signal transitions		NAND output signal transition
Input a	Input b	Output q
$0 \rightarrow 0$	$0 \rightarrow 0$	$1 \rightarrow 1$
$0 \rightarrow 0$	$0 \rightarrow 1$	$1 \rightarrow 1$
$0 \rightarrow 0$	$1 \rightarrow 1$	$1 \rightarrow 1$
$0 \rightarrow 1$	$1 \rightarrow 1$	$1 \rightarrow 0$
$1 \rightarrow 1$	$1 \rightarrow 1$	$0 \rightarrow 0$
$1 \rightarrow 1$	$1 \rightarrow 0$	$0 \rightarrow 1$
$1 \rightarrow 0$	$0 \rightarrow 1$	$1 \rightarrow 1$
$0 \rightarrow 0$	$1 \rightarrow 0$	$1 \rightarrow 1$
$0 \rightarrow 1$	$0 \rightarrow 0$	$1 \rightarrow 1$
$1 \rightarrow 1$	$0 \rightarrow 1$	$1 \rightarrow 0$
$1 \rightarrow 0$	$1 \rightarrow 1$	$0 \rightarrow 1$
$0 \rightarrow 1$	$1 \rightarrow 0$	$1 \rightarrow 1$
$1 \rightarrow 1$	$0 \rightarrow 0$	$1 \rightarrow 1$
$1 \rightarrow 0$	$0 \rightarrow 0$	$1 \rightarrow 1$
$0 \rightarrow 1$	$0 \rightarrow 1$	$1 \rightarrow 0$
$1 \rightarrow 0$	$1 \rightarrow 0$	$0 \rightarrow 1$

the power consumption of the evaluation phase and the subsequent precharge phase. The variations in the power traces of the CMOS NAND cell are due to the fact that for some input transitions, the output capacitance is charged, while for other input transitions, this is not the case. For the DRP NAND cells, all input transitions lead to a charging event for one output capacitance. Since the output capacitances are balanced, the variation of the power consumption is much smaller than for the CMOS NAND cell.

Variance and standard deviation of the energy consumption of the different NAND implementations for balanced complementary wires are shown in Table 8.2. The variance of the energy consumption of the SABL NAND cell is approximately four orders of magnitude smaller compared to the CMOS NAND cell. The variance of the WDDL NAND cell is around three orders of magnitude smaller than the one of the CMOS NAND cell. The reason for the higher variance is that combinational WDDL cells are built in a simple way from basic SR cells like AND cells and OR cells. Therefore, the possibilities for fine-tuning the functionality and the layouts of combinational WDDL cells to achieve a high DPA resistance are limited.

While the variance for the SABL NAND cell is also representative for the SABL D-flip-flop due to the similar cell structure, this is not the case for the WDDL NAND cell and the WDDL D-flip-flop. The variance of the energy consumption of the WDDL D-flip-flop is typically much lower than the one of

Table 8.2. Variance and standard deviation of the energy consumption of the CMOS NAND cell, the SABL NAND cell, and the WDDL NAND cell for balanced complementary wires.

Logic style	CMOS	SABL	WDDL
$Var(E_{NAND})$	$22\,469 \cdot 10^{-29}\,\mathrm{J}^2$	$1.6954 \cdot 10^{-29}\,\mathrm{J}^2$	$26.853 \cdot 10^{-29}\,\mathrm{J}^2$
$Std(E_{NAND})$	$474\,\mathrm{fJ}$	$4.12\,\mathrm{fJ}$	$16.4\,\mathrm{fJ}$

the WDDL NAND cell. The reason is that identical SR D-flip-flops are used in both datapaths of the WDDL D-flip-flop. This leads to a more balanced power consumption.

8.3.2 Unbalanced Complementary Wires

As explained in Section 8.3.1, the DPA resistance of DRP cells is very high if the capacitances at the complementary outputs are balanced. However, this balancing is never perfect in practice, and thus, the DPA resistance of DRP cells is reduced. Figure 8.4 shows how the variance of the energy consumption of an SABL NAND cell and a WDDL NAND cell increases when the complementary output wires of the cells become less balanced.

The graphs show that the DPA resistance of the SABL NAND cell is maximal in the balanced case (the energy variation at this point is minimal). The DPA resistance decreases quadratically with the difference of the capacitances at the complementary outputs q and $\bar{q}$. The graph of the WDDL NAND cell shows basically the same behavior as the graph of the SABL NAND cell. An interesting point is that the WDDL NAND cell does not reach its maximum DPA resistance when the capacitances at the outputs q and $\bar{q}$ are perfectly balanced. The graph shows that the maximum DPA resistance is reached when the capacitance at output q is around 2 fF lower than the capacitance at output $\bar{q}$. This indicates an unbalanced internal structure of the WDDL NAND cell. The reason is that the SR AND and the SR OR cells within the WDDL NAND cell (see Figure 7.9) do not consume exactly the same power when switching their outputs in the same manner. During the design of the cell layout, it is possible to correct the unbalanced internal structure of the WDDL NAND cell to some degree. However, this increases the design and implementation effort of WDDL cells significantly. Note that the minimum energy variance of the WDDL NAND cell (*i.e.* for -2 fF capacitance difference) is still slightly higher than the minimum energy variance of the SABL NAND cell.

8.4 Notes and Further Reading

Effectiveness of Hiding Countermeasures in General. The effectiveness of the randomization of the execution of cryptographic algorithms was analyzed

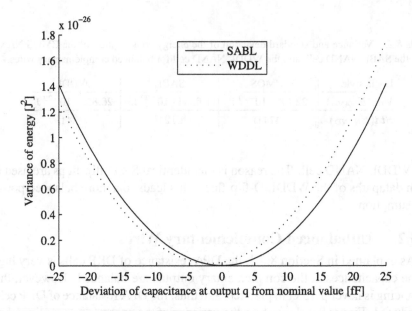

Figure 8.4. Variance of the energy consumption of an SABL NAND cell and a WDDL NAND cell as a function of the difference of the capacitances at the complementary cell outputs q and $\bar{q}$. The nominal capacitance at the outputs is 100 fF.

by Chari *et al.* in [CJRR99b]. In this article, they pointed out that the effect of the randomization can be undone by aligning the power traces with the help of signal processing techniques. Furthermore, they also mentioned that the number of needed power traces grows quadratically with the number of shuffled operations. In [CCD00], Clavier *et al.* introduced the sliding-window DPA attack that makes the number of power traces grow only linearly with the number of shuffled operations. In this article, the attack was presented based on the difference-of-means method. Mangard analyzed the effect of hiding countermeasures in general for DPA attacks that are done based on the correlation coefficient in [Man04].

Effectiveness of DPA-Resistant Logic Styles.   Simulation results comparing the DPA resistance of SABL and CMOS circuits were for example presented by Tiri *et al.* in [TAV02] and [TV03]. Similar simulation results for WDDL circuits were published in [TV04a]. In these articles, the so-called normalized energy deviation (NED) was used to quantify the effectiveness of SABL and WDDL. However, this metric favors logic styles with a high power consumption. In [TV05c], Tiri and Verbauwhede analyzed logic styles based on the formulas presented in [Man04].

Kulikowski *et al.* [KKT06] analyzed the problem of early propagation of logic cells, which causes a decrease of the DPA resistance. Sundström and Alvandpour [SA05] compared the variance of the power consumption of complementary CMOS logic, dynamic CMOS logic, dynamic differential logic, SABL, and DyCML. The security evaluation of an asynchronous test chip implementing a 16-bit microcontroller was presented in [FML$^+$03]. Results of DPA attacks on a test chip implemented using WDDL were presented in [THH$^+$05].

Asymmetric Cryptography. Attacks were published for many of the hiding techniques that we sketched for asymmetric cryptosystems in Section 7.5. Now, we survey these attacks using the same notation as in Section 7.5.

The first hiding technique was to randomize the representation of the exponent. Walter [Wal02b, Wal03] discussed security issues for MIST (the algorithm that was published in [Wal02a]) and Oswald [Osw03] discussed security issues for randomized addition-subtraction chains. Markov theory (Markov chains and hidden Markov models) turned out to be useful in this context. Walter and Oswald focused on SPA attacks in which the attacker can record one power trace only. However, researchers also looked at SPA attacks in which an attacker has several traces for the same parameters (*i.e.* base and exponent in RSA, and scalar and point in ECC). Walter [Wal02b, Wal02a] considered this scenario for MIST. Karlof and Wagner [KW03] and later on Green *et al.* [GNS05] discussed this scenario for randomized addition-subtraction chains. They also worked out nicely how to use hidden Markov models in this context. Fouque *et al.* [FMPV04] discussed an attack on [HM02]. Nobody so far published a thorough investigation of the security against DPA attacks of these hiding techniques.

The second hiding technique was to fix the sequence of operations in the exponentiation algorithm. It is clear that a square-and-multiply-always algorithm as well as a double-and-add-always algorithm are secure against SPA attacks that are based on a visual inspection of the trace. However, it should be clear that security against more advanced SPA techniques is not guaranteed. In particular, the reuse of operands might be detectable, e.g. the attacker might search for collisions within a trace or computes the correlation between parts of a trace. This often allows determining the key if a simple (binary) algorithm is used. For instance, in the binary double-and-add-always algorithm [Cor99], if and only if the bit of the key is zero in step i, the result of the double operation of step i is used in the double operation in step $i + 1$. Hence, the detection of a reuse of operands allows determining the bits of the key. The same observation was made by Fouque and Valette [FV03] in their *doubling attack*. This attack was described for a binary double-and-add-always algorithm that scans the bits from the MSB down to the LSB and requires the attacker to obtain power traces from two scalar multiplications (one for a point P and one for its double $[2]P$). Okeya and Sakurai [OS02] observed that the detection of the reuse of operands also poses a threat to exponentiation algorithms such as [Möl01]) that use window methods to preserve a double-and-add-always (or square-and-multiply-always) structure. Naturally, if more than one trace is available for a fixed key (scalar or exponent) the attacks can be improved. Note that algorithms that fix the sequence of operations do not increase the resistance against DPA attacks.

The third hiding technique was to make the operation (square and multiply, double and add) indistinguishable, either by inserting dummy operations or by rewriting the formulas of the operations. Like before, this prevents SPA attacks based on a visual inspection of the power trace but it does not provide security against advanced SPA techniques. Walter [Wal04] discussed an attack on a particular implementation of the unified code for double-and-add operations on Weierstrass elliptic curves [BJ02]. He assumed that the unified formulas, the double-and-add algorithm, and Montgomery multiplication using the conditional subtraction are used in an implementation. It is well known that this conditional subtraction leaks information about the processed intermediate values. Walter observed that in the unified code a specific intermediate value is computed twice, if and only if a double operation is performed. Hence, if one can detect whether or not these two intermediate values are equal, one can distinguish addition operations from double operations. Since the double-and-add algorithm was assumed to be used, this enables to recover the key. Akishita and Takagi [AT06] made a similar observation as Walter [Wal04] and provided some empirical evidence based on simulated attacks that they could turn their observation into a real attack. As before, if more than one trace is available, the attack can be improved. Note that algorithms that make the operations look uniform do not increase the resistance against DPA attacks.

It is apparent that hiding countermeasures for asymmetric cryptographic algorithms typically do not provide resistance against DPA attacks. Consequently, hiding countermeasures must be combined with masking (blinding) countermeasures in order to achieve resistance against SPA and DPA attacks.

Chapter 9

MASKING

The goal of every countermeasure is to make the power consumption of a cryptographic device independent of the intermediate values of the cryptographic algorithm. Masking achieves this by randomizing the intermediate values that are processed by the cryptographic device. An advantage of this approach is that it can be implemented at the algorithm level without changing the power consumption characteristics of the cryptographic device. In other words, masking allows making the power consumption independent of the intermediate values, even if the device has a data-dependent power consumption. Masking is one of the countermeasures that has been extensively discussed in the scientific community. Numerous articles have been published that explain different types of masking schemes. Even security proofs have been delivered for some of the schemes. Recently, masking has also been applied to the cell level.

In this chapter, we discuss how masking works and under which assumptions it leads to secure implementations. We also discuss how to implement masking at the architecture and the cell level.

9.1   General Description

In a masked implementation, each intermediate value v is concealed by a random value m that is called mask: $v_m = v * m$. The mask m is generated internally, $i.e.$ inside the cryptographic device, and varies from execution to execution. Hence, it is not known by the attacker.

> A masked intermediate value v_m is an intermediate value v that is concealed by a random value m: $v_m = v * m$. The attacker does not know the random value.

The operation $*$ is typically defined according to the operations that are used in the cryptographic algorithm. Hence, the operation $*$ is most often the Boolean exclusive-or function $\oplus$, the modular addition $+$, or the modular multiplication $\times$. In case of modular addition and modular multiplication, the modulus is chosen according to the cryptographic algorithm.

Typically, the masks are directly applied to the plaintext or the key. The implementation of the algorithm needs to be slightly changed in order to process the masked intermediate values and in order to keep track of the masks. The result of the encryption is also masked. Hence, the masks need to be removed at the end of the computation in order to obtain the ciphertext. A typical *masking scheme* specifies how all intermediate values are masked and how to apply, remove, and change the masks throughout the algorithm.

It is important that every intermediate value is masked all the time. This must be guaranteed also for intermediate values that are calculated based on previous intermediate values. For instance, if two masked intermediate values are exclusive-ored, we need to ensure that the result is masked as well. For this reason, we typically use several masks. Hence, different intermediate values are concealed by different masks. It turns out that it is not advisable to use a new mask for each intermediate value because the number of masks decreases the performance. Consequently, the number of masks needs to be chosen carefully in order to achieve a reasonable performance.

In the remainder of this section we review several important concepts that occur in the context of masking. In particular, we discuss different types of masking (Boolean vs. arithmetic), and we explain how secret sharing relates to masking. Furthermore, we explain the meaning of blinding and discuss the security of masking.

9.1.1   Boolean vs. Arithmetic Masking

We distinguish between Boolean and arithmetic masking. In Boolean masking, the intermediate value is concealed by exclusive-oring it with the mask: $v_m = v \oplus m$. In arithmetic masking the intermediate value is concealed by an arithmetic operation (addition or multiplication). Often, one uses the modular addition: $v_m = v + m \pmod{n}$. The modulus n is defined according to the cryptographic algorithm. The other arithmetic operation that is frequently used is the modular multiplication: $v_m = v \times m \pmod{n}$.

Some algorithms are based on Boolean and arithmetic operations. Therefore, they require both types of masking. This is problematic because switching from one type of masking to another type often requires a significant amount of additional operations, see [CG00], and [Gou01].

In addition, cryptographic algorithms use linear and non-linear functions. A linear function f has the property that $f(x * y) = f(x) * f(y)$. For example, if the operation $*$ is the exclusive-or operation $\oplus$, then a linear function has the property that $f(x \oplus m) = f(x) \oplus f(m)$. Hence, in a Boolean masking scheme linear operations change the mask m in a way that is easily computable. This means, a linear operation is easy to mask with Boolean masking. The AES S-box is a non-linear operation: $S(x \oplus m) \neq S(x) \oplus S(m)$. Since the Boolean masks are changed in a more complicated way, a lot of effort has to be spent on computing how the masks are changed. Thus, it cannot be easily concealed by Boolean masking. However, the S-box is based on computing the multiplicative inverse of a finite field element: $f(x) = x^{-1}$. It is compatible to multiplicative masking because $f(x \times m) = (x \times m)^{-1} = f(x) \times f(m)$. The authors of [AG01] show an efficient scheme to switch between Boolean masks and multiplicative masks. However, multiplicative masks have one major disadvantage. They cannot conceal the intermediate value 0. In Section 10.2 we show how this observation can be turned into a DPA attack.

In this book, we focus on Boolean masking. We always use the letter m to refer to a mask. If we want to point out that a particular mask is used to conceal a particular intermediate value v, then we refer to this mask by m_v.

9.1.2 Secret Sharing

Recall that masking means to conceal an intermediate value with a mask. For instance, in Boolean masking, a masked intermediate value v_m is equal to $v_m = v \oplus m$. The intermediate value v can be computed given v_m and m. In other words, the intermediate value v is represented by the two shares (v_m, m). Given only one of the two shares, no information about v is revealed. Knowing both shares allows determining v. Consequently, masking corresponds to a secret-sharing scheme that uses two shares.

Secret sharing with several masks is therefore a general way of masking. It means that several masks are applied to one intermediate value. It has been shown in [CJRR99b] that n masks can prevent up to an n-th order DPA attack. Higher-order DPA attacks are one of the subjects of Chapter 10.

Applying several masks to one intermediate value, and consequently keeping track of several shares, increases the cost of an implementation. Such an implementation needs more memory (storage elements) in order to store the shares and more computing time in order to compute the shares. Hence, in practice mostly secret sharing based on two shares, *i.e.* masking, is implemented. Resistance against higher-order DPA attacks is achieved by combining hiding and masking.

9.1.3 Blinding

In typical asymmetric cryptographic algorithms, additive or multiplicative masking is a good choice. The application of arithmetic masks in the context of asymmetric cryptographic schemes is called blinding. For example, in RSA decryption we can apply multiplicative masking to the input message v: $v_m = v \times m^e$. The result v^d can be easily recovered at the end of the algorithm because $(v_m)^d \equiv (v^d \times m) \pmod{n}$. This technique is called message blinding. Another slightly different masking technique can be applied to the exponent d: $d_m = d + m \times \phi(n)$. The result is automatically unmasked because $v^{d_m} \equiv v^d \pmod{n}$. This scheme is called exponent blinding. The exponent is concealed by an additive mask in this case.

9.1.4 Provable Security

DPA attacks work because the instantaneous power consumption of a cryptographic device depends on the intermediate values that it processes. Masking schemes try to destroy this dependency by masking the intermediate values. If an intermediate value v is masked, the corresponding masked intermediate value $v_m = v * m$ should be independent of v. The theory is that if v_m is independent of v, also the power consumption of v_m is independent of v.

> Masking provides security against first-order DPA attacks, if each masked intermediate value v_m is pairwise independent of v and m.

Hence, in typical proofs for masking schemes, it is shown that every masked intermediate value induces a distribution which does not depend (in a statistical sense) on the unmasked intermediate value. Then, the implementation will resist first-order DPA attacks. For instance, the distribution of $v \oplus m$ is always the same, no matter which value v has. Hence, it is independent of v.

For Boolean masking schemes, security proofs of this type have been given. In particular, the authors of [BGK05] and [OMPR05] have shown that $v \oplus m$, $(v \oplus m_v) \times (w \oplus m_w)$, $(v \oplus m_v) \times m_w$, $(v \oplus m_v)^2$ and $(v \oplus m_v)^2 \times p$ are pairwise independent of v (and w). Furthermore, by using another independent mask m' we can also add masked intermediate results: if the v_{m_i} are arbitrary and m' is pairwise independent of all v_{m_i}, then the distribution of $\sum v_{m_i} \oplus m'$ is pairwise independent of the v_{m_i}.

9.2 Architecture Level

The first papers that have discussed masking mainly have looked at software implementations. A lot of research has been devoted to 8-bit smart card implementations. In particular, the AES selection process has stimulated research that has investigated how to secure the AES finalists against power analysis attacks on smart cards, see for instance [CJRR99a]. Most of the recent research on masking focuses more or less exclusively on AES implementations. Recently, masking schemes have also been designed for implementations in dedicated hardware. Most papers in this context also deal with implementations of AES.

9.2.1 Software

A typical implementation of a Boolean masking scheme in software works in a straightforward manner. As explained before, we exclusive-or the mask(s) with the plaintext (or key), make sure that all intermediate values are masked throughout the computation, keep track of how the masks are changed, and at the end, remove the masks from the output. If all operations of the algorithm are (linear) Boolean operations, then Boolean masking fits nicely and is easy to implement. This is different in case of non-linear operations. These operations are more difficult to cope with. In the remainder of this section, we first discuss the masking of non-linear operations. Subsequently, we describe how random precharging can make implementations of masking easier, and we discuss pitfalls for masking. Finally, we give an example of how to efficiently mask an AES implementation on an 8-bit microcontroller.

Masking Table Look-Ups

In addition to simple operations, cryptographic algorithms also use complex operations including non-linear operations. These complex operations require more than plain Boolean masking. Consequently, special attention needs to be paid to their secure and efficient implementation. Most modern block ciphers allow implementing the non-linear operations as table look-ups. This means, for each input v of the non-linear operation, the output is stored at the corresponding index in a table T. The table is stored in memory where it can be accessed fast. This method is actually the most popular method for software implementations of block ciphers on smart cards. In a masked implementation, such a table needs to be masked. Consequently, one needs to produce a table T_m with the property $T_m(v \oplus m) = T(v) \oplus m$. Generating such a table is a simple process. However, in order to generate such a masked table, it is necessary to run through all inputs v, look up $T(v)$ and store $T(v) \oplus m$ for all m in the masked table. This process needs to be done for all masks m that are used in context with this operation. Consequently, the computational effort and the amount of memory increases with the number of masks that are used to mask the table look-up.

Random Precharging

A simple trick to implicitly mask intermediate values is called random precharging. Implicit masking means that random precharging masks the power consumption of intermediate values rather than the intermediate values themselves.

It works for operations that leak the Hamming distance. For instance, assume that the power consumption is related to the Hamming distance of two intermediate values that are transferred over a bus, or that are stored in the same register consecutively. In this case, it can be useful to simply load or store a random value m before the actual intermediate value v occurs. The device hence leaks $HD(v, m) = HW(v \oplus m)$. This prevents DPA attacks because the attacker cannot predict the Hamming distance between the intermediate value and the random value.

Pitfalls

As discussed in Section 9.1.4, security proofs for masking are usually based on the following assumption. They assume that the power consumption of masked intermediate values is independent of the unmasked intermediate values, if the masked and the unmasked values are independent of each other. However, this assumption does not always hold in practice. Often, the instantaneous power consumption of a device depends not only on one, but on several values. Such a combined power consumption of two or more intermediate values can make implementations insecure, even if all intermediate values are provably secure.

For example, assume a device that leaks the Hamming distance of two intermediate values. Hence, if two masked intermediate values, which have the same mask, are processed consecutively, the power consumption is related to the Hamming distance of the unmasked values because $HD(v_m, w_m) = HW(v_m \oplus w_m) = HW(v \oplus w)$. Consequently, in such a device special attention must be paid that no two values that are concealed by the same mask are transferred consecutively over the bus or stored consecutively in the same register. This example shows that the power consumption characteristics of a device can both help (see random precharging) and make implementations difficult.

Example for Masked AES

We now discuss an example of a masked AES smart card implementation. Our masking scheme uses Boolean masks only, and it is tailored to the AES software implementation that is described in Appendix B. The focus of our discussion is on the round transformation of this masking scheme. However, the key schedule is masked as well. This means that at the beginning of an encryption some masks are exclusive-ored with the plaintext, and some other masks are exclusive-ored with the first round key.

In the following, we first discuss how the four operations of the AES round transformation are masked. Subsequently, we describe the masking scheme as a whole. Finally, some performance figures are provided.

AddRoundKey: Since the round key bytes k are masked with m in our scheme, performing AddRoundKey automatically masks the bytes d of the state: $d \oplus (k \oplus m) = (d \oplus k) \oplus m$. Masking the round keys is important to prevent SPA attacks on the key schedule. We have discussed such an attack in Section 5.3.5.

SubBytes: The only non-linear operation of AES is SubBytes. In software implementations on a microcontroller, SubBytes is typically implemented as table look-up. Hence, we use a masked S-box table for this operation.

ShiftRows: The ShiftRows operation moves the bytes of the state to different positions. In our scheme, all bytes of the state are masked with the same mask at this point of the algorithm. Therefore, this operation does not affect the masking.

MixColumns: The MixColumns operation requires more attention because MixColumns mixes the bytes from different rows of a column. Hence, MixColumns requires at least two masks. If only two masks are used for the bytes of a column, then MixColumns needs to be done very carefully to make sure that all intermediate values stay masked. This leads to an inefficient implementation. Instead, it is better to make sure that each row is masked with a separate mask at this point of the algorithm. In this case, it is an advantage if the same masks are used in every round. Then, also the output masks of MixColumns are the same in all rounds. Hence, they need to be determined only once. In our scheme, we pursue this approach.

Now, we put these observations together and describe the masked AES implementation. We use six independent masks in our scheme. The first two masks, m and m', are the input and output masks for the masked SubBytes operation. The remaining four masks m_1, m_2, m_3, and m_4 are the input masks of the MixColumns operation. At the beginning of each AES encryption, two precomputations take place. First, we compute a masked S-box table S_m such that $S_m(x \oplus m) = S(x) \oplus m'$. Second, we compute the output masks for the MixColumns operation by applying this operation to (m_1, m_2, m_3, m_4). We denote the resulting output masks of MixColumns by (m_1', m_2', m_3', m_4').

A masked AES round works as follows. At the beginning of each round, the plaintext is masked with m_1', m_2', m_3', and m_4'. Then, the AddRoundKey operation is performed. The round key is masked such that the masks change from m_1', m_2', m_3', and m_4' to m. Next, the table look-up with the S-box table S_m is performed. This changes the masks to m'. ShiftRows has no influence on the masks because all bytes of the state are masked with m' at this point. Before MixColumns, we change the mask from m' to m_1 in the first row, to m_2 in the second row, to m_3 in the third row, and to m_4 in the fourth row. MixColumns changes the masks m_i to m_i' for $i = 1, \ldots, 4$. Note that these are the masks that we also had at the beginning of the round. Consequently, we can mask an arbitrary number of rounds in this way. At the end of the last encryption round,

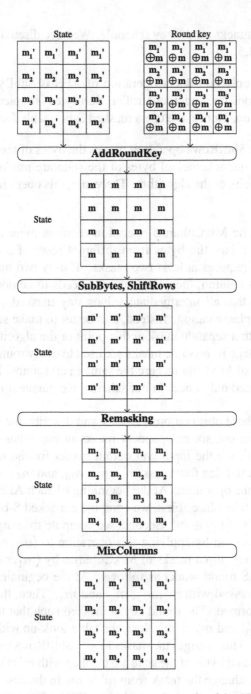

Figure 9.1. The AES round functions change the masks of the AES state bytes.

the masks are removed by the final AddRoundKey operation. Figure 9.1 shows a graphical representation of this scheme.

The costs that are imposed by masking are typically quite high in terms of computation time. However, the high effort does not come from the additional operations that have to be performed within a round. It comes from precomputing the masked S-box table. We now provide some performance figures for an implementation of our masking scheme on an 8-bit microcontroller. This implementation and the corresponding performance figures can also be found in [HOM06].

The overall number of clock cycles for this masked AES is 8 420 clock cycles. An unmasked AES implementation (in the same style) on the same platform takes 4 427 clock cycles. Hence, the masked implementation requires about twice as many clock cycles. Out of the 8 420 clock cycles, about 2 800 are spent on precomputations (masked S-box, masked MixColumns output, preparation of masks). Hence, one third of the runtime goes into precomputations. The additional effort that is induced by masking in one round of AES (including one round of the key schedule) is only about 78 clock cycles. This is not surprising since the masking does not require altering most AES steps. Furthermore, we use the round keys to change the masks. There is one important consequence of these figures. For this implementation, it does not make sense to remove the masking for the inner rounds. It is a widely spread believe that doing so improves the performance of masked implementations significantly. However, our example proves this believe to be untrue for a rather typical software implementation. Removing the masking from the six inner AES rounds in our scheme only gives an advantage of about 468 clock cycles. Hence the performance gain is only about 5.6%.

9.2.2 Hardware

Implementations of masking schemes in hardware require similar considerations as implementations in software. Boolean masking schemes fit well to many block ciphers. Hence, only for those parts of the round function that require different types of masking more effort has to be spent. In contrast to software implementations, more trade-offs between size and speed are possible.

In this section, we discuss how to mask multipliers, how to use random precharging, and how to mask buses. Furthermore, we discuss pitfalls of masking and explain how to mask an AES S-box implementation using composite field arithmetic.

Masking Multipliers

In hardware, adders and multipliers are among the basic building blocks when implementing cryptographic algorithms. For example, hardware implementations of the AES S-box usually decompose the S-box into a sequence of

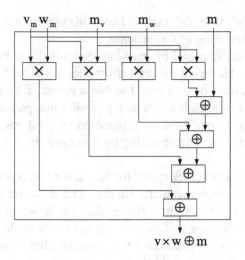

Figure 9.2. A masked multiplier *MM* consists of four standard multipliers and four standard adders.

$$MM(v_m, w_m, m_v, m_w, m) = (v_m \times w_m)$$
$$\oplus (w_m \times m_v) \oplus (v_m \times m_w) \oplus (m_v \times m_w) \oplus m \quad (9.1)$$

additions and multiplications. Because additions are typically easier to mask than multiplications, we focus on how to mask multiplications.

Our goal is to define a masked multiplier *MM*. Hence, we need a circuit that computes the product of two masked inputs $v_m = v \oplus m_v$, $w_m = w \oplus m_w$ and some masks m_v, m_w and m such that $MM(v_m, w_m, m_v, m_w, m) = (v \times w) \oplus m$. We have already stated in Section 9.1.4 that values such as $v_m \times m_w = (v \oplus m_v) \times m_w$ and $w_m \times m_v = (w \oplus m_w) \times m_v$ are secure. This observation can be used to build a masked multiplier *MM* as shown in (9.1). Figure 9.2 shows a block diagram of this multiplier. It can be observed that the masked multiplier *MM* requires four standard multipliers and four standard adders.

Random Precharging

Random precharging can also be applied to hardware. This means, random values are sent through the circuit in order to randomly precharge all combinational and sequential cells of the circuit. In a typical implementation, such as the one reported in [BGLT04], random precharging requires duplicating the sequential cells, *i.e.* the number of registers is doubled. The duplicates of the registers are inserted in between the original registers and the combinational cells of the circuit. Hence, random precharging is achieved as follows.

In the first clock cycle, the duplicates of the registers contain random values. As these registers are connected to the combinational cells, the outputs of all combinational cells are randomly precharged. When switching from the first to the second clock cycle, the result of the combinational cells (notice that this a random result) is stored in the original registers that contain the intermediate values of the executed algorithm. At the same time, the intermediate values are moved from these registers to the duplicates. This means that the role of the registers is switched. Therefore, in the second clock cycle the combinational cells are connected to registers that contain the intermediate values of the algorithm. Hence, the execution of the algorithm is continued in this cycle. When switching from the second to the third clock cycle, the role of the registers is switched again and the combinational cells are precharged again.

When implementing random precharging like this, all combinational and sequential cells process random data in one cycle and intermediate values in the next clock cycle. Hence, assuming the device leaks the Hamming weight of the data it processes, resistance against power analysis attacks is achieved. The power consumption is masked implicitly. Notice that random precharging is implemented in a very similar way as the random insertion of dummy cycles, see Section 7.2.2. Hence, these countermeasures can be combined easily.

Another approach to implement random precharging has been presented in [MMS01b]. The basic idea of this approach to randomize the register usage. Hence, the intermediate values of the algorithm are stored in different registers with potentially different data during each execution. Consequently, also a kind of implicit masking is achieved.

Masking Buses

Bus encryption has a long tradition in small devices. Bus encryption refers to the concept of encrypting the data and address buses that connect the processor of a smart card to memory and cryptographic co-processors. The purpose of bus encryption is to prevent eavesdropping on the bus. Buses are particularly vulnerable to power analysis attacks because of their large capacitance.

The encryption algorithms that are used for bus encryption are often quite simple. A pseudo-random key is generated and used in a simple scrambling algorithm (simple means mainly using exclusive-or operations). Hence, the simplest version of bus encryption, where a random value is exclusive-ored to the value on the bus, corresponds to masking the bus. There are only few articles available that discuss bus encryption techniques. Some recent articles are [BGM+03], [Gol03], and [ETS+05].

Pitfalls

Similar to software implementations, special attention needs to be paid when a masked value and its corresponding mask (or two intermediate values that are concealed by the same mask) are processed consecutively. For instance, a masked value and its corresponding mask should not be stored consecutively in the same register, if the register leaks Hamming-distance information.

Differently to software implementations, one also needs to pay attention to the optimizations that synthesis tools, which are typically used in hardware design, apply to a design. These tools have the property that they remove redundancies in the circuit description of the designer. Masked implementations have a lot of redundancies because masks are exclusive-ored at some point in the algorithm and removed or changed at another point. The tools recognize this and remove the parts of the circuit that correspond to the masking. This is of course absolutely undesired. Hence, the circuit designer needs to define which parts of the circuit description the tools may not touch.

Example for Masked AES S-Box

The most challenging part of a masked AES implementation in hardware is the masking of the S-box. In this example, we summarize formulas to compute the S-box in a masked way. It is important to observe that all terms of these formulas are "provably secure". At the end of this example, we provide performance figures for implementations of this masking scheme of the S-box.

The presented masking scheme is based on the S-box architecture described in [WOL02], which uses composite field arithmetic. In this approach, the S-box input is seen as an element of a finite field with 256 elements. Mathematicians often use the abbreviation GF(256) to refer to this finite field. There are several representations of finite field elements. Naturally, one chooses a representation that leads to an efficient implementation. In case of the AES S-box, it turns out that representing each byte of the state as a linear polynomial $v_h x + v_l$ over a finite field with 16 elements is efficient. In other words, an element of GF(256) can be represented by a combination of two elements of GF(16). Therefore, the finite field GF(256) is a quadratic extension of GF(16).

The inversion of the element $v_h x + v_l$ can be computed using operations in GF(16) only:

$$(v_h x \oplus v_l)^{-1} = v_h' x \oplus v_l' \tag{9.2}$$

$$v_h' = v_h \times w' \tag{9.3}$$

$$v_l' = (v_h \oplus v_l) \times w' \tag{9.4}$$

$$w' = w^{-1} \tag{9.5}$$

$$w = (v_h^2 \times p_0) \oplus (v_h \times v_l) \oplus v_l^2 \tag{9.6}$$

All operations are done modulo a field polynomial that is fixed when the quadratic extension is defined. The element p_0 is defined in accordance with this field polynomial.

In order to calculate the inversion of a masked input value, we first map the value as well as the mask to the composite field representation. Such a mapping has been defined in [WOL02]. The mapping is a linear operation and therefore it is easy to mask. After the mapping, the value that needs to be inverted is represented by $(v_h \oplus m_h)x \oplus (v_l \oplus m_l)$. Note that both elements in the composite field representation are masked additively.

Our goal is that all input and output values in the computation of the inverse are masked, see (9.7).

$$((v_h \oplus m_h)x \oplus (v_l \oplus m_l))^{-1} = (v'_h \oplus m'_h)x \oplus (v'_l \oplus m'_l) \qquad (9.7)$$

Hence, we replace each addition and multiplication in (9.3) to (9.5) with a masked addition and a masked multiplication. It turns out that we can do this in such a way that (9.8) to (9.10) hold.

$$v'_h \oplus m'_h = v_h \times w' \oplus m'_h \qquad (9.8)$$
$$v'_l \oplus m'_l = (v_h \oplus v_l) \times w' \oplus m'_l \qquad (9.9)$$
$$w' \oplus m'_w = w^{-1} \oplus m'_w \qquad (9.10)$$
$$w \oplus m_w = (v_h^2 \times p_0) \oplus (v_h \times v_l) \oplus v_l^2 \oplus m_w \qquad (9.11)$$

We still have to master one more difficulty. In (9.10) we need to calculate the inverse in GF(16). Calculating the inverse in GF(16) can be reduced to calculating the inverse in GF(4) by representing GF(16) as quadratic extension of GF(4). Like before, we can express an element of GF(16) as a linear polynomial, but now the coefficients are elements of GF(4). Hence, the same formulas as given in (9.8) to (9.11) can be used to calculate the masked inverse in the quadratic extension of GF(4). In GF(4), the inversion operation is equivalent to squaring: $x^{-1} = x^2 \ \forall x$. Hence, in GF(4) we have that $(x \oplus m)^{-1} = (x \oplus m)^2 = x^2 \oplus m^2$. The inversion operation preserves the masking in this field.

Now we discuss the performance of this masking scheme. When looking at the formulas it should be clear that an implementation of this scheme is considerably slower and larger than an implementation of an unmasked S-box in composite field arithmetic. For instance, the most efficient implementation of this idea so far, which has been published in [OMPR05], requires nine multiplications, two multiplications with a constant and two square operations in GF(16). Note that for the sake of simplicity we only count the expensive operations in the bigger field and do not consider the operations in GF(4). An efficient implementation of an unmasked S-box in composite field arithmetic, which has been reported in [WOL02] requires only three multiplications, one multiplication with a constant and two squaring operations in GF(16). This is considerably less. In addition, the length of the critical path of the masked S-box increases significantly. It has been reported in [POM+04] that an implementation of this scheme is about two to three times larger and slower than an implementation of a corresponding unmasked S-box in composite field arithmetic.

9.3   Cell Level

The first DPA-resistant logic styles that have been proposed to counteract power analysis attacks have all been based on the concept of hiding, see Section 7.3. Masking has mainly been implemented at the architecture level. Recently, also several DPA-resistant logic styles have been proposed that use masking. Such DPA-resistant logic styles are usually referred to as *masked logic styles*.

In this section, we give a general description of masked logic styles and discuss the approaches that exist to build masked circuits. Furthermore, we discuss how masked logic styles can be used for semi-custom circuit design.

9.3.1   General Description of Masked Logic Styles

Applying masking to the cell level means that the logic cells in a circuit only work on masked values and the corresponding masks. Cells that are used in such circuits are called *masked cells*. The circuits themselves are called *masked circuits*. The theory of these circuits is the following. Since the masked values are independent of the unmasked values, the power consumption of the masked cells should also independent of the unmasked values. As a result, the total power consumption of a cryptographic device should be independent of the processed data and the performed operations. However, like in case of all other countermeasures, complete independence cannot be achieved in practice. It is only possible to make the power consumption largely independent of the corresponding unmasked values.

> When using masked logic styles, the logic cells only work on masked values and masks. Therefore, the power consumption of these cells is largely independent of the corresponding unmasked values.

Usually, Boolean masking is used for masked circuits. Figure 9.3 shows a 2-input unmasked cell and a corresponding 2-input masked cell. The input signals a, b and the output signal q of the unmasked cell are carried on single wires. In case of the masked cell, the input signals and the output signal are split into masked values and the corresponding masks.

Note that glitches in masked circuits can lead to a strong dependency between the power consumption and the unmasked values [MPG05]. As a result, masked logic styles are usually built in a way such that glitches are completely avoided.

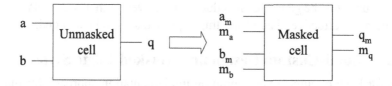

Figure 9.3. A 2-input unmasked cell and a corresponding 2-input masked cell.

The Number of Different Masks in a Circuit

There exist three approaches to build masked circuits. First, *one distinct mask for each signal* can be used. As a result, all masked values are pairwise independent of each other, no matter whether the corresponding unmasked values are independent or not. Using this approach, the functionality of the logic cells is typically very complex, because every input and output signal is masked separately. The number of necessary masks for such a circuit is huge. Therefore, this approach is not practical.

The second possibility is to partition the signals of a circuit in several groups and to use *the same mask for a group of signals*. This decreases the number of required masks significantly. Furthermore, the complexity of cells working on input signals that are masked with the same mask is typically reduced. For every masked signal that passes over from one signal group $G1$ to another signal group $G2$, some additional circuitry is necessary to change the mask from m_{G1} to m_{G2}. Defining the number of signal groups with different masks is a non-trivial task.

The third approach is to use *the same mask for all signals* in a circuit. Hence, there is no overhead for handling different masks. If unmasked values depend on each other, also the corresponding masked values depend on each other. A drawback of this approach is that changing the single mask value can typically be detected via the power consumption of the huge mask net. The mask net is responsible for distributing the mask to every cell in the circuit. Possible solutions to this problem are to resort to the second approach discussed above or to implement the mask net in a DRP-like manner.

Mask Value Changing Frequency

When specifying how often the mask values in a circuit are changed, the following considerations are important. If the *mask values are changed in each clock cycle*, the rate at which new mask values must be generated is very high. This is especially true if different masks are used in a circuit. A main advantage of changing the mask values in each clock cycle is that higher-order DPA attacks as presented in Section 10.1.1 are more difficult.

In order to reduce the rate at which new mask values are required, the *mask values can be used in several clock cycles*. As a result, the rate at which new mask values must be generated is reduced. However, higher-order DPA attacks become more effective. Therefore, this approach should be avoided.

9.3.2   Semi-Custom Design and Masked Logic Styles

As for DRP circuits, a very common implementation approach for masked circuits is semi-custom design based on standard cells. Most of the issues that have been discussed for DRP logic styles in Section 7.3.3 are also valid for masked logic styles.

The main differences are that during the logic style conversion the unmasked single-rail cells in the synthesized circuit are replaced by the corresponding masked cells, and that the balancing of complementary wires is not necessary. The latter difference is a main advantage of masked logic styles over DRP logic styles, because it significantly reduces the complexity of the floorplanning, the placement, and the routing steps. Another difference in the semi-custom design flow is that one or more mask nets must be added to the circuit and connected appropriately to the masked cells. The additional mask nets increase the complexity of the routing process.

9.4   Examples of Masked Logic Styles

In the following section, the masked logic style *Masked Dual-Rail Precharge Logic (MDPL)* is presented. The functionality and the properties of MDPL cells and MDPL circuits are discussed. Furthermore, one combinational MDPL cell (NAND) and one sequential MDPL cell (D-flip-flop) are presented in detail.

9.4.1   Masked Dual-Rail Precharge Logic

Masked Dual-Rail Precharge Logic has been introduced in [PM05]. MDPL uses the same mask m for all signals in the circuit. Each masked signal d_m that is processed in an MDPL circuit corresponds to an unmasked value $d = d_m \oplus m$. MDPL circuits are implemented as dual-rail precharge circuits, which are discussed in more detail in Section 7.3.1. This is solely done to avoid glitches. Therefore, it is not necessary to balance complementary wires in an MDPL circuit. MDPL cells are built based on SR cells that are available in existing standard-cell libraries.

The general architecture of an MDPL circuit is shown in Figure 9.4. In MDPL circuits, only the sequential cells are connected to the clock signal. Therefore, only these cells precharge at the same time. Furthermore, they also evaluate at the same time. Combinational MDPL cells precharge when their inputs have been set to the precharge value. They evaluate, when their inputs have been set to complementary values. MDPL D-flip-flops perform three operations. In the precharge phase, they start the precharge wave. In the evaluation phase, they provide the stored complementary values that are masked with the mask value

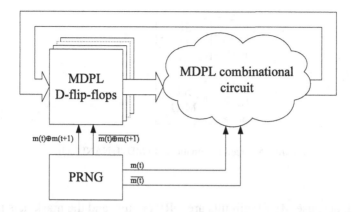

Figure 9.4. Architecture of an MDPL circuit.

$m(t)$ of the current clock cycle. Third and last, the MDPL D-flip-flops perform the mask change from $m(t)$ to $m(t + 1)$. The mask $m(t + 1)$ is the mask value of the next clock cycle.

During the precharge phase, the precharge wave is started at the outputs of the MDPL D-flip-flops. The precharge value that is used in MDPL circuits is 0. In the precharge phase, also the mask signals $m(t)$, $\overline{m(t)}$, $m(t) \oplus m(t + 1)$, and $\overline{m(t) \oplus m(t + 1)}$ are precharged. At the end of the precharge phase, all MDPL cells in the combinational circuit have been precharged. Furthermore, the MDPL D-flip-flops store the masked values for the subsequent evaluation phase. Let us assume that these values are masked with the mask value $m(t)$. In the subsequent evaluation phase, the MDPL D-flip-flops provide their stored masked values to the combinational MDPL cells. Since these stored values are masked with mask $m(t)$, this mask must also be provided to the combinational MDPL cells. The MDPL D-flip-flops are provided with the mask $m(t) \oplus m(t + 1)$ and its inverse. The flip-flops change the masks of the input values from $m(t)$ to $m(t + 1)$ before these re-masked values are stored at the beginning of the next precharge phase.

The transitions of the mask values cause significant current peaks in MDPL circuits because the mask nets are usually quite big. The situation is made worse by the fact that the transitions of the mask values coincide with the transition of the clock signal. Note that the mask nets in MDPL circuits are complementary wires that are precharged. This allows balancing the power consumption of the mask nets to a degree that prevents SPA attacks, *i.e.* it is not possible to determine the mask values by looking at the power traces of the circuit.

The area requirements of MDPL circuits are at least doubled compared to corresponding unmasked CMOS circuits, while the maximum clock rates are typically halved. The power consumption of MDPL circuits is significantly

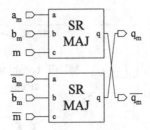

Figure 9.5.   Cell schematic of an MDPL NAND cell.

increased, because MDPL circuits are DRP circuits and the mask nets must be switched. The properties of AES modules implemented in MDPL are presented in [PM05] and [PM06]. The latter article also includes details about using MDPL in a semi-custom design flow.

General Description of MDPL Cells

The generic structure of combinational MDPL cells is quite similar to the one of the combinational WDDL cells shown in Figure 7.8. The main difference is that the inputs to the positive monotonic Boolean functions F_1 and F_2 are masked values and the corresponding masks. As output values, only masked values are calculated. All input signals are precharged, and hence, no glitches occur.

The DPA resistance of MDPL cells is limited by two factors. First, not all internal nodes of the SR cells on which MDPL cells are based are perfectly masked. Fortunately, the power consumption caused by charging and discharging these internal nodes is typically very small. Second, the time-of-evaluation of MDPL cells is data dependent. Avoiding these two problems would significantly increase the size and the complexity of MDPL cells.

Examples of MDPL Cells

In the following, a combinational MDPL cell implementing the NAND function is presented. Furthermore, as an example of a sequential MDPL cell, an MDPL D-flip-flop is shown.

MDPL NAND Cell.   The cell schematic of an MDPL NAND cell is shown in Figure 9.5. It consists of two 3-input SR MAJ cells. A MAJ cell calculates the majority function. If more input values of a MAJ cell are 1 than 0, the output is 1. Otherwise, the output is 0. The majority function fulfills (7.1) for complementary input values and it is a positive monotonic Boolean function. Therefore, no glitches are produced by the MDPL NAND cell and the outputs are precharged correctly when the inputs are set to the precharge value. The

Table 9.1. Truth table of an MDPL NAND cell for complementary input values.

a_m	b_m	m	$\overline{a_m}$	$\overline{b_m}$	$\overline{m}$	$q_m = MAJ(\overline{a_m}, \overline{b_m}, \overline{m})$	$\overline{q_m} = MAJ(a_m, b_m, m)$
0	0	0	1	1	1	1	0
0	1	0	1	0	1	1	0
1	0	0	0	1	1	1	0
1	1	0	0	0	1	0	1
0	0	1	1	1	0	1	0
0	1	1	1	0	0	0	1
1	0	1	0	1	0	0	1
1	1	1	0	0	0	0	1

truth table of an MDPL NAND cell for complementary input values is shown in Table 9.1.

MDPL D-Flip-Flop. Figure 9.6 shows the cell schematic of an MDPL D-flip-flop. It is a complex cell that has to perform three operations. First, the mask of the input signal must be changed from the current mask $m(t)$ to the mask $m(t + 1)$ that is used in the next clock cycle. This change of the mask is performed by the two SR AND cells, the two SR OR cells, and the SR MAJ cell at the input of the MDPL D-flip-flop. The circuit built by these cells takes as inputs $d_{m(t)} = d \oplus m(t)$ and $m(t) \oplus m(t+1)$ (both non-inverted and inverted). It calculates as output $d_{m(t+1)} = d \oplus m(t + 1)$ and its inverse. The SR AND and SR OR cells realize positive monotonic functions, and hence, the circuit responsible for mask changing is precharged correctly during the precharge phase and produces no glitches.

The second operation of the MDPL D-flip-flop is realized by the SR D-flip-flop. This flip-flop stores the value $d_{m(t+1)}$ that is masked with the new mask $m(t + 1)$ of the next clock cycle. An MDPL D-flip-flop stores its input value at the positive clock edge. Note that no precharge value needs to be stored in the SR D-flip-flop as it was the case for the WDDL D-flip-flop shown in Figure 7.10. This is because the MDPL D-flip-flip stores masked values only.

The third operation is realized by the two SR NOR cells at the output of the MDPL D-flip-flop. During the precharge phase ($clk = 1$) the outputs of the SR NOR cells and thus the outputs $q_{m(t+1)}$ and $\overline{q_{m(t+1)}}$ of the MDPL D-flip-flop are set to the precharge value 0. This starts the precharge wave for the combinational MDPL cells connected to the outputs of the MDPL D-flip-flop. In the subsequent evaluation phase ($clk = 0$), the complementary values at the outputs of the SR D-flip-flop are just passed through the SR NOR cells.

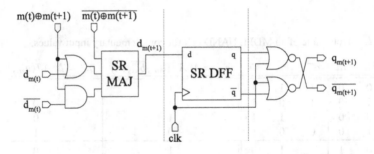

Figure 9.6. Cell schematic of an MDPL D-flip-flop.

9.5   Notes and Further Reading

Basic Masking Techniques and Secret Sharing.   The idea of using secret-sharing schemes was proposed independently by Goubin and Patarin [GP99] and Chari *et al.* [CJRR99b]. The latter paper was also the first to put forward a concept for proving resistance against DPA attacks. Messerges [Mes00a] discussed masking techniques for the AES candidates. In this article, algorithms to switch between Boolean and arithmetic masking were presented. However, it was pointed out by Coron [CG00] that those algorithms are not secure. Later, Goubin [Gou01] and Coron and Tchulkine [CT03] described secure and more efficient algorithms to switch between Boolean and arithmetic masking. Akkar and Goubin discussed a masking scheme for DES in [AG03]. Naturally, secret sharing can also be applied in the context of asymmetric cryptography. For instance, one can imagine that the secret RSA exponent d is represented by two shares d_m and m with $m < d$ such that $d = d_m + m$. Consequently, the two exponentiations v^{d_m} and v^m do not reveal information about d.

Provable Security.   Security proofs that were given in the context of masking schemes work according to the following idea. By masking the intermediate values, the masked intermediate values become independent of the unmasked intermediate values and the masks. The theory is, that then the unmasked intermediate values and power consumption of the masked intermediate values is independent. This way of thinking is fine in theory. However, there are pitfalls in practice. If a device leaks the Hamming distance, then the power consumption is a function of two intermediate values. Hence, even if all intermediate values fulfill the independence property, an implementation using them is insecure if two intermediate values that are concealed by the same mask are processed consecutively. This shows the limits of these security proofs.

Masking AES.   The first proposal to mask the AES S-box was presented by Akkar and Giraud [AG01]. It introduced multiplicative masking, which later turned out to be insecure. Simplified multiplicative masking by Trichina *et al.* [TSG03] inherits this security problem. Golić and Tymen [GT03] presented an idea to overcome this problem by mapping the finite field elements to elements of a larger ring. In this way, the zero value is mapped to different non-zero values. The security of this scheme was not investigated in detail.

Itoh *et al.* reported on their masking scheme in [ITT02]. The masking scheme that we gave as example in Section 9.2.1 was reported by Herbst *et al.* [HOM06].

Independently, several groups developed masking schemes based on masked arithmetic. For example, the papers by Pramstaller *et al.* [POM+04, PGH+04] report on the practical realization of the masking scheme by Oswald *et al.* [OMPR05]. This approach exploits that the AES S-box can be implemented nicely with composite field arithmetic, see also Section 9.2.2. The practical realization of this scheme has led to the discovery of the glitch problem, which we discuss in Section 10.2. A variant of this scheme that has been adapted for smart card implementations with flexible masks has been reported by Oswald and Schramm in [OS06]. Trichina *et al.* [TKL05], Blömer *et al.* [BGK05], Carlier *et al.* [CCD04], and Morioka and Akishita [MA04] reported on similar masking schemes. It is imperative that actual implementations of these schemes take the glitch problem into account. Another masking scheme for AES was presented by Schramm and Paar in [SP06]. Since they focused on security against higher-order DPA attacks, they investigated recomputation techniques for masked S-boxes. In particular, they concentrated on techniques to efficiently compute a masked S-box table based on another masked S-box table.

Blinding. Kocher [Koc96] listed blinding techniques for RSA already in his article on timing attacks. We have discussed these techniques in Section 9.1.3. Note that these techniques do not necessarily protect against SPA attacks, see Section 10.7.

Blinding techniques for ECC were discussed by Coron in [Cor99]. Those techniques are similar to the techniques for RSA. We can blind the secret scalar d by adding a value m to it that is a multiple of the group order $ord(P)$: $d_m = d + m \times ord(E)$. Then, we have that $[d_m]P = [d]P \bmod ord(E)$ and the result is automatically unmasked. We can blind the base point P by adding a random point M to the base point: $P_m = P + M$, such that $[d]M$ is known. Then we can derive $[d]P$ from $[d]P_m = [d](P + M)$ by subtracting $[d]M$. The third blinding technique is genuine for implementations of ECC in projective coordinates. It exploits that coordinates of a point in the projective plane are not unique: $(X, Y, Z) = (mX, mY, mZ)$ for every $m \neq 0$ in the finite field. Hence, we can randomize the representation of a point by choosing m randomly.

Joye and Tymen [JT01] proposed to randomize an elliptic curve point by using a random isomorphism to map a given point to a point on an isomorphic curve. In the same article, they showed how to randomize a point by randomizing the finite field that underlies the elliptic curve and they gave another method for randomizing the scalar in case of Koblitz curves.

Masking Cells. The approach of masked logic cells was also independently developed by several researchers. For instance, Trichina *et al.* [TKL05] proposed a masked AND gate. Golić and Menicocci [GM04] proposed another masked AND gate that has a shorter critical path than the one proposed by Trichina. Ishai *et al.* [ISW03] also discuss masking (secret sharing) at the cell level. Their approach comes with a definition of security. However, the model that is used for this definition does not take glitches into account. In practice, glitches usually lead to a data-dependent power consumption, *i.e.* they make power analysis attacks possible.

There exist several masked logic styles which take the glitch problem into account. MDPL for example, which we discussed as example in Section 9.4.1, completely avoids glitches. Another masked logic style that avoids glitches was proposed by Suzuki *et al.* [SSI04]. Fischer and Gammel [FG05] proposed a masked logic style which overcomes the glitch problem if each masked input value of a cell arrives at the same time as the corresponding mask. A masked logic style that takes the glitch problem and also the early propagation problem into account was proposed by Chen and Zhou [CZ06].

Chapter 10

ATTACKS ON MASKING

The use of masking schemes to counteract power analysis attacks is popular for several reasons. For instance, masking can be implemented in software on processors without altering their power consumption characteristics. Probably because of their popularity, many researchers have studied the security of masking schemes and their implementations. It has turned out that virtually every masking scheme can be attacked.

In this chapter, we discuss different types of power analysis attacks on masking schemes, including second-order DPA attacks and template-based DPA attacks. We start by discussing DPA attacks on masking schemes in general. Next, we discuss implementation issues of masking schemes that can be exploited by DPA attacks. Subsequently, we focus on second-order DPA attacks on software implementations. In addition, we explain second-order DPA attacks using templates and template-based DPA attacks on software implementations. Last, we discuss second-order DPA attacks on hardware implementations.

10.1 General Description

Masking provides security against DPA attacks if each masked intermediate value v_m is pairwise independent of the unmasked intermediate value v and the mask m. Hence, only if this pairwise independence does not hold for some reason, a masking scheme is vulnerable to DPA attacks. The DPA attacks, which we have discussed so far, have the property that one intermediate value is predicted and used in the attack. Because only one intermediate value is used, these DPA attacks are also referred to as *first-order* DPA attacks. If several intermediate values are used to formulate the hypotheses, then the corresponding DPA attacks are called *higher-order* DPA attacks. We continue to write DPA attacks in order to refer to first-order DPA attacks in the remainder of this book.

Higher-order DPA attacks exploit the joint leakage of several intermediate values that occur inside the cryptographic device. Remember that due to performance reasons, typical implementations of masking schemes conceal several intermediate values by the same mask. However, even if several masks are used throughout the algorithm, they are generated before the algorithm starts, they are applied to the data and (or) the key, and they are altered by the operations of the algorithm. Consequently, in an implementation where efficiency (memory, speed) is needed, it is always the case that a mask (or a combination of masks) and an intermediate value that is concealed by this mask (or a combination of masks) occur in the device. Hence, in practice it is typically not necessary to study higher-order DPA attacks in general. In practice, it is sufficient to concentrate on higher-order DPA attacks that exploit the leakage that is related to *two* intermediate values. These attacks are called *second-order* DPA attacks. The two intermediate values can either be two values that are concealed by the same mask or a masked value and the corresponding mask.

> Second-order DPA attacks exploit the joint leakage of two intermediate values that are processed by the cryptographic device.

10.1.1   Second-Order DPA Attacks

Second-order DPA attacks exploit the leakage of two intermediate values that are related to the same mask. In general, this leakage cannot be exploited directly because the two intermediate values often occur in different operations of the algorithm. Hence, they might be computed subsequently and contribute to the power consumption at different times. In this case, it is necessary to preprocess the power traces in order to obtain power consumption values that depend on both intermediate values.

However, even if the intermediate values contribute to the power consumption at the same time, it is possible that the distribution of the power consumption has the same mean but different variances for all hypotheses. In this case, a DPA attack using the statistical methods that we have used in Chapters 4 and 6 do not succeed because these methods work with the mean value. In order to mount successful DPA attacks in this case it is necessary to either use other statistical methods that exploit the variance, or to preprocess the traces in such a way that the mean-based methods work. The preprocessing is typically done in step 2 of a DPA attack, which consists of recording the power consumption of the device.

> Second-order DPA attacks work in the same way as first-order DPA attacks except that they sometimes require preprocessing the power traces.

Preprocessing

The preprocessing prepares the power traces for the DPA attack. There are three cases that occur in practice. In the first case, the targeted intermediate values occur in different clock cycles. In this case, the preprocessing combines two points within a trace. This first case typically occurs in software implementations of masking schemes. Second, the targeted intermediate values occur within one clock cycle. In this case, the preprocessing function is applied to single points in the trace. Third, the targeted intermediate values occur within a clock cycle and the power consumption characteristics allow exploiting the leakage directly. In this case, the preprocessing step can even be omitted. The two latter cases typically occur in hardware implementations of masking schemes.

DPA Attacks on Preprocessed Traces

A second-order DPA attack simply applies a DPA attack to the preprocessed traces. This means, in step 1 of the second-order DPA attack, we choose two intermediate values u and v. These values do not occur as such in the device because we study a masked implementation. Recall that in an implementation that uses Boolean masking, only the masked intermediate values $u_m = u \oplus m$ and $v_m = v \oplus m$ are present in the device. In step 2, we record the power traces and we actually do the preprocessing. In step 3, we calculate hypothetical values that are a combination of u and v: $w = comb(u, v)$. In attacks on Boolean masking, this combination function typically is the exclusive-or function:

$$w = comb(u, v) = u \oplus v = u_m \oplus v_m \qquad (10.1)$$

Note that we can calculate the value of the combination of two masked intermediate values without having to know the mask! In step 4, we map w to hypothetical power consumption values h. In step 5, we compare the hypothetical power consumption with the preprocessed traces.

10.2 DPA Attacks

Before we discuss second-order DPA attacks in detail, we discuss when and why even first-order DPA attacks can break masked implementations. Recall from Sections 9.2.1 and 9.2.2 that DPA attacks only work if the masking scheme does not fulfill the independence property or if some mistake has been made during the implementation. In this section, we survey attacks that exploit common problems, such as multiplicative masks, reused masks, and biased masks.

10.2.1 Multiplicative Masking

We have distinguished between Boolean and arithmetic masking in Section 9.1.1 and we have outlined that depending on the algorithm, either one or

Table 10.1. Multiplicative masking for the AES S-box.

Intermediate value	Applied operation
$v \oplus m$	$\times m'$
$(v \oplus m) \times m'$	$\oplus m \times m'$
$(v \times m')$	-1
$(v \times m')^{-1}$	$\oplus m \times m'^{-1}$
$(v \times m')^{-1} \oplus m \times m'^{-1}$	$\times m'$
$v^{-1} \oplus m$	

both types of masking are used. In addition, we have pointed out that multiplicative masking does not satisfy the independence condition: $v \times m$ is not statistically independent of v. This is because if $v = 0$, then $v \times m = 0$ regardless of m. Consequently, multiplicative masking is vulnerable to DPA attacks. In particular, it is vulnerable to zero-value DPA attacks. Zero-value DPA attacks exploit that processing an intermediate value that is zero requires a different amount of power than processing an intermediate value that is not zero. These attacks essentially use the ZV power model that has been introduced in Section 6.2.2.

Example for Software

In case that it is not possible to mask table look-ups, the authors of [AG01] have suggested using multiplicative masking for the AES S-box. The idea works as follows. The AES S-box function is actually defined by two operations. These are a finite field inversion and a linear mapping, see B.1. It is easy to determine how the finite field inversion changes a multiplicative mask because $(v \times m)^{-1} = v^{-1} \times m^{-1}$. Hence, it is only necessary to convert the masking of the intermediate values from Boolean masking to multiplicative masking and vice versa. Table 10.1 shows the conversion and the masked inversion that has been suggested in [AG01].

The conversion consists of five steps. Steps three and four work with the masked intermediate values $v \times m'$ and $(v \times m')^{-1}$. Hence, they are ideal candidates for a zero-value DPA attack. In this example, we focus on the input of the inversion which is $v \times m'$. Let us assume that we attack the first round of AES. For a plaintext d_i and a key hypothesis k_j the corresponding hypothetical intermediate value is $v_{i,j} = d_i \oplus k_j$. In order to get the hypothetical power consumption $h_{i,j}$ we apply the ZV power model to $v_{i,j}$:

$$h_{i,j} = ZV(v_{i,j}) = ZV(d_i \oplus k_j)$$

This power model describes the power consumption well because the power consumption for $v = 0$ is independent of the mask. In the last step of a

DPA attack, we compare the hypothetical power values $h_{i,j}$ with the measured traces. As usual, the correct key is indicated by the highest correlation peak in the resulting correlation traces.

In order to compare this zero-value DPA attack with a DPA attack that uses the HW model, we determine the correlation coefficient for the correct key k_{ck}, and for the correct moment of time t_{ct} of both attacks on an unprotected implementation. We have already discussed in Section 6.3 how to calculate (or simulate) correlation coefficients. Now we use this technique to obtain the correlation for a zero-value DPA attack. This means we simulate an attack which is based on the ZV power model. According to Section 6.3, the correlation coefficient that we are interested in is defined as $\rho(\mathbf{h}_{ck}, \mathbf{s}_{ct})$. The hypotheses $\mathbf{h}_{ck}$ are given by $h_{i,ck} = ZV(d_i \oplus k_{ck})$ and the simulated traces $\mathbf{s}_{ct}$ are given by $s_{i,ct} = HW((d_i \oplus k_{ck}) \times m_{d_i})$. It turns out that $\rho(\mathbf{h}_{ck}, \mathbf{s}_{ct})$ is 0.17. Summarizing, the correlation coefficient for the correct key hypothesis is 0.17 in this zero-value DPA attack. In contrast to this, the correlation coefficient for the simulated DPA attack on the software implementation using the Hamming-weight model is 1 according to Section 6.3.1.

10.2.2 Mask Reuse Attacks

Masks can be reused in several ways. First, the same masks can be used for different intermediate values. Second, the masks can be used in several encryption runs, and third, the same masks can be used for Boolean and arithmetic operations.

As we have pointed out in Section 9.1, it is advisable to use several masks in a masking scheme. Otherwise, the result of an exclusive-or of two intermediate values that are concealed by the same Boolean mask would be unmasked. This is not desired. Hence, one must use two masks if such an exclusive-or operation is part of the cryptographic algorithm. We have also pointed out that one of the pitfalls of masking is to accidentally unmask intermediate values. For instance, if a device leaks the Hamming distance of the intermediate values, then it is not advisable to consecutively move two values that are concealed by the same mask over the bus.

Another problem is the reuse of masks in several encryption runs. For instance, within a masking scheme several masks might be used. However, in order to minimize the overhead for masking table look-ups, one might decide to reuse the masks for the tables in subsequent encryption runs. The policy could be to reuse the mask for a table for instance ten times. This would mean that a new mask is used to recompute a the masked table every ten encryptions. For an infinite sequence of encryption runs, this policy would not change the distribution of the masks. However, in practice an attacker only measures a limited number of power traces. Within this limited set, the masks are then likely to be biased. As a result, DPA attacks work.

There also occurs a problem if the same additive and multiplicative masks are used such as suggested in [TSG03]. This article pursues the idea to simplify the multiplicative masking scheme by Akkar *et al.* by setting $m = m'$, *i.e.* the Boolean mask m is equal to the arithmetic mask m'. This introduces another vulnerability into the multiplicative masking scheme for the AES S-box. The reason is that the intermediate value $(v \oplus m) \times m$ is computed. This intermediate value is not independent of v. For instance, if $v = 0$ then $m \times m$ can only be zero if $m = 0$. However, if $v = 1$ then $(1 \oplus m) \times m$ gives zero if either $m = 0$, or $m = 1$. Apparently, the number of solutions of $(v \oplus m) \times m = 0$ depends on v. Consequently, the value $(v \oplus m) \times m$ depends in a statistical sense on v. Hence, it is vulnerable to DPA attacks.

10.2.3   Biased Mask Attacks

We know already that uniformly distributed masks are essential for the security of a masking scheme. Hence, the strategy of an attacker could be to force some sort of bias into the masks. This can either be done by actively manipulating a device (fault attacks) or by selecting a subset of the measured traces that corresponds to a subset of masks. The latter strategy can be implemented in the following way. The attacker encrypts one plaintext a large number of times. If masking is used, then the intermediate values will vary. More precisely, the intermediate values that are masked will be different in subsequent encryption runs. In this way, the attacker can determine when the masks are generated and when the masks are applied to the plaintext. If the attacker is able to determine some information (for instance the Hamming weight) about the masks, then it is possible to select a subset of traces based on this information. Because in this subset of traces only a subset of masks has been used, the masks are no longer uniformly distributed and a DPA attack works. This type of attack can also be realized as a variant of a second-order DPA attack.

10.3   Second-Order DPA Attacks on Software Implementations

In this section, we study second-order DPA attacks in the context of software implementations on microcontrollers. In particular, we study how to apply them to the masked AES implementation that we have described in Section 9.2.1. This implies that we focus on Boolean masking only. First, we discuss preprocessing functions. Second, we discuss how certain preprocessing functions influence the correlation coefficient in the DPA attack. Third, we give an example for a practical second-order DPA attack on a masked AES software implementation and fourth, we give an example for the same attack on an AES software implementation which uses masking and shuffling.

10.3.1 Preprocessing

In the preprocessing step, we apply a preprocessing function *pre* to each power trace. The result of the preprocessing step is a preprocessed trace that we refer to as $\tilde{t}$.

In an attack, we are interested in the two points of a power trace that correspond to the calculation of u_m and v_m. Typically, we do not know exactly when these two masked intermediate values are computed. Consequently, we can at best guess an interval $I = t_{r+1}, \ldots, t_{r+l}$ of the power trace, which likely contains u_m and v_m. Hence in practice, we apply the preprocessing function to all combinations of points in this interval. If the preprocessing function is symmetric, and only pairs of points (t_x, t_y) with $(x \neq y)$ are considered, then the preprocessed traces consist of $l - 1$ segments of decreasing length, *i.e.* the length of the preprocessed traces is $(l - 1) + (l - 2) + \ldots 2 + 1 = l \cdot (l - 1)/2$. Consequently, a preprocessed trace $\tilde{t}$ is typically given by $(pre(t_{r+1}, t_{r+2}), pre(t_{r+1}, t_{r+3}), \ldots, pre(t_{r+2}, t_{r+3}), \ldots pre(t_{r+l-1}, t_{r+l}))$.

Several types of preprocessing functions have been discussed in the scientific literature so far. The first function has been suggested by [CJRR99b]. This function computes the product of two points: $pre(t_x, t_y) = t_x \cdot t_y$. In [Mes00b], the author has proposed the absolute value of the difference of two points: $pre(t_x, t_y) = |t_x - t_y|$. Because we use this preprocessing function frequently, we refer to it as "*absolute-difference*" function. In [WW04], the authors have proposed to use the square of the sum of two points: $pre(t_x, t_y) = (t_x + t_y)^2$. In addition, they have pointed out that an application of the FFT also leads to a combination of points.

A preprocessing function that selects traces depending on the height of a certain point has been described in [Jaf06b]. An example for such a choice is:

$$pre(t_x, t_y) = \begin{cases} t_y & \text{if } t_x > c \\ - & \text{otherwise} \end{cases}$$

This preprocessing function is not symmetric. For a fixed point t_x the preprocessed trace $\tilde{t}$ is given by $(pre(t_x, t_{r+1}), pre(t_x, t_{r+2}), \ldots, pre(t_x, t_{r+l}))$. This means, only if a certain point t_x of a trace is higher than a certain threshold c, the trace (or a part of it) is used for the DPA step. Otherwise the trace is discarded. This preprocessing corresponds to biasing the masks. Note that there are two ways to use this preprocessing function in practice. First, the point t_x can be chosen such that it corresponds to the processing of a mask. Then, by selecting only a subset of traces, the mask is biased. Second, the point t_x can be chosen such that it corresponds to the processing of a masked intermediate value. Then, the attacker needs to bias the unmasked intermediate value as well in order to bias the masks. Note that an attacker can easily bias the plaintexts in a chosen plaintext attack.

10.3.2   DPA Attacks on the Preprocessed Traces

Now we investigate the effect of different preprocessing functions on the correlation coefficient. First, we discuss how the maximum correlation coefficient in a second-order DPA attack is defined. Then, we provide an intuitive discussion for a 1-bit scenario. This scenario shows that different preprocessing functions can lead to very different correlation coefficients. Based on the 1-bit scenario, we discuss how the correlation coefficient is influenced by different preprocessing functions in the case of a multiple-bit scenario.

For all our discussions we assume that the attacker exploits the joint distribution of two masked intermediate values u_m and v_m. This means the attacker computes hypothetical intermediate values u and v and combines them in order to get a joint hypothetical intermediate value w. Since we deal with Boolean masking only, our combination function is the exclusive-or function: $w = u \oplus v$. We also assume that the attacked device leaks the Hamming weight. Hence, the hypothetical intermediate values w are mapped to hypothetical power consumption values using the Hamming-weight model $h = HW(w) = HW(u \oplus v)$.

In step 5 of the DPA attack, we compare the hypothesis $HW(u \oplus v)$ with the preprocessed traces. This means, we estimate $\rho(\mathbf{H}, \tilde{\mathbf{T}}) = \rho(HW(\mathbf{V} \oplus \mathbf{W}), \tilde{\mathbf{T}})$. The maximum of this correlation (see (4.14) for the definition) is reached for the correct key hypothesis k_{ck} at the point $\tilde{t}_{ct}$. Hence, $\rho_{ck,ct}$ determines the number of traces that are needed for an attack, see also Section 6.4. The correlation $\rho_{ck,ct}$ is essentially determined by the preprocessing function, because we have fixed the combination function before. The point $\tilde{t}_{ct}$ corresponds to the processing of the two targeted intermediate values: $\tilde{t}_{ct} = pre(HW(u_m), HW(v_m))$. Consequently, the goal is to determine a preprocessing function pre that maximizes (10.2).

$$\rho(HW(u \oplus v), pre(HW(u_m), HW(v_m)) \tag{10.2}$$

One-Bit Scenario

As a first step to determine a good preprocessing function, we study (10.2) in a simplified scenario. We assume that u_m, v_m, and m are one-bit values. Table 10.2 shows the results of (10.2) for different preprocessing functions. The first column lists the masked intermediate values and the preprocessing functions. The next four columns list the four possible combinations of the values u_m and v_m as well as the corresponding results of the preprocessing functions. The last column gives the correlation coefficient that is calculated according to (10.2). Hence, each value in this column is the correlation between the four values in the row $HW(u \oplus v)$ and the four values in the row of the corresponding preprocessing function.

Table 10.2 shows that taking the absolute value of the difference leads to the highest correlation. Thus, it is the best choice in this case. Table 10.2 also

Table 10.2. The effect of different preprocessing functions on the correlation coefficient in the DPA attack on the preprocessed traces.

	Value				Correlation according to (10.2)
u_m	0	0	1	1	
v_m	0	1	0	1	
$HW(u \oplus v)$	0	1	1	0	
$HW(u_m) \cdot HW(v_m)$	0	0	0	1	$\rho = -0.57$
$|HW(u_m) - HW(v_m)|$	0	1	1	0	$\rho = 1$
$(HW(u_m) + HW(v_m))^2$	0	1	1	4	$\rho = -0.33$
$HW(u_m) + HW(v_m)$	0	1	1	2	$\rho = 0$
$HW(u_m) - HW(v_m)$	0	-1	1	0	$\rho = 0$

shows that two other preprocessing functions, which have been listed in the previous paragraph, lead to non-zero correlation coefficients as well. Hence, attacks based on them work as well. The last two rows contain variants of the preprocessing functions. Since the correlation is zero, attacks based on them do not work in our scenario. Note that the preprocessing function that uses the square of the sum of points leads to a non-zero correlation while taking just the sum of points leads to a zero correlation. This can be explained in the following way. The sum of two points is one if $HW(u \oplus v) = 1$. The sum of two points is either zero or two if $HW(u \oplus v) = 0$. Hence, on average the power consumption in the case $HW(u \oplus v) = 1$ is the same as in the case $HW(u \oplus v) = 0$ although the distributions of the two cases are different. If we take the square of the sum, then the power consumption in the case $HW(u \oplus v) = 0$ is on average smaller than the power consumption in the case $HW(u \oplus v) = 1$. This has the effect that the two distributions can now be distinguished by looking at their means.

Multiple-Bit Scenario

In practice, we are interested in the correlation for larger values of u_m and v_m. Therefore, we have produced Table 10.3 that lists the correlations for different numbers of bits of u_m and v_m. They decrease for all preprocessing functions for an increasing number of bits of u_m and v_m. However, the preprocessing function that computes the absolute value of the difference of two points is still the best function. It leads to a correlation coefficient of 0.24 in the case of 8-bit values.

Similar to the simplified (single bit) scenario, using the square of the sum of two points leads to a correlation coefficient that is non-zero. In contrast to this, taking the sum of two points leads to a correlation coefficient that is zero. It appears that also in the general case, using a non-linear function during preprocessing changes the correlation coefficient. This should not come as a

Table 10.3. Correlation coefficients for different preprocessing functions and different numbers of bits of the operands.

	Number of bits of u_m and v_m			
	1	2	4	8
$HW(u_m) \cdot HW(v_m)$	-0.58	0.32	-0.17	-0.09
$\lvert HW(u_m) - HW(v_m) \rvert$	1.00	0.53	0.34	0.24
$(HW(u_m) + HW(v_m))^2$	-0.33	-0.16	0.08	-0.04
$HW(u_m) + HW(v_m)$	0.00	0.00	0.00	0.00
$HW(u_m) - HW(v_m)$	0.00	0.00	0.00	0.00

surprise. Recall that the correlation coefficient is a linear measure. Hence, linear functions do not change the correlation, but non-linear functions do. A simple non-linear extension to the absolute-difference function is to combine it with an exponentiation: $\rho(HW(u \oplus v), \lvert HW(u_m) - HW(v_m) \rvert^{\beta})$. Table 10.4 shows results for some values of β. The number of bits has been fixed to eight in this table. The conclusion from Table 10.4 is that one can slightly increase the correlation coefficient, but no significant improvement can be achieved in this way. More complicated preprocessing methods seem to be required.

Table 10.4. Correlation coefficients for the absolute-difference function raised to the power β.

	Different values of β					
	1	2	3	4	5	6
$\lvert HW(u_m) - HW(v_m) \rvert^{\beta}$	0.24	0.26	0.25	0.23	0.20	0.18

Summarizing, we can see that the different preprocessing functions lead to different correlation coefficients in the DPA step. Which preprocessing function is the best depends on the power model of the device. If a device leaks the Hamming weight, taking the absolute-difference function is a good choice.

> The absolute-difference preprocessing function is a good choice for second-order DPA attacks on devices that leak the Hamming weight.

10.3.3   Example for Masked AES

In this example, we show how to attack the masked software implementation of AES that is described in Section 9.2.1. In the implementation of this masking scheme have set $m' = m$, which means that the input and the output of the SubBytes operation are concealed by the same mask m. We have also verified that the implementation is secure against DPA attacks. Note that setting $m = m'$

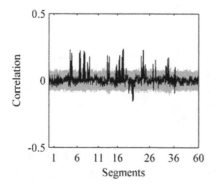 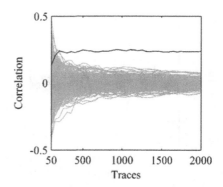

Figure 10.1. Result of a second-order DPA attack on a masked AES implementation in software.

Figure 10.2. Evolution of the correlation coefficient over an increasing number of traces.

does not make attacks easier in general. It is just one of the many scenarios that can occur in practice and we restrict our discussion to this scenario for the sake of conciseness.

We have targeted the S-box input and the S-box output in our second-order DPA attack. For the second-order DPA attack we have measured the power consumption of our masked AES software implementation during the first encryption round. In order to reduce the number of points in the measured traces, we have compressed them. We have identified the first round of AES with a visual inspection of the compressed power traces. From the first round, we have only taken the first 61 points of the compressed traces. These points are within the interval that likely contains the first S-box look-up. Consequently, we have applied a preprocessing function to this interval only. According to the conclusion from the previous section, we have used the absolute-difference preprocessing function. Since the interval contains 61 points, the preprocessing step, which requires looking at all combinations of two out of 61 points, leads to preprocessed traces with 60 segments having in total $61 \cdot 60/2 = 1\,830$ points.

Then, we have computed the hypothetical intermediate values $u_{i,j} = d_i \oplus k_j$ and $v_{i,j} = S(d_i \oplus k_j)$. We have combined them with the exclusive-or function to derive $w_{i,j} = u_{i,j} \oplus v_{i,j} = (d_i \oplus k_j) \oplus S(d_i \oplus k_j)$, and we have mapped them to hypothetical power consumption values $h_{i,j}$ using the Hamming-weight model.

$$h_{i,j} = HW(w_{i,j}) = HW((d_i \oplus k_j) \oplus S(d_i \oplus k_j)) \tag{10.3}$$

Finally, we have compared the hypothetical power consumption with the preprocessed traces. Figures 10.1 and 10.2 show the result of this attack. Figure 10.1 depicts the correlation trace for the correct key hypothesis, which is plotted in

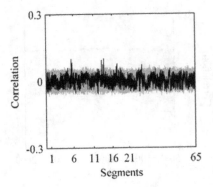

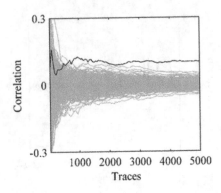

Figure 10.3. Result of a second-order DPA attack on an AES implementation in software that uses masking and shuffling.

Figure 10.4. Evolution of the correlation coefficient over an increasing number of traces.

black, versus the correlation traces for the incorrect key hypotheses, which are plotted in gray. Note that high correlation peaks occur in all segments that are related to the processing of the two attacked intermediate values. The highest correlation that occurs in this trace is about 0.23. This is very close to the theoretical correlation coefficient $\rho_{ck,ct}$ for 8-bit values, which is 0.24 according to Table 10.3.

According to (6.8), about 460 traces are needed for a successful attack with $\rho = 0.24$. Figure 10.2 shows that it is easily possible to distinguish the correct key hypothesis from the incorrect key hypotheses with about 500 traces.

10.3.4   Example for Masked and Shuffled AES

In this example, we discuss an attack on an AES software implementation that implements the masking scheme of Section 9.2.1 and that uses shuffling such as described in Section 7.1.1.

Remember that the effect of shuffling is that the height of the correlation peak is reduced. For example, if a shuffling scheme randomizes the occurrence of the AES state bytes 1 and 2, then the height of the correlation peak is reduced by a factor of 2 if bytes 1 or 2 are attacked and no windowing is used. The same reasoning also holds for second-order DPA attacks. This means, even if shuffling occurs in addition to masking, we can still perform a second-order DPA attack as before. The only difference is that the correlation coefficient for the correct key hypothesis is smaller.

We provide evidence that this reasoning holds by showing the results of a concrete experiment. Our implementation only shuffles state bytes 1 and 2 and uses the same masking scheme as in the previous example. We have performed a second-order DPA attack with the same hypotheses as before. Since the correlation coefficient for the correct key hypothesis was about 0.23 in the previous example, we now expect to see a correlation coefficient of about 0.11.

Figure 10.3 shows the result of this attack. The correlation trace of the correct key hypothesis, which is plotted in black, shows indeed correlation peaks with a value of about 0.11. The traces that correspond to the incorrect key hypotheses are firmly below the trace that corresponds to the correct key hypothesis. Figure 10.4 indicates that it is possible to distinguish the correct key hypothesis from the incorrect key hypotheses with about 3 000 traces. Notice though that there is already a clear difference after 1 000 traces. However, this difference becomes smaller again as the number of traces is increased. This is a nice example why determining the number of traces by inspecting estimated correlation coefficients for one single experiment is not very precise, see Section 6.7. On average the attack presented in this section requires four times more traces than the attack presented in Section 10.3.3.

10.4   Second-Order DPA Attacks on Software Implementations Using Templates

In this section we discuss how second-order DPA attacks can be conducted, and even be improved, by using templates. Since we typically preprocess traces in a second-order DPA attack, we can use templates before, during, or after the preprocessing of the traces. We discuss all these three scenarios. First, we explain how templates can be used before preprocessing. In this scenario, templates are used in order to extract information directly from the measured power traces. Second, we discuss how templates can be used as preprocessing. In this scenario, we use templates in order to discard some power traces. This scenario is an example of an attack that biases the masks by discarding traces. Third, we discuss how templates can be used after preprocessing. In this scenario, templates are used on the preprocessed traces.

For all scenarios, we make the same assumptions as in the previous section. This means, we assume software implementations on 8-bit microcontrollers that leak the Hamming weight.

10.4.1   Templates Before Preprocessing the Traces

In this scenario we use templates in order to improve the preprocessing function. Recall that the second-order DPA attacks that we have shown in the previous examples have worked because the preprocessing function $|HW(u_m) - HW(v_m)|$ correlates reasonably well to the hypotheses $HW(u \oplus v)$. As indicated before, we can improve second-order DPA attacks by using other preprocessing functions.

Therefore, we would like to find preprocessing functions that maximize $\rho(HW(u \oplus v), pre(HW(u_m), HW(v_m))$. The function $HW(u \oplus v)$ has a complicated structure. When approximating functions with a complicated structure, theory suggests using trigonometric functions. Our experiments have shown that using higher degree polynomials based on the sine function improves the correlation coefficient dramatically. In addition, instead of using $HW(u \oplus v)$ we can use more complex combination functions. For example, it turns out that the correlation between the two following sine-based functions is about 0.83:

$$comb(u, v) = - 89.95 \cdot \sin \left(HW(u \oplus v)^3 \right) -$$
$$- 7.82 \cdot \sin \left(HW(u \oplus v)^2 \right) + 67.66$$
$$pre(HW(u_m), HW(v_m)) = \sin \left(HW(u_m) - HW(v_m) \right)^2$$

$$\rho(comb(u, v), pre(HW(u_m), HW(v_m))) = 0.83$$

There is one significant problem when using these improved preprocessing and combination functions in practice. We have derived them based on noise-free data (*i.e.* the Hamming weights). In practice however, we do not have noise free measurements. Hence, the improved preprocessing does not work immediately.

In the light of this, templates can be useful. The idea is to build templates that allow identifying the Hamming weight of the processed data. If such templates are available, a second-order DPA attack works as follows. The templates are used to deduce $HW(u_m)$ and $HW(v_m)$. Instead of doing the preprocessing on the traces, we do the improved preprocessing on the extracted Hamming weights. Because we now work directly with the Hamming weights, we can use an improved preprocessing function and thereby increase the correlation coefficient.

Example for Masked AES

In this example we sketch how to apply this idea to attack the masked AES software implementation. In the attack, we use templates to extract the Hamming weights of the masked intermediate values before and after SubBytes. Hence, we obtain $HW((d_i \oplus k_j) \oplus m)$ and $HW(S(d_i \oplus k_j) \oplus m)$ with the templates and apply the improved preprocessing function to these values. The hypotheses are $u = HW((d_i \oplus k_j)$ and $v = S(d_i \oplus k_j)$. We calculate the improved combination function using them. Figure 10.5 shows the result of a second-order DPA attack using templates before preprocessing. Since we extract the Hamming weights of the two targeted intermediate values $(d_i \oplus k_j) \oplus m$ and $S(d_i \oplus k_j) \oplus m$ with templates, we get the correlation coefficients for t_{ct} only. This is why Figure 10.5 shows only 256 values (one for each key hypothesis). It is clearly visible that the correct key hypothesis has a correlation coefficient of about 0.83. Figure 10.6 shows that with about 30 traces, the correct key hypothesis can be distinguished from the incorrect key hypotheses.

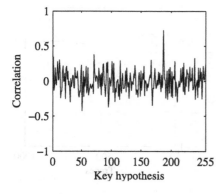

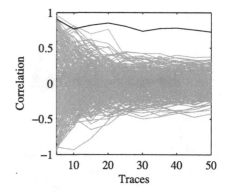

Figure 10.5. Result of a second-order DPA attack that uses templates before preprocessing.

Figure 10.6. Evolution of the correlation coefficient over an increasing number of traces.

10.4.2   Templates for Preprocessing the Traces

In this scenario we use templates for preprocessing. This means, we use templates in order to bias the masks by discarding a subset of the power traces. Recall that this corresponds to using the last preprocessing function that we have introduced in Section 10.3.1.

Templates can be naturally applied to such a second-order DPA attack. For instance, we simply build templates that can identify the Hamming weight of intermediate values. During the attack, the attacker uses these templates to identify the Hamming weight of the processed masks. All traces that do not belong to a chosen subset, e.g. which have Hamming weight smaller than some constant c, are discarded.

Example for Masked AES

In this example, we have used templates to identify the traces in which the mask has a Hamming weight that is smaller than six. In the preprocessing step we have discarded all the traces with this property. Then, we have performed a DPA attack on the remaining traces. Figure 10.7 shows the result of such an attack. It shows the correlation for all 256 key hypotheses. The line that is plotted in black indicates the correct key hypothesis. Figure 10.8 shows that with about 450 traces the correct key hypothesis can be identified.

10.4.3   Templates After Preprocessing the Traces

In the last scenario, we use templates after preprocessing. This means, the attacker builds templates for $HW(u \oplus v)$ (the hypothetical values that are used during the DPA attack) with the preprocessed traces. During the DPA attack,

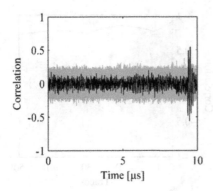

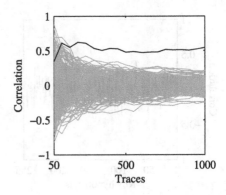

Figure 10.7. Result of a second-order DPA attack that uses templates for preprocessing.

Figure 10.8. Evolution of the correlation coefficient over an increasing number of traces.

the attacker uses these templates to extract the Hamming weights $HW(u \oplus v)$ and performs a DPA attack on these values.

The preprocessing removes quite an amount of information from the power traces. Recall that the correlation coefficient goes down from 1 to 0.24. Hence, it cannot be expected that this strategy leads to better results than the attacks described before.

Example for Masked AES

We have built templates for $HW((d_i \oplus k_j) \oplus S(d_i \oplus k_j))$ based on the pre-processed traces. In the second-order DPA attack we have used these templates to extract $HW((d_i \oplus k_j) \oplus S(d_i \oplus k_j))$. Then, we have mounted a DPA attack on these values. Figure 10.9 shows the result of such an attack. The figure shows the correlations for all 256 key hypotheses. The correct key hypothesis is indicated by the highest correlation coefficient. Figure 10.10 shows that about 2 000 traces are necessary to identify the correct key hypothesis. Apparently, this attack requires a large number of power traces.

10.5   Template-Based DPA Attacks

We have already introduced the concept of template-based DPA attacks in Section 6.6. Recall also from Section 5.3.2 that there are several ways to build templates. For example, the attacker can build templates for pairs of data and key or the attacker can build templates for intermediate values. In addition, the attacker can take the power model of the device into account when building templates. Most importantly, the attacker can decide which points of interest to include in the templates. For instance, if templates are built for pairs of data

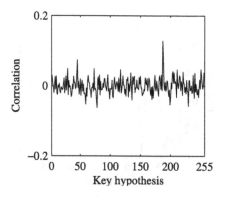
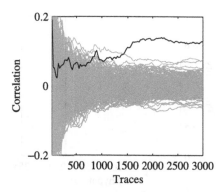

Figure 10.9. Result of a second-order DPA attack that uses templates after preprocessing.

Figure 10.10. Evolution of the correlation coefficient over an increasing number of traces.

and key, the attacker can actually include all the points in the trace that depend on this pair.

The ability to include several points of interest, which correspond to several intermediate values, makes template-based DPA attacks directly applicable to attack masked implementations. Remember that a second-order DPA attack works because it exploits the joint leakage of two intermediate values. By building templates with interesting points that correspond to two intermediate values concealed by the same mask, we exploit the joint leakage of these two intermediate values. Consequently, template-based DPA attacks can be directly applied to break masked implementations. This is the strongest attack that can be mounted on a masked implementation in the sense that it requires the smallest number of traces.

10.5.1 General Description

We assume that the attacker has the ability to build templates for the targeted cryptographic device. These templates are built in such a way that the interesting points correspond to at least two intermediate values that are concealed by the same mask.

During the attack, the templates and traces are matched. Recall that we attack a masked intermediate value and we do not know the value of the mask m in a certain encryption run. This implies that we have to perform the template matching for all values that the mask m can take. Consequently, the template matching gives the probabilities $p(\mathbf{t}'_i | k_j \wedge m)$ and we have to derive $p(\mathbf{t}'_i | k_j)$

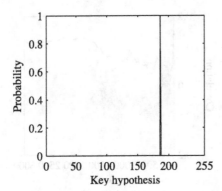
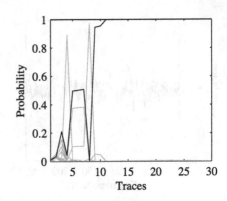

Figure 10.11. Result of a template-based DPA attack. The correct key hypothesis has probability one. The incorrect key hypotheses all have probability zero.

Figure 10.12. Evolution of the probability over an increasing number of traces. The correct key hypothesis is plotted in black. The incorrect key hypotheses are plotted in gray.

by calculating (10.4).

$$p(\mathbf{t}_i'|k_j) = \sum_{m=0}^{M-1} p(\mathbf{t}_i'|k_j \wedge m) \cdot p(m) \tag{10.4}$$

With $p(\mathbf{t}_i'|k_j)$ we can calculate (6.18). Hence, except for the extra calculation of (10.4), a template-based DPA attack on a masked implementation works in exactly the same manner as a template-based DPA attack on an unmasked implementation.

10.5.2 Example for Masked AES

In this example, we have built templates that contain the joint leakage of two instructions. The first instruction is related to the mask m that is used to conceal the S-box output $S(d_i \oplus k_j)$ and the second instruction is related to the masked S-box output $S(d_i \oplus k_j) \oplus m$. Our templates take the power model of the microcontroller into account. Hence, we have built 81 templates, one for each pair of $HW(m)$ and $HW(S(d_i \oplus k_j) \oplus m)$:

$$h_{HW(m),HW(S(d_i \oplus k_j) \oplus m)}$$

The template matching then gives the probabilities for $p(\mathbf{t}_i'|k_j \wedge m)$:

$$p(\mathbf{t}_i'|k_j \wedge m) = p(\mathbf{t}_i'; h_{HW(m),HW(S(d_i \oplus k_j) \oplus m)})$$

With these probabilities, and by assuming that $p(m) = 1/M$, we have calculated (10.4) and subsequently we have derived $p(k_j|\mathbf{T})$ with (6.18).

The result of this attack is depicted in Figure 10.11. It shows that one key has probability one whereas all other key hypotheses have probability zero. Figure 10.12 shows that with about 15 traces the correct key can be identified. This shows that this template-based DPA attack on a masked AES implementation in software works in the same way, and leads to similar results, as the template-based DPA attack on the unmasked AES implementation in software that we have discussed in Section 6.6.2.

> Template-based DPA attacks can be applied to unmasked and masked implementations in the same way.

This example demonstrates that template-based DPA attacks require about the same number of traces in order to break an unmasked and a masked implementation on the same device.

10.6 Second-Order DPA Attacks on Hardware Implementations

In software implementations of masking schemes, the intermediate values are computed sequentially. Therefore, it is necessary to combine points of power traces in order to perform second-order DPA attacks. In hardware implementations, usually several intermediate values are computed in parallel. It often happens that masks and correspondingly masked values are processed in the same clock cycle. Hence, second-order DPA attacks on hardware implementations can typically be performed based on the power consumption of a single clock cycle. This clock cycle contains information about the joint distribution of masked values and masks.

The degree of information that is available in a single clock cycle strongly depends on the power consumption characteristics of the cryptographic device and on the way the masked values and the masks are processed in this clock cycle. Essentially, there are two scenarios that need to be distinguished. The masked values and the masks can either be processed in parallel in the clock cycle, or they can be processed jointly.

- **Parallel Processing:** Parallel processing means that masked values and masks are processed separately. There is no computation that takes masked values and the corresponding masks as input, *i.e.* there is no connection between the modules working on the masks and the modules working on the masked values.

- **Joint Processing:** Joint processing means that masked values and masks are used as input to the same function, *i.e.* a part of the circuit performs computations that depend on masked values and masks.

We now discuss attacks on hardware implementations that perform parallel or joint processing. In particular, we describe suitable preprocessing techniques, and we provide an example of an attack on a masked hardware implementation of an AES S-box.

10.6.1   Preprocessing

The preprocessing step of a second-order DPA attack depends on whether the cryptographic device performs parallel or joint processing. In case of parallel processing, the power consumption of the device is usually the sum of the power that is needed to process the masks and the power that is needed to process the masked values. For example, if the device leaks the Hamming weight of the intermediate values, the power consumption for the processing of a masked value v_m and the corresponding mask m is proportional to $HW(v_m) + HW(m)$. According to Section 10.3.2, squaring the power traces is a suitable preprocessing technique in this case.

In general, also other non-linear preprocessing functions can be used. For the selection of the preprocessing function, the same considerations have to be made as in Sections 10.3.1 and 10.3.2. In some special cases, even no preprocessing is necessary when attacking hardware implementations. It can happen that a kind of preprocessing is already performed by non-linear effects in the cryptographic device. The power consumption of a device is not always the sum of the power consumption of the masked values and the masks. For example, cross coupling between the wires that process the masked values and the masks can lead to non-linear effects in the power consumption. In this case, second-order DPA attacks can be performed without the need to preprocess the power traces.

Furthermore, there is usually no need to preprocess power traces, if the cryptographic device performs a joint processing of masked values and masks. In the case of joint processing, not only the cross coupling of wires, but mainly glitches lead to non-linear effects in the power consumption. As explained in Section 3.1.3, the power consumption of CMOS circuits strongly depends on the number of glitches that occur in the circuit. If there is a part in the circuit that calculates intermediate results based on masks and masked values, the glitches that are caused by this calculation also depend on the masks and the masked values. In fact, the number of glitches depends on the joint distribution of the masks and the masked values. In this case, the power consumption is not proportional to the sum of the power consumption of the masks and the masked values. It is rather a non-linear function of both. Hence, the power consumption can be exploited in a second-order DPA attack without the need to preprocess the power traces.

10.6.2 DPA Attacks on Preprocessed Traces

In case of glitches and other non-linear effects in the cryptographic device, the power traces do not need to be preprocessed. In all other cases, a non-linear function (e.g. squaring) needs to be applied to the traces during the preprocessing. Hence, the (preprocessed) power traces that are used for the actual DPA attack have the following property: There is a point in the traces that is a non-linear function of the power consumption of a masked value v_m and the corresponding mask m. Therefore, hypotheses about the unmasked intermediate value v need to be formulated. This task is straightforward. The hypotheses can be formulated just like in all other DPA attacks.

The most difficult part of a DPA attack on a masked hardware implementation is to find a suitable power model. Recall that the power consumption depends in a non-linear way on v_m and m, while the hypotheses are formulated for $v = v_m \oplus m$. The goal of the attacker is to find a power model for v that maximizes the correlation to the power consumption that depends on v_m and m. We provide an example of such a power model in the following section.

The fact that hypotheses about v are formulated, makes second-order DPA attacks often appear like a first-order DPA attack. However, it is important to realize that these attacks exploit the joint leakage of two intermediate results. The attacker uses a power model for an unmasked value in order to describe the power consumption that is a function of a masked value and a mask.

10.6.3 Example for a Masked S-Box

In this example, we attack a hardware implementation of the masking scheme for the AES S-box that we have described in the example in Section 9.2.2. The masking scheme is based on composite field arithmetic, and it uses masked multipliers similar to the ones described in Section 9.2.2. The computation of the S-box output is performed within one clock cycle. Furthermore, a joint processing of masked values and masks is performed, *i.e.* the implementation of the S-box calculates intermediate results that are functions of masks and masked values. The S-box is a non-linear function. Therefore, it is not possible to calculate the masked output and the corresponding mask separately and in parallel. Due to the joint processing, the number of glitches that occur in the S-box is a non-linear function of the masked input value of the S-box and the corresponding mask. Therefore, no preprocessing of the power traces is necessary for a second-order DPA attack.

The challenging task for the attacker is to find out how the number of glitches is related to the unmasked input value of the S-box. In case of the concrete implementation, we have had access to the netlist of the S-box. Therefore, we have been able to simulate the number of transitions that occur in the S-box. Figure 10.13 shows the average toggle count of the masked S-box as a function

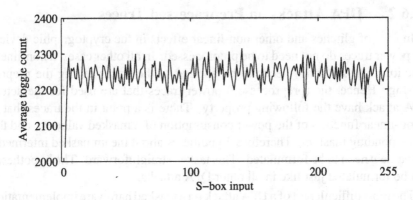

Figure 10.13. Average number of transitions occurring in the masked S-box for the 256 input values.

of the unmasked S-box input. It is clearly visible that the input value zero leads to significantly less activity than all other input values. The masked implementation of the AES S-box can therefore be attacked using the zero-value model that we have already used to attack the unmasked implementation of the S-box in Section 6.2.2, *i.e.* $h_{i,j} = ZV(S(d_i \oplus k_j))$.

10.6.4   Example for MDPL

We now discuss the effectiveness of the masked logic style MDPL. We do this by analyzing the DPA resistance of an MDPL NAND cell, see Figure 9.5. As for the DRP NAND cells in Section 8.3.1, the DPA resistance of the MDPL NAND cell has been determined by simulating the variance of its energy consumption for different input transitions. The simulation environment has been the same as for the DRP NAND cells. Figure 10.14 shows the 16 different power traces of an MDPL NAND cell when the 16 combinations of signal transitions given in Table 8.1 have been applied to the inputs of the cell. Every combination of input signal transitions has been simulated for the four possible mask signal transitions. Afterwards, the mean of these four traces has been calculated. Since an attacker does not know the values of the mask, he implicitly does the same during a DPA attack. The power traces show the power consumption of the evaluation phase and the subsequent precharge phase.

Variance and standard deviation of the energy consumption of the MDPL NAND cell for balanced complementary wires are shown in Table 10.5. The variance of the energy consumption of the MDPL NAND cell decreases approximately by two orders of magnitude compared to the CMOS NAND cell. The values for the CMOS NAND cell have been taken from Table 8.2. The values in Table 10.5 indicate that the DPA resistance of an MDPL NAND cell is lower than the one of the DRP NAND cells for balanced complementary wires.

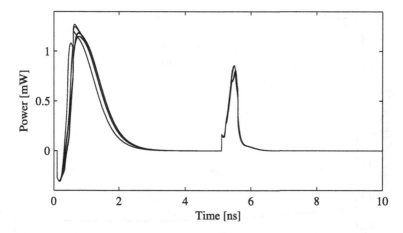

Figure 10.14. Simulated power traces of an MDPL NAND cell for different input transitions.

Table 10.5. Variance and standard deviation of the energy consumption of the CMOS NAND cell and the MDPL NAND cell for balanced complementary wires.

Logic style	CMOS	MDPL
$Var(E_{NAND})$	$224.69 \cdot 10^{-27}$ J^2	$1.7048 \cdot 10^{-27}$ J^2
$Std(E_{NAND})$	474 fJ	41.3 fJ

However, in contrast to the DRP NAND cells, the DPA resistance of the MDPL NAND cell does not depend on the balancing of the complementary wires at the output of the cell. This is a major advantage of MDPL.

Figure 10.15 shows how the variance of the energy consumption of an MDPL NAND cell and an SABL NAND cell depends on the balancing of the complementary output wires of the cells. The result for the SABL NAND cell has been taken from Figure 8.4. While the DPA resistance of the SABL NAND cell decreases quadratically with the difference of the capacitances at the complementary outputs, the DPA resistance of the MDPL NAND cell is more or less independent of the size of the difference. In the particular case, if the difference of the capacitances is approximately 10 fF or more, the DPA resistance of the MDPL NAND cell is higher than the one of the SABL NAND cell. More simulation results indicating the DPA resistance of MDPL circuits are presented in [PM05] and in [PM06].

10.7 Notes and Further Reading

SPA Attacks. Blinding techniques do not necessarily protect against SPA attacks. For instance, even if we use exponent blinding for RSA decryption,

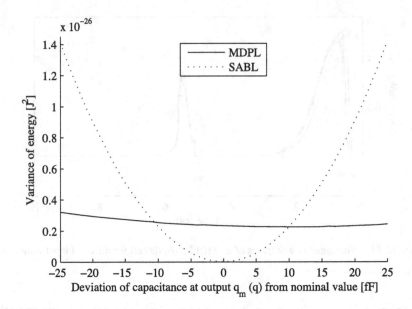

Figure 10.15. Variance of the energy consumption of an MDPL NAND cell and an SABL NAND cell as a function of the difference of the capacitances at the complementary cell outputs q_m (q) and $\overline{q_m}$ $(\overline{q})$. The nominal capacitance at the outputs is 100 fF.

i.e. one uses d_m instead of d, then an SPA attack might allow recovering d_m. Since $d_m \equiv d \pmod{\phi(n)}$, the attacker can use d_m to decipher messages. If we use message blinding, *i.e.* we use v_m instead of v, then an SPA attack might allow recovering d.

An issue that arises in the context of masked logic styles is the secure distribution of the masks to all cells in the circuit. This distribution typically requires large nets of wires that transport the masks to all cells. For example, in MDPL only one mask is used for all cells in the circuit. Since the mask network is typically large, there is a threat that an attacker can determine changes of the mask with an SPA attack. Hence, it is vital that the mask net is protected. In MDPL, SPA resistance is ensured by implementing the mask net using complementary wires that are precharged.

DPA Attacks. Several researchers observed that multiplicative masking can be attacked using zero-value DPA attacks, see Akkar and Goubin [AG03], and Golić and Tymen [GT03]. Akkar *et al.* [ABG04] made the observation that the simplified multiplicative masking scheme of Trichina *et al.* [TSG03] leads to even more insecure intermediate values than their original multiplicative masking scheme.

As discussed in Chapter 9, there exists the opinion that removing the masking of the inner rounds of an algorithm leads to a significant improvement in performance. Although our example in Section 9.2.1 provides good arguments why this is wrong for a typical software implementation of a masking scheme on an 8-bit microcontroller, the opinion still exists. There are two more reasons why the masking should not be removed from the inner encryption rounds. Handschuh and Preneel [HP06] discussed attacks that make use of the Hamming-weight information of some intermediate values that occur in the inner encryption rounds. Another attack that targets intermediate values in inner encryption rounds was presented by Kunz-Jacques *et al.* [KJMV04].

Goubin [Gou03] further reported on a zero-value attack on implementations of ECC that apply multiplicative masking (*i.e.* blinding) to projective coordinates. He pointed out that blinding leaks information if one of the coordinates is zero. Akishita and Takagi [AT03] extended this idea to cases where some intermediate value becomes zero during the calculation that involves a blinded point on an elliptic curve.

Second-Order DPA Attacks on Software Implementations. Although the idea of "high-order DPA functions" was mentioned already in the article by Kocher *et al.* [KJJ99], the article by Messerges [Mes00b] was the first to discuss an actual implementation of this idea. He made the observation that using the absolute-difference function for preprocessing leads to good results if the cryptographic device leaks the Hamming weight.

However, the article by Messerges [Mes00b] brought up several issues. Besides that it was not clear whether his suggested preprocessing function leads to the best attack, it was also unclear how to find the two points in the trace that need to be subtracted. Akkar and Goubin [AG03] made the observation that if one attacks the first and the last round of a cipher and assumes that the same masks are used in these rounds, then the points to be subtracted will be at the same position within the respective rounds. This can simplify second-order DPA attacks in practice.

Waddle and Wagner [WW04] were the next to work on this topic. In their article, they assumed some power model and showed that in their model, multiplication works as preprocessing. In addition they made two observations. First, they observed that whenever two masked intermediate values are computed in parallel, the preprocessing simply consists of squaring the points in the traces. Second, they observed that computing an FFT over (a part of) a trace can also be seen as a preprocessing function.

The important question of how the preprocessing influences the effectiveness of a second-order DPA attack was first investigated by Joye *et al.* [JPS05]. They studied the absolute-difference function in the Hamming-weight and in the Hamming-distance model. It turned out that this preprocessing function can be used in both models. This observation was also made in a practical attack by Herbst *et al.* in [HOM06]. Furthermore, Joye *et al.* studied how the absolute-difference function behaves when it is combined with the power function. It turns out that the correlation increases only slightly. Oswald *et al.* [OMHT06] described a strategy for conducting second-order DPA attacks on software implementations of block ciphers. Yoo *et al.* [YHM$^+$06] used the same strategy to analyze a masked implementation of the block cipher ARIA. The second-order DPA attacks that we have described in this chapter follow the strategy described by Oswald *et al.* They demonstrated (in theory as well by providing practical results) that different combinations of intermediate results in the first and last encryption round are typically good targets. They also calculated the correlation coefficients for different attack scenarios using the absolute-difference function and assuming the Hamming-weight model. Their results coincide with those of Joye *et al.* [JPS05]. Furthermore, Oswald *et al.* showed that using multiplication as preprocessing leads to lower correlations in the Hamming-weight model. Schramm and Paar [SP06] provided another study for multiplication as preprocessing in the Hamming-weight model. Their results confirm the results of Oswald *et al.* Hence, it is clear from all these articles that using the absolute-difference function is the best known preprocessing method, if the cryptographic device leaks either the Hamming weight or the Hamming distance.

We want to point out that the attack described in Section 10.4.2 can be modified in such a way that it also works without using templates. The idea behind this attack was presented by Jaffe [Jaf06b].

Second-Order DPA Attacks on Hardware Implementations. There are significantly less articles that discuss second-order DPA attacks on hardware implementations. The first articles in this direction were Mangard *et al.* [MPG05] and Suzuki *et al.* [SSI05]. They showed that masked CMOS cells leak in practice. This observation was refined and applied to a masked AES implementation by Mangard *et al.* [MPO05]. It turned out that the leakage of a masked AES S-box can be efficiently exploited with the zero-value power model. Mangard and Schramm [MS06] provided an explanation for this leakage.

Higher-Order DPA Attacks. Chari *et al.* [CJRR99b] gave a proof that a masking scheme using n different masks can protect at most against an n-th order DPA attack. Akkar and Goubin [AG03] discussed the application of higher-order DPA attacks to the masking scheme that they presented in the same article. They concluded that a second-order DPA attack (and therefore any higher-order DPA attack) is not possible, because they use different masks in the first and the last DES round. However, they assumed that the attacker can only mount a second-order DPA attack by using the first and the last round.

Hence, under a more general assumption, second-order DPA attacks are possible on their scheme.

Schramm and Paar [SP06] also discussed higher-order attacks using multiplication as preprocessing.

For asymmetric cryptography, there is almost no literature available. Muller and Valette [MV06] were the only ones who investigated how to attack RSA implementations that apply a secret-sharing scheme to split the exponent.

Compression vs. Preprocessing. Recall that we have introduced different techniques for compressing power traces in Section 4.5. Of course, these techniques can be seen as preprocessing techniques, such as introduced in this chapter or in Section 8.2.3. However, compression techniques such as described in Section 4.5 have quite a different goal than the preprocessing techniques described in this chapter or in Chapter 8. The goal of compression is to reduce the length of the recorded power traces in order to make DPA attacks more efficient. The goal of the preprocessing techniques presented in this chapter is to produce a joint leakage that can be exploited in second-order DPA attacks. In case of attacks on misaligned traces, the goal is to increase the correlation coefficient by removing the misalignment at the cost of a lowered SNR.

Template Attacks. Agrawal *et al.* [ARRS05] discussed another way to use templates within second-order DPA attacks. They made the assumption that they have a test device with an imperfect random number generator. Since in this case their test device had biased masks, they could mount a DPA attack to identify the masked output bits of the S-box operation. For each output bit, they built a template. During the actual attack, they made hypotheses for the unmasked S-box output bits and derived the actual masked S-box output bits by using their templates. Then, they exclusive-ored their hypothesis with the masked bit (which they derived from the trace). This gave them the hypotheses on m. In the last step, they made a DPA attack with the hypotheses on m. Only for the correct key the hypotheses on the unmasked S-box output bits were true. Hence, only in this case, the correct mask bit m was derived and peaks occurred in the final DPA attack.

Peeters *et al.* [PSDQ05] discussed template-based DPA attacks on hardware implementations of block cipher primitives. Oswald and Mangard [OM07] discussed different strategies to apply template attacks to masked implementations, see also Sections 10.4 and 10.5.

DPA vs. Second-Order DPA vs. Template Attacks. DPA attacks exploit the leakage that is caused by one intermediate value. Second-order DPA attacks exploit the joint leakage that is caused by two intermediate values. In order to calculate a point that contains this joint leakage, second-order DPA attacks typically require preprocessing the traces. However, in case of hardware implementations we have seen that, depending on the leakage of the device, sometimes no preprocessing is required. Furthermore, if mask and masked value are processed in parallel, the hypothesis that the attacker uses in the second-order DPA attack is actually the unmasked value. Hence, in these cases the attack appears to be a DPA attack. Nevertheless, since the leakage is a joint leakage that is caused by two intermediate values, the attack is actually a second-order DPA attack.

Templates can be used in two ways in second-order DPA attacks. First, templates can be used to enhance a conventional second-order DPA attack. For example, the preprocessing can be improved by templates. Second, one can make a template-based DPA attack. The latter case is particularly interesting. Recall that in a template-based DPA attack on a masked implementation, the attacker builds templates in such a way that several intermediate results (that are concealed by the same mask) contribute to the templates. Note that this is analogous to a template-based DPA attack on an unmasked implementation, where templates can also be built with several intermediate values contributing to the templates. It follows that template-based DPA attacks can be applied in the same manner to both masked and unmasked implementations. This means, template-based DPA attacks should neither be categorized as first-order nor second-order DPA attacks.

Chapter 11

CONCLUSIONS

When we started writing this book, we thought that the final manuscript would have about 200 pages. However, after having finished the first couple of chapters, we realized that the final manuscript would get significantly longer. Power analysis attacks are a very interdisciplinary topic. Hence, these attacks have attracted the attention of people with very different backgrounds. This has lead to a large number and a great variety of publications that discuss power analysis attacks from many different points of view.

The different views on power analysis attacks can essentially be categorized into two groups. On the one hand, power analysis attacks can be viewed as a mathematical problem. The goal is to find mathematical models that describe the leakage of cryptographic devices in order to build secure systems based on these models. This line of thinking has lead to countermeasures like masking. On the other hand, power analysis attacks can be viewed as an engineering problem that can be solved by decreasing the leaking signal or by increasing the noise. This line of thinking has lead to the different hardware counter-measures including DPA-resistant logic styles. It has been very fruitful during the last years that there is more than one approach to model and to counteract power analysis attacks. The two approaches have continuously stimulated new research on attacks and countermeasures.

The motivation for writing this book was to provide a comprehensive over-view of the research on this topic. Furthermore, our goal was also to provide an introduction to power analysis attacks for people who start working in this field. Obviously, it is not possible to do an in-depth discussion of every aspect of power analysis attacks in a single book. However, we have tried to put together the aspects that we consider to be the most important ones. We now provide specific conclusions for the attacks and countermeasures that have been explained in this book. Subsequently, we provide some general conclusions.

11.1   Specific Conclusions

In the introduction of this book we have pointed out that cryptographic devices have become essential building blocks of many security-sensitive systems. Consequently, it is important to study their resistance against power analysis attacks. The example provided in Chapter 1 has shown that conducting power analysis attacks is simple, although expertise from many different fields is required. Using an off-the-shelf oscilloscope and following the original paper of Kocher *et al.* allows performing DPA attacks on unprotected implementations of AES without actually knowing the details of the implementation.

However, in order to improve power analysis attacks and in order to develop countermeasures it is necessary to understand how cryptographic devices work, how they are built, and how they consume power. This is why Chapters 2 and 3 have introduced these topics. Based on these chapters we have discussed many aspects of power analysis attacks and countermeasures in this book. The following paragraphs summarize the most important issues concerning measurement setups, characteristics of power traces, SPA attacks, DPA attacks, template attacks, software countermeasures, hardware countermeasures, and DPA-resistant logic styles.

Measurement Setups.   Building a measurement setup is the first task that is needed in order to perform power analysis attacks in practice. However, although this task is crucial, there exist almost no publications on this topic. In Chapter 3, we have put together the experiences we have gained about measurement setups during the last years. We hope that this contribution stimulates a broader discussion of this topic. Such a discussion would help to make fair comparisons of attack results that have been obtained using different measurement setups.

In Chapter 3, we have distinguished two kinds of noise that account for the quality of a measurement setup. There is electronic noise on the one hand and switching noise on the other hand. The main part of electronic noise is typically caused by conducted and radiated emissions of other devices than the attacked one, e.g. clock generators, PCs, LCDs, *etc.* Hence, this kind of noise can be reduced by shielding. Switching noise is caused by cells of the attacked circuit that are not relevant for the attack. It can be reduced by using low clock frequencies and by using small probes that only measure the power consumption of a part of the attacked device.

As a general strategy to perform power analysis attacks, we recommend to characterize the noise characteristics of the measurement setup before performing power analysis attacks. The electronic noise should be analyzed right after having built the setup. Knowing the electronic noise of the setup is necessary for decisions like the number of traces that are used to build templates and for the calculation of the correlation coefficient occurring in DPA attacks.

Characteristics of Power Traces. The analysis of power traces is the foundation of all power analysis attacks. Hence, it is important to have methods and concepts for understanding and analyzing power traces.

In Chapter 4 we discussed such methods and concepts. We have first identified components of the power consumption, and then we have studied the statistical behavior of these components. It has turned out that we can describe the components reasonably well using normal distributions. Notice though that it is always advisable to determine the distributions empirically. Thus, whenever possible, one should conduct experiments in order to determine the statistical characteristics of the power consumption of the attacked cryptographic device.

We have also shown how standard statistical methods can be used to describe and analyze properties of power traces. It has turned out that these standard methods directly lead to the methods that are typically used in DPA attacks. Therefore, using standard methods from statistics is the right way to conduct power analysis attacks.

SPA Attacks. SPA attacks exploit key-dependent differences within a power trace. Chapter 5 has shown that there are different types of SPA attacks such as visual inspection of power traces, template-based SPA attacks, and collision attacks. Although these variants are quite different from each other, they have two common properties. In order to reveal the key, they all require that the attacker has some knowledge about the cryptographic device and the implementation of the cryptographic algorithm. The attacker can obtain this knowledge from data sheets or other documents about the attacked device. Alternatively, this knowledge can be derived directly from power traces of the device. In particular the visual inspection of power traces is not only useful to determine the key of a cryptographic device, but also to reveal implementation details. The second property of SPA attacks is that they are only based on a small number of power traces. Consequently, SPA attacks are typically used in scenarios where the attacker has only a small number of power traces, but knowledge about the device and the implementation.

SPA attacks work especially well if the attacked algorithm repeatedly uses the same sequence of operations. This is the case in almost all cryptographic algorithms. Hence, SPA attacks allow detecting the rounds and inner-round features in case of symmetric cryptographic algorithms, and multiplications (square operations) or point additions (point double operations) in case of asymmetric cryptographic algorithms. Especially in naïve implementations of asymmetric cryptographic algorithms, the occurrence of these operations depends on the key. Hence, in order to prevent SPA attacks it is vital that the operations of the executed algorithm occur independently of the key.

DPA Attacks. DPA attacks are the most popular power analysis attacks. Hence, there exists a large number of publications on this topic. However, although the published DPA attacks look rather different at first sight, they all consist of the five steps that have been explained in Chapter 6. This means that the attacker measures power traces, chooses an intermediate result, formulates key hypotheses to calculate hypothetical intermediate values, maps these intermediate values to hypothetical power consumption values, and finally performs a statistical test to compare the hypothetical power consumption values with the measured traces. These five steps make up a DPA attack.

It is important to realize that DPA attacks often differ in more than one step. In particular, the power models and the statistical tests are often different. There are attacks that use simple power models like the Hamming-weight model, and there are attacks that use more sophisticated models. It is obvious that the better the power model describes the power consumption of the attacked device, the better the DPA attack works.

For the statistical test that is performed in DPA attacks, usually the correlation coefficient, the difference-of-means, or the distance-of-means method are used. As the correlation coefficient is the most flexible and powerful of these three methods, the description in Chapter 6 is done for this method. The most favorable properties of the correlation coefficient are that it can be used for all power models and that it uses a normalized scale. This makes it easy to compare power analysis attacks based on the correlation coefficient. Furthermore, the number of needed power traces can be calculated easily.

In Chapter 6, we have provided several examples of DPA attacks on a hardware and on a software implementation of AES. We have also explained methods to simulate and calculate the correlation coefficients that occur in these DPA attacks. The attacks as well as the simulations have shown that DPA attacks are very powerful against software implementations of AES. The effort to break such implementations is minimal. However, also hardware implementations can be broken with reasonable effort.

In general, DPA attacks can be applied whenever a cryptographic device encrypts or decrypts data with a fixed key. Therefore, they can be used for attacks on implementations of symmetric and asymmetric ciphers. However, as there are many more ways to implement asymmetric ciphers than symmetric ciphers, DPA attacks on asymmetric ciphers are less popular.

Template Attacks. Template attacks typically require the least amount of traces among all types of power analysis attacks. Consequently, template attacks are the strongest type of power analysis attacks known today. They exploit the characteristics of the power consumption of the cryptographic device by using advanced statistical methods. Chapters 5, 6, and 10 have shown that we can use template attacks in quite different scenarios. First, we can use them in SPA attacks. There, they allow the extraction of information from a trace that is not directly visible. Second, we can use templates in DPA attacks. There, they are the most efficient way to extract information and they lead to the best attacks. Third, we can use templates in second-order DPA attacks. There, they can be applied in just the same way as in DPA attacks.

So far, template attacks have mainly been applied to software implementations of symmetric algorithms. More research is needed for other types of implementations.

Although we use templates in different ways in SPA attacks, DPA attacks, and second-order DPA attacks, they have one thing in common. They require that the attacker characterizes the cryptographic device. Consequently, template attacks can only be used if such a characterization is possible.

Software Countermeasures. Implementing countermeasures to counteract power analysis attacks in software has one significant advantage: The underlying hardware does not need to be changed. Another perspective on this is the following. If a cryptographic device has already been designed and manufactured, but has not been secured against power analysis attacks, implementing countermeasures in software is the only option. Most of the countermeasures that we have discussed in Chapters 7 and 9 can be implemented in software and in hardware.

In software, the execution of a cryptographic algorithm can be randomized by shuffling or by randomly inserting dummy operations. These hiding techniques can easily be applied to implementations of symmetric and asymmetric cryptographic algorithms. Today, there seem to be more implementation options available for asymmetric cryptographic algorithms than for symmetric cryptographic algorithms. Note that these countermeasures are only effective if the attacker is not able to align the power traces. If the attacker is able to align the traces, the effect of such countermeasures can be completely removed by the attacker. Hence, it must be ensured that neither the shuffling, nor the random insertion of dummy operations can be detected by an attacker.

Hiding in the amplitude dimension is also easy to implement in software. However, it is typically not very effective because cryptographic devices have a fixed set of instructions. Hence, the choice of instructions is also limited. This implies that only a certain reduction of the signal can be achieved.

Masking is probably the most popular software countermeasure—at least when looking at the huge number of articles that have been published. Masking can be implemented efficiently in software. However, the number of masks that are used needs to be chosen carefully. Blinding is also frequently used to secure asymmetric cryptographic algorithms. It is even easier to implement than masking and typically comes at a reasonable cost. There is one thing that most countermeasures (hiding and masking) have in common. They require random numbers.

Chapters 8 and 10 have listed numerous attacks on hiding and masking schemes. Second-order DPA attacks and template-based DPA attacks break implementations of masking schemes in software and in hardware. It has also turned out that biasing random numbers renders countermeasures ineffective. This holds for random number that are used as masks and for random numbers that are used for hiding. Counteracting the attacks that we have described in Chapters 8 and 10 requires "help" from the cryptographic device. In other words, securing a device such as the microcontroller that we have used in many of the examples in this book is very difficult. In security-sensitive applications, cryptographic devices should be used that have already some hardware countermeasures implemented.

Hardware Countermeasures.   In hardware, there are more ways to counteract power analysis attacks than in software. We now provide conclusions for hardware countermeasures that can be implemented at the architecture level. Both, hiding as well as masking countermeasures can be implemented at this level. Hence, it is possible to randomize the execution, to change the SNR of the executed operations, and to mask the implemented algorithm. These countermeasures are not perfect though. Nevertheless, hardware countermeasures can push the security of cryptographic devices significantly beyond levels that can be reached by software countermeasures.

In hardware, the execution of cryptographic algorithms can be randomized by shuffling, by the random insertion of dummy operations, by the random insertion of dummy cycles, and by randomly varying the frequency of the clock signal. Just like in the case of software, these countermeasures are only effective if the attacker is not able to align the power traces. In hardware, it is possible to prevent attackers from aligning traces by taking care that the power consumption of all clock cycles is identically distributed. In case of randomly varying the frequency of the clock signal, it is crucial to have multiple clock domains on the chip. If there is only one clock signal, the attacker can easily remove the effect of the varying clock signal because the beginning of a clock cycle can easily be detected in a power trace.

The second way to protect cryptographic devices in hardware is to reduce the SNR of the performed operations. This can be done by filtering the power consumption or by increasing the noise. Filtering can be done by switched capacitors or by active filtering circuits. Noise can be generated by dedicated noise engines. Such noise engines essentially consist of random number generators that are connected to switching networks. When applying countermeasures like this, it is important to keep in mind that the SNR not only depends on the cryptographic device, but also on the measurement setup that is used by an attacker.

As a designer, it is therefore crucial to consider all possible measurement setups that can be used to attack the cryptographic device. Hardware countermeasures at the architecture level can sometimes be bypassed by sophisticated attack setups.

The third way to protect hardware implementations of cryptographic algorithms is masking. In this case, designers need to ensure that attackers cannot easily perform second-order DPA attacks. Glitches and coupling effects can lead to a power consumption that depends on the mask and the masked values in a non-linear way. This allows second-order DPA attacks that do not even require a preprocessing of the power traces. Designers also need to take care that attackers cannot detect the power consumption that is caused by the masks. Otherwise, the attacker can select power traces in such a way that the masks in these traces are biased.

DPA-Resistant Logic Styles. DPA-resistant logic styles are a class of hardware countermeasures that is implemented at the cell level. Counteracting DPA attacks at the cell level is a very intuitive approach. The basic idea is to build logic cells with a power consumption that is independent of the processed logic values. The two approaches to achieve this goal are to make the power consumption of the cells equal or random in each clock cycle. The first approach requires that every cell charges an equal capacitance over a path with an equal resistance in each clock cycle. The second approach can be implemented by masking all signals in the circuit.

Logic cells with an equal power consumption in each clock cycle are typically built using DRP logic styles. The most important thing when using such logic styles is to balance the capacitances and resistances of the complementary wires between the DRP cells. The more these wires are balanced, the higher is the DPA resistance of the circuit. However, building sufficiently balanced wires is a challenging task in practice.

In case of masking, there is no balancing problem. However, if not carefully implemented, masked circuits are vulnerable to second-order DPA attacks. During the design of such circuits, it must be ensured that neither glitches nor the parallel processing of the masks and the masked values causes a power consumption that is related to the unmasked values. Especially in the context of the latter issue, more research is necessary on how this problem can be efficiently avoided. Another issue for masked logic styles is the power consumption of the mask network(s). It must be ensured that an attacker cannot determine the mask values via an SPA attack.

DPA-resistant logic styles have the advantage that they can be applied automatically within a semi-custom design flow. The implementation of cryptographic devices is usually done using such a design flow. Effective DPA-resistant logic styles are typically very costly in terms of area requirements and power consumption. However, if they are implemented correctly, they provide a very good protection against power analysis attacks. Usually, a trade-off is possible between the level of protection, the size, and the power consumption of a circuit.

11.2 General Conclusions

Power analysis attacks always have to be considered when implementing cryptographic devices. They are a powerful cryptanalytic tool, and hence, they pose a serious threat to the security of cryptographic devices. SPA attacks are especially powerful when applied to implementations of asymmetric cryptographic algorithms. However, they also allow deriving useful information about implementations in general. DPA attacks can be applied to all kinds of cryptographic algorithms. They are the most popular power analysis attacks because they can be conducted without detailed knowledge about the attacked implementations. Template attacks are the most powerful attacks in terms of the number of needed power traces. However, they require a statistical characterization of the attacked device.

Preventing power analysis attacks is not trivial although there are several countermeasures. In this book we have distinguished hiding and masking countermeasures. These two types of countermeasures can be implemented at the architecture level (in software and in hardware) and at the cell level. In general, implementations in hardware increase the resistance of devices against power analysis attacks significantly more than implementations in software. However, although the resistance can be increased significantly, there is no practical way to achieve perfect security against power analysis attacks.

Each countermeasure has its weaknesses. For example, many countermeasures require random numbers. Obviously these countermeasures become ineffective if an attacker can bias the random numbers. Thereby it does not matter whether the random numbers are used for masking or for hiding countermeasures. Also countermeasures like dual-rail precharge logic styles have weaknesses. For perfect security, it would be necessary to perfectly balance all cells and wires of a cryptographic device. However, this is not possible in practice.

In practice, a reasonable compromise needs to be found between the resistance against power analysis attacks and the implementation costs of the countermeasures (design time, throughput, power consumption, area, *etc.*). Clearly, the more effort is spent on a countermeasure, the more resistance it provides. However, spending all effort on a single countermeasure is typically not the best strategy in practice. The costs for improving a countermeasure do not grow linearly. Therefore, it is the best strategy to counteract power analysis attacks by combining countermeasures. Implementing a combination of several cheap countermeasures typically leads to a much better protection than one expensive countermeasure.

Combining countermeasures is not only a good strategy in terms of the implementation costs. It is a also good strategy in terms of security. This is due to the fact that not all countermeasures provide the same level of protection against SPA and DPA attacks. For instance, blinding schemes are often used to protect implementations of asymmetric cryptographic algorithms against DPA attacks. However, these schemes do not necessarily also protect against SPA attacks. In contrast to this, there are hiding countermeasures for asymmetric algorithms that only protect against SPA attacks and not against DPA attacks. Hence, both types of countermeasures should be used. In addition to combining hiding and masking, we also recommend to limit the lifetime of cryptographic keys. This can be done using the protocol countermeasures sketched in Section 1.4.

When implementing a set of countermeasures for a cryptographic device, it is important to consider interactions that can occur between the countermeasures. Such interactions can lead to a lower degree of protection than expected. It is also important to understand that it is often sufficient for an attacker to find only a small part of the key. The remaining bits can be typically revealed by using other cryptanalytic methods. Hence, all bits of the key need to be protected adequately.

Summarizing, in practice resistance against power analysis attacks is best achieved by a combination of countermeasures. This means, the overall budget that is available to secure a device against power analysis attacks should not be spent on a single countermeasure. Instead, different types of countermeasures should be implemented in hardware and in software. Furthermore, the lifetime of keys that are used in cryptographic devices should be limited.

Appendix A

Differential Power Analysis

This paper appeared in *Advances in Cryptology – Proceedings of Crypto '99,* Lecture Notes in Computer Science, Vol. 1666, Springer-Verlag, 1999, pp. 388-397.

Paul Kocher, Joshua Jaffe, and Benjamin Jun

Cryptography Research, Inc.
575 Market Street, 21st Floor
San Francisco, CA 94105, USA.
http://www.cryptography.com
E-mail: {paul,josh,ben}@cryptography.com.

Abstract Cryptosystem designers frequently assume that secrets will be manipulated in closed, reliable computing environments. Unfortunately, actual computers and microchips leak information about the operations they process. This paper examines specific methods for analyzing power consumption measurements to find secret keys from tamper resistant devices. We also discuss approaches for building cryptosystems that can operate securely in existing hardware that leaks information.

Keywords: differential power analysis, DPA, SPA, cryptanalysis, DES

A.1   Background

Attacks that involve multiple parts of a security system are difficult to predict and model. If cipher designers, software developers, and hardware engineers do not understand or review each other's work, security assumptions made at each level of a system's design may be incomplete or unrealistic. As a result, security faults often involve unanticipated interactions between components designed by different people.

Many techniques have been designed for testing cryptographic algorithms in isolation. For example, differential cryptanalysis[3] and linear cryptanalysis [8] can exploit extremely small statistical characteristics in a cipher's inputs and outputs. These methods have been well studied because they can be applied by analyzing only one part of a system's architecture — an algorithm's mathematical structure.

A correct implementation of a strong protocol is not necessarily secure. For example, failures can be caused by defective computations[5, 4] and information leaked during secret key operations. Attacks using timing information[7, 11] as well as data collected using invasive measuring techniques[2, 1] have been demonstrated. The U.S. government has invested considerable resources in the classified TEMPEST program to prevent sensitive information from leaking through electromagnetic emanations.

A.2   Introduction to Power Analysis

Most modern cryptographic devices are implemented using semiconductor logic gates, which are constructed out of transistors. Electrons flow across the silicon substrate when charge is applied to (or removed from) a transistor's gate, consuming power and producing electromagnetic radiation.

To measure a circuit's power consumption, a small (e.g., 50 ohm) resistor is inserted in series with the power or ground input. The voltage difference across the resistor divided by the resistance yields the current. Well-equipped electronics labs have equipment that can digitally sample voltage differences at extraordinarily high rates (over 1GHz) with excellent accuracy (less than 1% error). Devices capable of sampling at 20MHz or faster and transferring the data to a PC can be bought for less than $400.[6]

Simple Power Analysis (SPA) is a technique that involves directly interpreting power consumption measurements collected during cryptographic operations. SPA can yield information about a device's operation as well as key material.

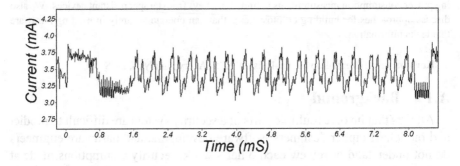

Figure A.1.   SPA trace showing an entire DES operation

A *trace* refers to a set of power consumption measurements taken across a cryptographic operation. For example, a 1 millisecond operation sampled at 5 MHz yields a trace containing 5000 points. Figure 1 shows an SPA trace from a typical smart card as it performs a DES operation. Note that the 16 DES rounds are clearly visible.

Figure 2 is a more detailed view of the same trace showing the second and third rounds of a DES encryption operation. Many details of the DES operation are now visible. For example, the 28-bit DES key registers C and D are rotated once in round 2 (left arrow) and twice in round 3 (right arrows). In Figure 2, small variations between the rounds just can be perceived. Many of these discernable features are SPA weaknesses caused by conditional jumps based on key bits and computational intermediates.

Figure 3 shows even higher resolution views of the trace showing power consumption through two regions, each of seven clock cycles at 3.5714 MHz.

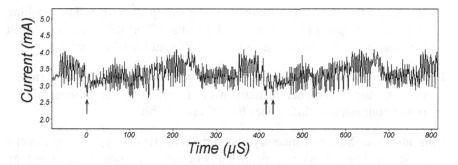

Figure A.2. SPA trace showing DES rounds 2 and 3.

The visible variations between clock cycles result primarily from differences in the power consumption of different microprocessor instructions. The upper trace in Figure 3 shows the execution path through an SPA feature where a jump instruction is performed, and the lower trace shows a case where the jump is not taken. The point of divergence is at clock cycle 6 and is clearly visible.

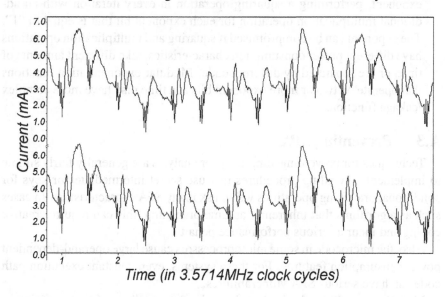

Figure A.3. SPA trace showing individual clock cycles.

Because SPA can reveal the sequence of instructions executed, it can be used to break cryptographic implementations in which the execution path depends on the data being processed. For example:

DES key schedule: The DES key schedule computation involves rotating 28-bit key registers. A conditional branch is commonly used to check the bit shifted off the end so that "1" bits can be wrapped around. The resulting power consumption traces for a "1" bit and a "0" bit will contain different SPA features if the execution paths take different branches for each.

DES permutations: DES implementations perform a variety of bit permutations. Conditional branching in software or microcode can cause significant power consumption differences for "0" and "1" bits.

Comparisons: String or memory comparison operations typically perform a conditional branch when a mismatch is found. This conditional branching causes large SPA (and sometimes timing) characteristics.

Multipliers: Modular multiplication circuits tend to leak a great deal of information about the data they process. The leakage functions depend on the multiplier design, but are often strongly correlated to operand values and Hamming weights.

Exponentiators: A simple modular exponentiation function scans across the exponent, performing a squaring operation in every iteration with an additional multiplication operation for each exponent bit that is equal to "1". The exponent can be compromised if squaring and multiplication operations have different power consumption characteristics, take different amounts of time, or are separated by different code. Modular exponentiation functions that operate on two or more exponent bits at a time may have more complex leakage functions.

A.3   Preventing SPA

Techniques for preventing simple power analysis are generally fairly simple to implement. Avoiding procedures that use secret intermediates or keys for conditional branching operations will mask many SPA characteristics. In cases such as algorithms that inherently assume branching, this can require creative coding and incur a serious performance penalty.

Also, the microcode in some microprocessors cause large operand-dependent power consumption features. For these systems, even constant execution path code can have serious SPA vulnerabilities.

Most (but not all) hard-wired hardware implementations of symmetric cryptographic algorithms have sufficiently small power consumption variations that SPA does not yield key material.

A.4   Differential Power Analysis of DES Implementations

In addition to large-scale power variations due to the instruction sequence, there are effects correlated to data values being manipulated. These variations tend to be smaller and are sometimes overshadowed by measurement errors and other noise. In such cases, it is still often possible to break the system using statistical functions tailored to the target algorithm.

Because of its widespread use, the Data Encryption Standard (DES) will be examined in detail. In each of the 16 rounds, the DES encryption algorithm performs eight S box lookup operations. The 8 S boxes each take as input six key bits exclusive-ORed with six bits of the R register and produce four output bits. The 32 S output bits are reordered and exclusive-ORed onto L. The halves L and R are then exchanged. (For a detailed description of the DES algorithm, see [9].)

The DPA selection function $D(C, b, K_s)$ is defined as computing the value of bit $0 \leq b < 32$ of the DES intermediate L at the beginning of the 16th round for ciphertext C, where the 6 key bits entering the S box corresponding to bit b are represented by $0 \leq K_s < 2^6$. Note that if K_s is incorrect, evaluating $D(C, b, K_s)$ will yield the correct value for bit b with probability $P \approx \frac{1}{2}$ for each ciphertext.

To implement the DPA attack, an attacker first observes m encryption operations and captures power traces $\mathbf{T}_{1..m}[1..k]$ containing k samples each. In addition, the attacker records the ciphertexts $C_{1..m}$. No knowledge of the plaintext is required.

DPA analysis uses power consumption measurements to determine whether a key block guess K_s is correct. The attacker computes a k-sample differential trace $\Delta_D[1..k]$ by finding the difference between the average of the traces for which $D(C, b, K_s)$ is one and the average of the traces for which $D(C, b, K_s)$ is zero. Thus $\Delta_D[j]$ is the average over $C_{1..m}$ of the effect due to the value represented by the selection function D on the power consumption measurements at point j. In particular,

$$
\begin{aligned}
\Delta_D[j] &= \frac{\sum_{i=1}^m D(C_i, b, K_s)\mathbf{T}_i[j]}{\sum_{i=1}^m D(C_i, b, K_s)} - \frac{\sum_{i=1}^m (1 - D(C_i, b, K_s))\,\mathbf{T}_i[j]}{\sum_{i=1}^m (1 - D(C_i, b, K_s))} \\
&\approx 2 \left(\frac{\sum_{i=1}^m D(C_i, b, K_s)\mathbf{T}_i[j]}{\sum_{i=1}^m D(C_i, b, K_s)} - \frac{\sum_{i=1}^m \mathbf{T}_i[j]}{m} \right).
\end{aligned}
$$

If K_s is incorrect, the bit computed using D will differ from the actual target bit for about half of the ciphertexts C_i. The selection function $D(C_i, b, K_s)$ is thus effectively *uncorrelated* to what was actually computed by the target device. If a random function is used to divide a set into two subsets, the difference in the averages of the subsets should approach zero as the subset sizes approach infinity. Thus, if K_s is incorrect,

$$
\lim_{m \to \infty} \Delta_D[j] \approx 0
$$

because trace components uncorrelated to D will diminish with $\frac{1}{\sqrt{m}}$, causing the differential trace to become flat. (The actual trace may not be completely flat, as D with K_s incorrect may have a weak correlation to D with the correct K_s.)

If K_s is correct, however, the computed value for $D(C_i, b, K_s)$ will equal the actual value of target bit b with probability 1. The selection function is thus *correlated* to the value of the bit manipulated in the 16th round. As a result, the $\Delta_D[j]$ approaches the effect of the target bit on the power consumption as $m \to \infty$. Other data values, measurement errors, etc. that are not correlated to D approach zero. Because power consumption is correlated to data bit values, the plot of Δ_D will be flat with spikes in regions where D is correlated to the values being processed.

The correct value of K_s can thus be identified from the spikes in its differential trace. Four values of b correspond to each S box, providing confirmation of key block guesses. Finding all eight K_s yields the entire 48-bit round subkey. The remaining 8 key bits can be found easily using exhaustive search or by analyzing one additional round. Triple DES keys can be found by analyzing an outer DES operation first, using the resulting key to decrypt the ciphertexts, and attacking the next DES key. DPA can use known plaintext or known ciphertext and can find encryption or decryption keys.

Figure 4 shows four traces prepared using known plaintexts entering a DES encryption function on another smart card. On top is the reference power trace showing the average power consumption during DES operations. Below are three differential traces, where the first was produced using a correct guess for K_s. The lower two traces were produced using incorrect values for K_s. These traces were prepared using 1000 samples ($m = 10^3$). Although the signal is clearly visible in the differential trace, there is a modest amount of noise.

Figure 5 shows the average effect of a single bit on detailed power consumption measurements. On top is a reference power consumption trace. The center trace shows the standard deviation in the power consumption measurements. Finally, the lower trace shows a differential trace prepared with $m = 10^4$. Note that regions that are not correlated to the bit are more than an order of magnitude closer to zero, indicating that little noise or error remains.

The size of the DPA characteristic is about $40\mu A$, which is several times less than the standard deviation observed at that point. The rise in the standard deviation at clock cycle 6 coinciding with a strong characteristic indicates that the operand value has a significant effect on the instruction power consumption and that there is considerable variation in the operand values being manipulated. Because low-level instructions often manipulate several bits, a selection function can simultaneously select for values of multiple bits. The resulting DPA characteristics tend to have larger peaks, but do not necessarily have better signal-to-noise ratios because fewer samples are included in the averaging.

Several sources introduce noise into DPA measurements, including electromagnetic radiation and thermal noise. Quantization errors due to mismatching

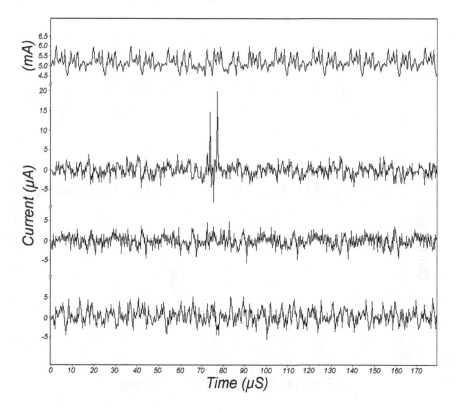

Figure A.4. DPA traces, one correct and two incorrect, with power reference.

of device clocks and sample clocks can cause additional errors. Finally, uncorrected temporal misalignment of traces can introduce a large amount of noise into measurements.

Several improvements can be applied to the data collection and DPA analysis processes to reduce the number of samples required or to circumvent countermeasures. For example, it is helpful to correct for the measurement variance, yielding the significance of the variations instead of their magnitude. One variant of this approach, automated template DPA, can find DES keys using fewer than 15 traces from most smart cards.

More sophisticated selection functions may also be used. Of particular importance are high-order DPA functions that combine multiple samples from within a trace. Selection functions can also assign different weights to different traces or divide traces into more than two categories. Such selection functions can defeat many countermeasures, or attack systems where partial or no information is available about plaintexts or ciphertexts. Data analysis using functions other than ordinary averaging are useful with data sets that have unusual statistical distributions.

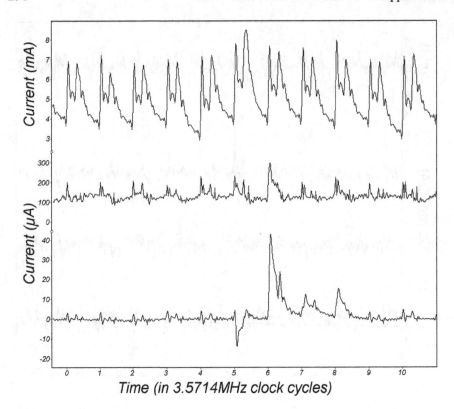

Figure A.5.   Quantitative DPA measurements

A.5   Differential Power Analysis of Other Algorithms

Public key algorithms can be analyzed using DPA by correlating candidate values for computation intermediates with power consumption measurements. For modular exponentiation operations, it is possible to test exponent bit guesses by testing whether predicted intermediate values are correlated to the actual computation. Chinese Remainder Theorem RSA implementations can also be analyzed, for example by defining selection functions over the CRT reduction or recombination processes.

In general, signals leaking during asymmetric operations tend to be much stronger than those from many symmetric algorithms, for example because of the relatively high computational complexity of multiplication operations. As a result, implementing effective SPA and DPA countermeasures can be challenging.

DPA can be used to break implementations of almost any symmetric or asymmetric algorithm. We have even used the technique to reverse-engineer unknown algorithms and protocols by using DPA data to test hypotheses about a device's computational processes. (It may even be possible to automate this reverse-engineering process.)

A.6   Preventing DPA

Techniques for preventing DPA and related attacks fall roughly into three categories.

A first approach is to reduce signal sizes, such as by using constant execution path code, choosing operations that leak less information in their power consumption, balancing Hamming Weights and state transitions, and by physically shielding the device. Unfortunately such signal size reduction generally cannot reduce the signal size to zero, as an attacker with an infinite number of samples will still be able to perform DPA on the (heavily-degraded) signal. In practice, aggressive shielding can make attacks infeasible but adds significantly to a device's cost and size.

A second approach involves introducing noise into power consumption measurements. Like signal size reductions, adding noise increases the number of samples required for an attack, possibly to an infeasibly-large number. In addition, execution timing and order can be randomized. Designers and reviewers must approach temporal obfuscation with great caution, however, as many techniques can be used to bypass or compensate for these effects. Several vulnerable products have passed reviews that used naïve data processing methods. For safety, it should be possible to disable temporal obfuscation methods during review and certification testing.

A final approach involves designing cryptosystems with realistic assumptions about the underlying hardware. Nonlinear key update procedures can be employed to ensure that power traces cannot be correlated between transactions. As a simple example, hashing a 160-bit key with SHA[10] should effectively destroy partial information an attacker might have gathered about the key. Similarly, aggressive use of exponent and modulus modification processes in public key schemes can be used to prevent attackers from accumulating data across large numbers of operations. Key use counters can prevent attackers from gathering large numbers of samples.

Using a leak-tolerant design methodology, a cryptosystem designer must define what leakage rates and functions that the cryptography can survive. Leakage functions can be analyzed as oracles providing information about computational processes and data, where the leakage rate is the upper bound on the amount of information provided by the leakage function. Implementers can then use leak reduction and leak masking techniques as needed to meet the specified parameters. Finally, reviewers must verify that the design assumptions are appropriate and correspond to the physical characteristics of the completed device.

A.7   Related Attacks

Electromagnetic radiation is a particularly serious issue for devices that pass keys or secret intermediates across a bus. Even a simple A.M. radio can detect strong signals from many cryptographic devices. A wide variety of other signal measurement techniques (such as superconducting quantum imaging devices) also show promise. Statistical methods related to SPA and DPA can be used to find signals in noisy data.

A.8   Conclusions

Power analysis techniques are of great concern because a very large number of vulnerable products are deployed. The attacks are easy to implement, have a very low cost per device, and are non-invasive, making them difficult to detect. Because DPA automatically locates correlated regions in a device's power consumption, the attack can be automated and little or no information about the target implementation is required. Finally, these attacks are not theoretical or limited to smart cards; in our lab, we have used power analysis techniques to extract keys from almost 50 different products in a variety of physical form factors.

The only reliable solution to DPA involves designing cryptosystems with realistic assumptions about the underlying hardware. DPA highlights the need for people who design algorithms, protocols, software, and hardware to work closely together when producing security products.

References for Appendix A

[1] R. Anderson, M. Kuhn, "Low Cost Attacks on Tamper Resistant Devices," *Security Protocol Workshop*, April 1997, `http://www.cl.cam.ac.uk/ftp/users/rja14/tamper2.ps.gz`.

[2] R. Anderson and M. Kuhn, "Tamper Resistance – a Cautionary Note", *The Second USENIX Workshop on Electronic Commerce Proceedings*, November 1996, pp. 1-11.

[3] E. Biham and A. Shamir, *Differential Cryptanalysis of the Data Encryption Standard*, Springer-Verlag, 1993.

[4] E. Biham and A. Shamir, "Differential Fault Analysis of Secret Key Cryptosystems," *Advances in Cryptology: Proceedings of CRYPTO '97*, Springer-Verlag, August 1997, pp. 513-525.

[5] D. Boneh, R. DeMillo, and R. Lipton, "On the Importance of Checking Cryptographic Protocols for Faults," *Advances in Cryptology: Proceedings of EUROCRYPT '97*, Springer-Verlag, May 1997, pp. 37-51.

[6] Jameco Electronics, "PC-MultiScope (part #142834)," February 1999 Catalog, p. 103.

[7] P. Kocher, "Timing Attacks on Implementations of Diffie-Hellman, RSA, DSS, and Other Systems," *Advances in Cryptology: Proceedings of CRYPTO '96*, Springer-Verlag, August 1996, pp. 104-113.

[8] M. Matsui, "The First Experimental Cryptanalysis of the Data Encryption Standard," *Advances in Cryptology: Proceedings of CRYPTO '94*, Springer-Verlag, August 1994, pp. 1-11.

[9] National Bureau of Standards, "Data Encryption Standard," Federal Information Processing Standards Publication 46, January 1977.

[10] National Institute of Standards and Technology, "Secure Hash Standard," Federal Information Processing Standards Publication 180-1, April 1995.

[11] J. Dhem, F. Koeune, P. Leroux, P. Mestré, J. Quisquater, and J. Willems, "A practical implementation of the timing attack," *UCL Crypto Group Technical Report Series: CG-1998/1*, 1998.

[12] R.L. Rivest, A. Shamir, and L.M. Adleman, "A method for obtaining digital signatures and public-key cryptosystems," *Communications of the ACM*, **21**, 1978, pp. 120-126.

Appendix B

The Advanced Encryption Standard

At the beginning of 1997, the US National Institute of Standards and Technology (NIST) started an initiative to develop a new encryption standard. The new standard, called *Advanced Encryption Standard*, was intended to succeed, and after some time, to completely replace the old Data Encryption Standard and also triple-DES.

In contrast to other selection processes, like the selection of the Secure Hash Algorithm (SHA), NIST decided to have an open process. This means that anyone could submit a candidate cipher. Each of the submissions that fulfilled the submission requirements was considered. However, NIST did not perform the evaluation itself. Instead it coordinated a community process. Hence, the cryptologic community was invited to cryptanalyze the candidates and to evaluate their implementation costs. All results were published on the NIST website. In addition, NIST organized conferences in which researchers presented their findings.

The selection process was divided into several rounds. At the end of the first round in 1998, 15 algorithms were accepted as candidates. All accepted candidates were presented during the first AES candidate conference. In the next round, these algorithms were evaluated with regard to their security, cost, and implementation characteristics. In March 1999, the second AES conference was held and it brought a lot of results with respect to the candidates. In August 1999, five finalist algorithms were selected from the 15 candidates. After the choice of the five finalist algorithms, the community was asked to focus all their efforts on the analysis of the finalists. The third AES conference was held in April 2000. No breakthrough cryptanalytic results were presented and the previous results regarding the implementation costs were confirmed. During this conference a questionnaire was handed out, asking about the preference of the attendees. Rijndael turned out to be the favorite algorithm.

On October 2nd, 2000, NIST officially announced that Rijndael had been chosen as Advanced Encryption Standard [Nat01].

The official scope of a standard such as AES seems to be quite limited at a first glance. AES is a *Federal Information Processing Standard* (FIPS) of the

d_0	d_4	d_8	d_{12}
d_1	d_5	d_9	d_{13}
d_2	d_6	d_{10}	d_{14}
d_3	d_7	d_{11}	d_{15}

k_0	k_4	k_8	k_{12}
k_1	k_5	k_9	k_{13}
k_2	k_6	k_{10}	k_{14}
k_3	k_7	k_{11}	k_{15}

Figure B.1.   State and key layout for AES

United States. Therefore, it applies to the US Federal Administration, where it is used to secure documents that contain *sensitive but not classified* information. However, in June 2003, the *National Security Agency* (NSA) published a policy which allows using AES with a key size of 128 bits to secure even *classified* data up to the SECRET level. For TOP SECRET data, AES with key size above 128 bits can be used. The approval of the NSA has given AES additional credibility.

However, not only the use in governmental organizations contributes to the importance of AES. Because AES is the replacement for DES and triple-DES, which have been the de-facto standards for encryption, AES has been adopted by banks, industry and administrations all over the world.

Note that we have chosen implementations of AES as our running example because of its importance in practice. Attacking (unprotected) implementations of any other algorithm is as simple as attacking unprotected implementations of AES.

B.1   Algorithm

AES encrypts data of a fixed block length (128 bits) under a key, which can either have 128, 192, or 256 bits. It is common practice to use the abbreviation AES-128 in order to refer to AES with 128-bit keys (AES-192 refers to 192-bit keys, *etc.*). The differences between AES-128, AES-192 and AES-256 are small: the number of rounds increases and the key schedule changes accordingly. Because all the attacks that we describe in this book are explained based on implementations of AES-128 encryption, we only describe AES-128 encryption in this appendix. Because we only use AES-128, we refer to this version as AES from now on.

B.1.1   Structure of the AES Encryption

AES encrypts a 128-bit data block under a 128-bit key. The data and the key are represented as a rectangular array of bytes with four rows and four columns, see Figure B.1. The data array is also called the *state*. AES is a so-called key-

Algorithm 297

```
AES-128(byte in[16], byte out[16], word w[44])
    byte state[4,4];
    state = in;
    AddRoundKey(state, w[0,3])
    for round = 1 step 1 to 9
        SubBytes(state)
        ShiftRows(state)
        MixColumns(state)
        AddRoundKey(state, w[round*4,(round+1)*4-1])
    end
    SubBytes(state)
    ShiftRows(state)
    AddRoundKey(state, w[40,43])
    out = state;
```

Figure B.2.   Pseudo code for AES.

iterated cipher. This means that a round transformation is repeatedly applied to the state. In AES, ten repetitions have to be computed. For each round, a round key is used. The round keys are generated by the key scheduling algorithm. Decryption works similar to encryption. However, the round keys have to be applied in reverse order and the inverse of the round transformation needs to be applied.

B.1.2   Round Transformation

The round transformation consists of four different steps (operations) that are called AddRoundKey, SubBytes, ShiftRows, and MixColumns. Figure B.2 shows in which order the steps are applied to the state. The encryption starts with an initial AddRoundKey operation that adds the key to the state. Then, nine rounds are executed. Each of these nine rounds consists of the four operations. Thereafter, a tenth round is executed, which skips the MixColumns operation.

AddRoundKey

The key addition, which exclusive-ors the state with the round key (see Figure B.3), is called AddRoundKey. The round key in the initial AddRound-Key operation is just the cipher key. The round key length is equal to the data block length, which is 128 bits.

SubBytes

The byte substitution, which is applied to each byte of the state separately (see Figure B.4), is called SubBytes. It is the only non-linear step in the round transformation. This non-linear function is also called S-box (short S), and is

$a_{0,0}$	$a_{0,1}$	$a_{0,2}$	$a_{0,3}$
$a_{1,0}$	$a_{1,1}$	$a_{1,2}$	$a_{1,3}$
$a_{2,0}$	$a_{2,1}$	$a_{2,2}$	$a_{2,3}$
$a_{3,0}$	$a_{3,1}$	$a_{3,2}$	$a_{3,3}$

$\oplus$

$k_{0,0}$	$k_{0,1}$	$k_{0,2}$	$k_{0,3}$
$k_{1,0}$	$k_{1,1}$	$k_{1,2}$	$k_{1,3}$
$k_{2,0}$	$k_{2,1}$	$k_{2,2}$	$k_{2,3}$
$k_{3,0}$	$k_{3,1}$	$k_{3,2}$	$k_{3,3}$

$=$

$b_{0,0}$	$b_{0,1}$	$b_{0,2}$	$b_{0,3}$
$b_{1,0}$	$b_{1,1}$	$b_{1,2}$	$b_{1,3}$
$b_{2,0}$	$b_{2,1}$	$b_{2,2}$	$b_{2,3}$
$b_{3,0}$	$b_{3,1}$	$b_{3,2}$	$b_{3,3}$

Figure B.3.   In AddRoundKey, the round key is added to the state with an exclusive-or operation.

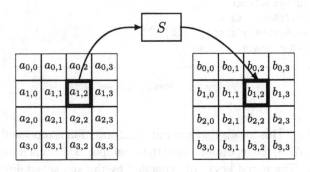

Figure B.4.   SubBytes works on the individual bytes of the state.

$x \backslash y$	0	1	2	3	4	5	6	7	8	9	a	b	c	d	e	f
0	63	7c	77	7b	f2	6b	6f	c5	30	01	67	2b	fe	d7	ab	76
1	ca	82	c9	7d	fa	59	47	f0	ad	d4	a2	af	9c	a4	72	c0
2	b7	fd	93	26	36	3f	f7	cc	34	a5	e5	f1	71	d8	31	15
3	04	c7	23	c3	18	96	05	9a	07	12	80	e2	eb	27	b2	75
4	09	83	2c	1a	1b	6e	5a	a0	52	3b	d6	b3	29	e3	2f	84
5	53	d1	00	ed	20	fc	b1	5b	6a	cb	be	39	4a	4c	58	cf
6	d0	ef	aa	fb	43	4d	33	85	45	f9	02	7f	50	3c	9f	a8
7	51	a3	40	8f	92	9d	38	f5	bc	b6	da	21	10	ff	f3	d2
8	cd	0c	13	ec	5f	97	44	17	c4	a7	7e	3d	64	5d	19	73
9	60	81	4f	dc	22	2a	90	88	46	ee	b8	14	de	5e	0b	db
a	e0	32	3a	0a	49	06	24	5c	c2	d3	ac	62	91	95	e4	79
b	e7	c8	37	6d	8d	d5	4e	a9	6c	56	f4	ea	65	7a	ae	08
c	ba	78	25	2e	1c	a6	b4	c6	e8	dd	74	1f	4b	bd	8b	8a
d	70	3e	b5	66	48	03	f6	0e	61	35	57	b9	86	c1	1d	9e
e	e1	f8	98	11	69	d9	8e	94	9b	1e	87	e9	ce	55	28	df
f	8c	a1	89	0d	bf	e6	42	68	41	99	2d	0f	b0	54	bb	16

Figure B.5.   The S-box table for inputs of the form xy in hexadecimal notation.

Algorithm 299

defined in (B.1).

$$S(x) = A \cdot x^{-1} + b \qquad (B.1)$$

In (B.1), the inverse of x is computed over a finite field with 256 elements. The variable A is a matrix. The variable b is a vector. Their definitions can be found in [Nat01]. The S-box has been chosen according to several criteria. The first criterion was non-linearity. This means that the maximum input-output correlation must be minimal. Furthermore, the maximum difference propagation probability must be minimal. The second criterion was algebraic complexity. This means that the algebraic expression of the S-box has to be complex. The two criteria have been motivated by linear and differential cryptanalysis. As a result, the function that defines the S-box requires computationally expensive operations such as a finite field inversion and a matrix multiplication. Therefore, the S-box is often precomputed and stored in a table, see Figure B.5.

ShiftRows

The byte transposition, which also works byte-wise, is called ShiftRows. It cyclically shifts the rows of the state by different offsets, see Figure B.6. In AES, the first row is not shifted, the second row is shifted by one position, the third row is shifted by two positions and the fourth row is shifted by three positions. The positions have been chosen according to two criteria that are both related to the diffusion. Diffusion means that the individual bytes of the state get dispersed all over the state. For optimal diffusion, the offsets of ShiftRows need to be chosen differently from each other.

MixColumns

The mixing of the elements, which works on the columns of the state, is called MixColumns. It takes a column of the state and performs a matrix multiplication with that column, see Figure B.7. As for SubBytes, the bytes of the state are considered as elements of a finite field with 256 elements. During the matrix multiplication, they are multiplied modulo $x^4 + 1$ with a fixed polynomial $c(x) = 03 \cdot x^3 + 01 \cdot x^2 + 01 \cdot x + 02$. The choice for this particular multiplication polynomial (and the resulting matrix) has been motivated by the so-called wide-trail design strategy, which provides high resistance against linear and differential cryptanalysis.

B.1.3 Key Schedule

The key schedule, which is often called key expansion, generates the round keys. In the first step, the key is expanded as shown in Figure B.8. Then, the round keys are extracted from the expanded key. The key expansion can be computed efficiently. The SubWord function applies the SubBytes function to the four bytes of a word. The RotWord function is a simple rotation. The

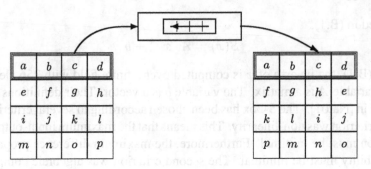

Figure B.6. ShiftRows operates on the rows of the state.

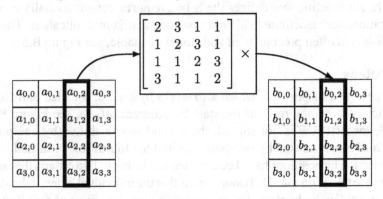

Figure B.7. MixColumns operates on the columns of the state.

```
RC[1..10] = ('01','02','04','08','10','20','40','80','1B','36')
Rcon[i]   = (RC[i],'00','00','00')

for(i = 0; i < 4; i++)
{
    W[i] = (key[4*i], key[4*i+1], key[4*i+2], key[4*i+3])
}
for(i = 4; i < 44; i++)
{
    temp = W[i-1]
    if (i mod 4 == 0)
        temp = SubWord(RotWord(temp)) xor Rcon[i/4]
    W[i] = W[i-4] xor temp
}
```

Figure B.8. Pseudo code for the AES key expansion.

addition (bit-wise exclusive-or) with the round constants $Rcon$ is supposed to eliminate symmetries. The result of the key expansion, the expanded key, has length $11 \cdot 128$ bits. Hence, for each round and the initial AddRoundKey, there is one 128-bit round key.

B.2 Software Implementation

It is possible to efficiently implement AES on 8-bit platforms such as 8051-compatible microcontrollers. Such microcontrollers are used in many devices, including smart cards. Therefore, they are an appropriate platform to study power analysis attacks.

Because we only deal with 8051-compatible microcontrollers, we simply refer to them as microcontrollers from now on. We assume that the reader has some basic understanding of how microcontrollers work. Hence, the subsequent section only briefly discusses several components and instructions of a microcontroller that we use in our implementation of AES.

B.2.1 Microcontrollers

Typical microcontrollers have different types of memories. The 8-bit internal address space is used to access 128 bytes of internal random access memory (RAM) and the special-function registers (SFRs). There is also external memory, which consists of RAM and code memory with 16-bit addresses each. Accessing internal memory is typically faster than accessing external memory.

Memory and Registers

Code memory is the memory in which the program code resides in. It is limited to 64 kB, however, it can be even less, depending on the type of the microcontroller. Hence, one has to take care not to exceed this limit.

A part of the internal RAM is divided into four register banks. Within a bank, the registers are referred to as R0, R1, ..., R7. The SFRs control specific functionality of the microcontroller. For instance, by manipulating the SFRs, one can send data over its serial communication port.

The accumulator (A) is a general register. It is used to accumulate the results of several instructions. Most of the instructions of a microcontroller make use of the accumulator. For example, all arithmetic operations take at least one operand from the accumulator and they also use it to store the result.

The data pointer (DPTR) is a 16-bit register. It can be used to access external memory (RAM and code memory).

Addressing Modes

A certain addressing mode specifies how memory locations are addressed. There are different addressing modes. Two of them are used in the code that we show in Figure B.10. The first of them is called direct addressing. In direct addressing, a value is directly retrieved from a given memory location. For instance, MOV A,30h moves the value that is stored in internal memory at the address hexadecimal 30 into the accumulator. The second one is called indirect addressing. In indirect addressing, a value is fetched from an address that is given in a register. For example, MOVC A,@A+DPTR fetches a value from a specific location in code memory. The address of this location is calculated by adding the value in A to the value of DPTR.

Subroutines

It is possible to structure code by putting it into subroutines that can be called during the execution of a program. In order to call a subroutine, the LCALL instruction is used. It stores the current value of the program counter (PC) on the stack and executes the code in the subroutine. At the end of the subroutine, the RET instruction is called. The RET instruction returns to the address that is stored on the stack.

Timing

Different instructions require a different amount of instruction cycles. The fastest instruction can be executed in only one instruction cycle, the slowest instruction takes up to four instruction cycles. One instruction cycle requires 12 clock cycles.

B.2.2   AES Assembly Implementation

Our assembly implementation of AES is optimized for speed. However, it also takes the limited memory into account. Because of the limited memory, the round keys are computed on the fly. Consequently, only 16 bytes of memory are used to store the current round key. The current round key is computed from the previous round key right before it is needed. This results in an AES implementation wherein the rounds and the round key generation are interleaved, see Figure B.9. Because AddRoundKey, SubBytes, and ShiftRows work byte-wise, they are grouped together in a subroutine which is called Round, see Figure B.10. After the subroutine Round has been executed, MixColumns is done. Thereafter, the round key for the next round is calculated by the subroutine NextRoundKey128. The round constants, which are used during the key schedule, are stored in R5. The table that is used in the SubBytes operation is stored in code memory, and the DPTR points to the start address of this part of the code memory. In order to optimize the performance, we store data (e.g. the S-box table) at fixed memory locations. This enables direct addressing. Furthermore, we unroll loops in order to avoid time consuming loop control instructions, and we use assembly as programming language.

```
ASM_AES128_Encrypt:
    MOV R5, #0x01
    MOV DPTR, #SBOX

    ; Round 1
    LCALL SET_NEW_INSTRUCTION
    LCALL Round
    LCALL MixColumns
    LCALL CLEAR_NEW_INSTRUCTION
    LCALL NextRoundKey128
    MOV R5, #0x02

    ; Round 2
    LCALL SET_NEW_INSTRUCTION
    LCALL Round
    LCALL MixColumns
    LCALL CLEAR_NEW_INSTRUCTION
    LCALL NextRoundKey128
    MOV R5, #0x04
```

Figure B.9.   Assembly code for two AES rounds.

In order to facilitate power analysis attacks, we call subroutines that set (or unset) a certain pin, which we call trigger pin, of the microcontroller. The trigger pin is used to trigger the oscilloscope. The subroutine called SET_NEW_INSTRUCTION sets the trigger pin of the microcontroller to 1. This indicates the start of an encryption round. The subroutine called CLEAR_ NEW_INSTRUCTION sets the trigger pin to 0. This indicates the end of an encryption round. The subroutines called SET_ROUND_TRIGGER and CLEAR_ROUND_TRIGGER are used to indicate the execution of AddRound-Key, SubBytes, and ShiftRows for the first state byte.

Because the examples that are given in this book only target intermediate results that occur before MixColumns, we omit details of the implementation of MixColumns and the key schedule.

B.3   Hardware Implementation

AES is not only suitable for software implementations. AES can also be effi-ciently implemented in hardware. In fact, there is a large number of publications that present hardware implementations of AES for different applications. The proposed implementations range from ultra low-power implementations for RFID devices to high-performance implementations for Internet servers. The AES ASIC that we use for the examples in this book features a 32-bit encryp-tion core. Such 32-bit architectures are very popular to implement AES for

```
Round:
    ; First row

    LCALL SET_ROUND_TRIGGER
    MOV A,ASM_input + 0      ; load a0
    XRL A,ASM_key + 0        ; add k0

    MOVC A,@A + DPTR         ; substitute a0
    MOV ASM_input,A          ; store a0
    LCALL CLEAR_ROUND_TRIGGER

    MOV A,ASM_input + 4      ; load a4
    XRL A,ASM_key + 4        ; add k4

    MOVC A,@A + DPTR         ; substitute a4
    MOV ASM_input + 4,A      ; store a4

    MOV A,ASM_input + 8      ; load a8
    XRL A,ASM_key + 8        ; add k8

    MOVC A,@A + DPTR         ; substitute a8
    MOV ASM_input + 8,A      ; store a8

    MOV A,ASM_input + 12     ; load a12
    XRL A,ASM_key + 12       ; add k12

    MOVC A,@A + DPTR         ; substitute a12
    MOV ASM_input + 12,A     ; store a12
```

Figure B.10.   Code for one row of the Round.

applications with medium throughput requirements. Our AES ASIC has been manufactured using a $0.25\,\mu$m CMOS process technology and it can perform AES-128 encryptions with a maximum clock frequency of 75 MHz. In the following sections, we briefly describe the architecture of the encryption core, and we present the way the S-boxes have been implemented. The S-boxes are the most critical part of implementations of AES in hardware.

B.3.1   Encryption Core

Figure B.11 shows a block diagram of the encryption core of our AES ASIC. We discuss the functionality of this core by describing how it encrypts one plaintext block.

- **Loading of Key and Data:** At the beginning, it is necessary to load the key and the plaintext into the encryption core. This is done in chunks of

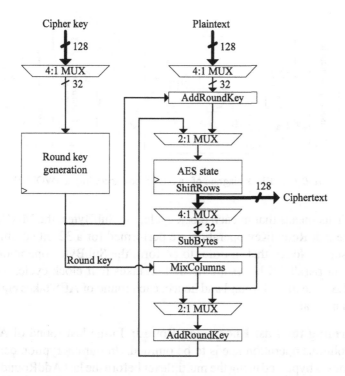

Figure B.11. Block diagram of the datapath of the AES ASIC that we have used for the examples in this book.

32 bits. Hence, it takes four clock cycles to load the key into the module labeled "round key generation". It takes another four clock cycles to load the plaintext into the module labeled "AES state". Note that during the loading of the plaintext, the initial AddRoundKey operation of AES is performed.

- **Performing a Normal Encryption Round:** After loading the key and the plaintext into the encryption core, nine normal encryption rounds are performed. In each round, essentially two things need to be done. It is necessary to perform the key expansion and it is necessary to update the AES state. In most encryption cores, these two operations are done in parallel. However, our encryption core performs the key expansion before the update of the AES state. This allows a separate analysis of the update of the AES state. The only operation that is done in parallel to the key expansion is the ShiftRows operation of the AES state. This operation is done during the first clock cycle of the key expansion. In total, the key expansion takes four clock cycles.

After having performed the key expansion and the ShiftRows operation, the AES state is updated. In case of our AES ASIC, this is done column by col-

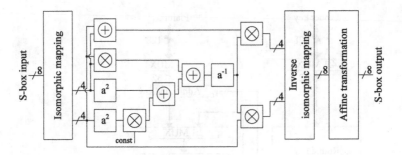

Figure B.12.   Block diagram of an AES S-box according to [WOL02].

umn. This means that in each clock cycle, the SubBytes, the MixColumns, and the AddRoundKey operation are performed for a 32-bit column of the AES state. Notice that in order to perform the SubBytes operation for 32 bits, four parallel S-boxes are needed. It takes four clock cycles to update the AES state in each round and hence, each round of AES takes eight clock cycles in total.

- **Performing the Last Encryption Round:** In the last round of AES, the MixColumns operation needs to be omitted. In our encryption core, Mix-Columns is bypassed using the multiplexer before the last AddRoundKey operation, see Figure B.11. Everything else is done exactly as in the normal encryption rounds. After the last round, the ciphertext is stored in the module "AES state".

B.3.2   S-Box

The S-boxes that are needed for the SubBytes operation are the most critical components of AES hardware implementations. This is due to the fact that the S-boxes require the most resources compared to the other operations of AES. In practice, there essentially exist three different ways to implement AES S-boxes. They can be implemented using a read-only memory (ROM), they can be synthesized as look-up table, or they can be implemented based on composite field arithmetic. In our encryption core, we have used the last approach. We have implemented the S-boxes as proposed in [WOL02]. Figure B.12 shows a block diagram of an S-box that is implemented according to this approach. As it can be observed, the S-box is calculated using arithmetic operations. This fact is exploited in the DPA attack presented in Section 6.2.2.

References

[AARR03]   Dakshi Agrawal, Bruce Archambeault, Josyula R. Rao, and Pankaj Rohatgi. The EM Side-channel(s). In Burton S. Kaliski Jr., Çetin Kaya Koç, and Christof Paar, editors, *Cryptographic Hardware and Embedded Systems – CHES 2002, 4th International Workshop, Redwood Shores, CA, USA, August 13-15, 2002, Revised Papers*, volume 2523 of *Lecture Notes in Computer Science*, pages 29–45. Springer, 2003.

[ABCS06]   Ross J. Anderson, Mike Bond, Jolyon Clulow, and Sergei P. Skorobogatov. Cryptographic Processors—A Survey. *Proceedings of the IEEE*, 94(2):357–369, February 2006. ISSN 0018-9219.

[ABDM00]   Mehdi-Laurent Akkar, Régis Bevan, Paul Dischamp, and Didier Moyart. Power Analysis, What Is Now Possible... In Tatsuaki Okamoto, editor, *Advances in Cryptology - ASIACRYPT 2000, 6th International Conference on the Theory and Application of Cryptology and Information Security, Kyoto, Japan, December 3-7, 2000, Proceedings*, volume 1976 of *Lecture Notes in Computer Science*, pages 489–502. Springer, 2000.

[ABG04]   Mehdi-Laurent Akkar, Régis Bevan, and Louis Goubin. Two Power Analysis Attacks against One-Mask Methods. In Bimal K. Roy and Willi Meier, editors, *Fast Software Encryption, 11th International Workshop, FSE 2004, Delhi, India, February 5-7, 2004, Revised Papers*, volume 3017 of *Lecture Notes in Computer Science*, pages 332–347. Springer, 2004.

[AE01]   Mohamed W. Allam and Mohamed I. Elmasry. Dynamic Current Mode Logic (DyCML): A New Low-Power High-Performance Logic Style. *IEEE Journal of Solid-State Circuits*, 36(3):550–558, March 2001. ISSN 0018-9200.

[AG01]   Mehdi-Laurent Akkar and Christophe Giraud. An Implementation of DES and AES, Secure against Some Attacks. In Çetin Kaya Koç, David Naccache, and Christof Paar, editors, *Cryptographic Hardware and Embedded Systems – CHES 2001, Third International Workshop, Paris, France, May 14-16, 2001, Proceedings*, volume 2162 of *Lecture Notes in Computer Science*, pages 309–318. Springer, 2001.

[AG03]   Mehdi-Laurent Akkar and Louis Goubin. A Generic Protection against High-Order Differential Power Analysis. In Thomas Johansson, editor, *Fast Software*

Encryption, 10th International Workshop, FSE 2003, Lund, Sweden, February 24-26, 2003, Revised Papers, volume 2887 of _Lecture Notes in Computer Science_, pages 192–205. Springer, 2003.

[AMM$^+$05] Manfred Aigner, Stefan Mangard, Renato Menicocci, Mauro Olivieri, Giuseppe Scotti, and Alessandro Trifiletti. A Novel CMOS Logic Style with Data Independent Power Consumption. In _International Symposium on Circuits and Systems (ISCAS 2005), Kobe, Japan, May 23-26, 2005, Proceedings_, volume 2, pages 1066–1069. IEEE, 2005.

[And01] Ross J. Anderson. _Security Engineering: A Guide to Building Dependable Distributed Systems_. Wiley, 2001. ISBN 0-471-38922-6.

[ARR03] Dakshi Agrawal, Josyula R. Rao, and Pankaj Rohatgi. Multi-channel Attacks. In Colin D. Walter, Çetin Kaya Koç, and Christof Paar, editors, _Cryptographic Hardware and Embedded Systems – CHES 2003, 5th International Workshop, Cologne, Germany, September 8-10, 2003, Proceedings_, volume 2779 of _Lecture Notes in Computer Science_, pages 2–16. Springer, 2003.

[ARRS05] Dakshi Agrawal, Josyula R. Rao, Pankaj Rohatgi, and Kai Schramm. Templates as Master Keys. In Josyula R. Rao and Berk Sunar, editors, _Cryptographic Hardware and Embedded Systems – CHES 2005, 7th International Workshop, Edinburgh, UK, August 29 - September 1, 2005, Proceedings_, volume 3659 of _Lecture Notes in Computer Science_, pages 15–29. Springer, 2005.

[AT03] Toru Akishita and Tsuyoshi Takagi. Zero-Value Point Attacks on Elliptic Curve Cryptosystem. In Colin Boyd and Wenbo Mao, editors, _Information Security, 6th International Conference, ISC 2003, Bristol, UK, October 1-3, 2003, Proceedings_, volume 2851 of _Lecture Notes in Computer Science_, pages 218–233. Springer, 2003.

[AT06] Toru Akishita and Tsuyoshi Takagi. Power Analysis to ECC Using Differential Power Between Multiplication and Squaring. In Josep Domingo-Ferrer, Joachim Posegga, and Daniel Schreckling, editors, _Smart Card Research and Advanced Applications, 7th IFIP WG 8.8/11.2 International Conference, CARDIS 2006, Tarragona, Spain, April 19-21, 2006, Proceedings_, volume 3928 of _Lecture Notes in Computer Science_, pages 151–164. Springer, April 2006.

[BCO04] Eric Brier, Christophe Clavier, and Francis Olivier. Correlation Power Analysis with a Leakage Model. In Marc Joye and Jean-Jacques Quisquater, editors, _Cryptographic Hardware and Embedded Systems – CHES 2004, 6th International Workshop, Cambridge, MA, USA, August 11-13, 2004, Proceedings_, volume 3156 of _Lecture Notes in Computer Science_, pages 16–29. Springer, 2004.

[BECN$^+$04] Hagai Bar-El, Hamid Choukri, David Naccache, Michael Tunstall, and Claire Whelan. The Sorcerer's Apprentice Guide to Fault Attacks. Cryptology ePrint Archive (http://eprint.iacr.org/), Report 2004/100, 2004.

[BGK05] Johannes Blömer, Jorge Guajardo, and Volker Krummel. Provably Secure Masking of AES. In Helena Handschuh and M. Anwar Hasan, editors, _Selected_

Areas in Cryptography, 11th International Workshop, SAC 2004, Waterloo, Canada, August 9-10, 2004, Revised Selected Papers, volume 3357 of *Lecture Notes in Computer Science*, pages 69–83. Springer, 2005.

[BGL+06]   Marco Bucci, Luca Giancane, Raimondo Luzzi, Giuseppe Scotti, and Alessandro Trifiletti. Enhancing Power Analysis Attacks Against Cryptographic Devices. In *International Symposium on Circuits and Systems (ISCAS 2006), Island of Kos, Greece, May 21-24, 2006, Proceedings*, pages 2905–2908. IEEE, May 2006.

[BGLT04]   Marco Bucci, Michele Guglielmo, Raimondo Luzzi, and Alessandro Trifiletti. A Power Consumption Randomization Countermeasure for DPA-Resistant Cryptographic Processors. In Enrico Macii, Odysseas G. Koufopavlou, and Vassilis Paliouras, editors, *14th International Workshop on Integrated Circuit and System Design, Power and Timing Modeling, Optimization and Simulation, PATMOS 2004, Santorini, Greece, September 15-17, 2004, Proceedings*, volume 3254 of *Lecture Notes in Computer Science*, pages 481–490. Springer, 2004.

[BGLT06]   Marco Bucci, Luca Giancane, Raimondo Luzzi, and Alessandro Trifiletti. Three-Phase Dual-Rail Pre-Charge Logic. In *Cryptographic Hardware and Embedded Systems – CHES 2006, 8th International Workshop, Yokohama, Japan, October 10-13, 2006, Proceedings*, Lecture Notes in Computer Science. Springer, 2006.

[BGM+03]   Luca Benini, Angelo Galati, Alberto Macii, Enrico Macii, and Massimo Poncino. Energy-Efficient Data Scrambling on Memory-Processor Interfaces. In Ingrid Verbauwhede and Hyung Roh, editors, *International Symposium on Low Power Electronics and Design, 2003, Seoul, Korea, August 25-27, 2003, Proceedings*, pages 26–29. ACM Press, 2003.

[BILT04]   Jean-Claude Bajard, Laurent Imbert, Pierre-Yvan Liardet, and Yannick Teglia. Leak Resistant Arithmetic. In Marc Joye and Jean-Jacques Quisquater, editors, *Cryptographic Hardware and Embedded Systems – CHES 2004, 6th International Workshop, Cambridge, MA, USA, August 11-13, 2004, Proceedings*, volume 3156 of *Lecture Notes in Computer Science*, pages 62–75. Springer, 2004.

[BJ02]   Eric Brier and Marc Joye. Weierstraß Elliptic Curves and Side-Channel Attacks. In David Naccache and Pascal Paillier, editors, *Public Key Cryptography, 5th International Workshop on Practice and Theory in Public Key Cryptosystems, PKC 2002, Paris, France, February 12-14, 2002, Proceedings*, volume 2274 of *Lecture Notes in Computer Science*, pages 335–345. Springer, 2002.

[BK03]   Régis Bevan and Erik Knudsen. Ways to Enhance Differential Power Analysis. In Pil Joong Lee and Chae Hoon Lim, editors, *Information Security and Cryptology - ICISC 2002, 5th International Conference Seoul, Korea, November 28-29, 2002, Revised Papers*, volume 2587 of *Lecture Notes in Computer Science*, pages 327–342. Springer, 2003.

[BMM+03a]   Luca Benini, Alberto Macii, Enrico Macii, Elvira Omerbegovic, Massimo Poncino, and Fabrizio Pro. A Novel Architecture for Power Maskable Arithmetic

Units. In *13th ACM Great Lakes Symposium on VLSI 2004, Washington, DC, USA, April 28-29, 2003, Proceedings*, pages 136–140. ACM Press, 2003.

[BMM⁺03b]   Luca Benini, Alberto Macii, Enrico Macii, Elvira Omerbegovic, Fabrizio Pro, and Massimo Poncino. Energy-Aware Design Techniques for Differential Power Analysis Protection. In *40th Design Automation Conference, DAC 2003, Anaheim, CA, USA, June 2-6, 2003, Proceedings*. ACM Press, 2003.

[BS99]   Eli Biham and Adi Shamir. Power Analysis of the Key Scheduling of the AES Candidates. In *Second Advanced Encryption Standard (AES) Candidate Conference*, Rome, Italy, 1999.

[BSYK03]   Alex Bystrov, Danil Sokolov, Alex Yakovlev, and Albert Koelmans. Balancing Power Signature in Secure Systems. In *14th UK Asynchronous Forum, Newcastle, June 2003*, 2003. Available online at http://www.staff.ncl.ac.uk/i.g.clark/async/ukasyncforum14/forum14-papers/forum14-bystrov.pdf.

[BZB⁺05]   Guido Bertoni, Vittorio Zaccaria, Luca Breveglieri, Matteo Monchiero, and Gianluca Palermo. AES Power Attack Based on Induced Cache Miss and Countermeasure. In *International Symposium on Information Technology: Coding and Computing (ITCC 2005), 4-6 April 2005, Las Vegas, Nevada, USA, Proceedings*, volume 1, pages 586–591. IEEE Computer Society, April 2005.

[Cad]   Cadence Design Systems. The Cadence Design Systems Website. http://www.cadence.com/.

[CCD00]   Christophe Clavier, Jean-Sébastien Coron, and Nora Dabbous. Differential Power Analysis in the Presence of Hardware Countermeasures. In Çetin Kaya Koç and Christof Paar, editors, *Cryptographic Hardware and Embedded Systems – CHES 2000, Second International Workshop, Worcester, MA, USA, August 17-18, 2000, Proceedings*, volume 1965 of *Lecture Notes in Computer Science*, pages 252–263. Springer, 2000.

[CCD04]   Vincent Carlier, Herve Chabanne, and Emmanuelle Dottax. A solution to protect AES against side channel attacks. Technical Report 0406 SEC 003, SAGEM SA, May 2004.

[CG00]   Jean-Sébastien Coron and Louis Goubin. On Boolean and Arithmetic Masking against Differential Power Analysis. In Çetin Kaya Koç and Christof Paar, editors, *Cryptographic Hardware and Embedded Systems – CHES 2000, Second International Workshop, Worcester, MA, USA, August 17-18, 2000, Proceedings*, volume 1965 of *Lecture Notes in Computer Science*, pages 231–237. Springer, 2000.

[Cha06]   Chair for Communication Security, Ruhr-Universität Bochum. Side Channel Cryptanalysis Lounge. http://www.crypto.ruhr-uni-bochum.de/en_sclounge.html, 2006.

[CJRR99a]   Suresh Chari, Charanjit S. Jutla, Josyula R. Rao, and Pankaj Rohatgi. A Cautionary Note Regarding Evaluation of AES Candidates on Smart-Cards. In

Second Advanced Encryption Standard (AES) Candidate Conference, Rome, Italy, 1999.

[CJRR99b]   Suresh Chari, Charanjit S. Jutla, Josyula R. Rao, and Pankaj Rohatgi. Towards Sound Approaches to Counteract Power-Analysis Attacks. In Michael J. Wiener, editor, *Advances in Cryptology - CRYPTO '99, 19th Annual International Cryptology Conference, Santa Barbara, California, USA, August 15-19, 1999, Proceedings*, volume 1666 of *Lecture Notes in Computer Science*, pages 398–412. Springer, 1999.

[CKN01]   Jean-Sébastien Coron, Paul C. Kocher, and David Naccache. Statistics and Secret Leakage. In Yair Frankel, editor, *Financial Cryptography, 4th International Conference, FC 2000 Anguilla, British West Indies, February 20-24, 2000, Proceedings*, volume 1962 of *Lecture Notes in Computer Science*, pages 157–173. Springer, 2001.

[CMCJ04]   Benoît Chevallier-Mames, Mathieu Ciet, and Marc Joye. Low-Cost Solutions for Preventing Simple Side-Channel Analysis: Side-Channel Atomicity. *IEEE Transactions on Computers*, 53(6):760–768, June 2004. ISSN 0018-9340.

[CNPQ03]   Mathieu Ciet, Michael Neve, Eric Peeters, and Jean-Jacques Quisquater. Parallel FPGA Implementation of RSA with Residue Number Systems. In *Proceedings of the 46th IEEE International Midwest Symposium on Circuits and Systems (MWSCAS '03)*, volume 2, pages 806–810. IEEE, 2003.

[Cor99]   Jean-Sébastien Coron. Resistance against Differential Power Analysis for Elliptic Curve Cryptosystems. In Çetin Kaya Koç and Christof Paar, editors, *Cryptographic Hardware and Embedded Systems – CHES'99, First International Workshop, Worcester, MA, USA, August 12-13, 1999, Proceedings*, volume 1717 of *Lecture Notes in Computer Science*, pages 292–302. Springer, 1999.

[CPM05]   Pasquale Corsonello, Stefania Perri, and Martin Margala. A New Charge-Pump Based Countermeasure Against Differential Power Analysis. In *Proceedings of the 6th International Conference on ASIC (ASICON 2005)*, volume 1, pages 66–69. IEEE, 2005.

[CRR03]   Suresh Chari, Josyula R. Rao, and Pankaj Rohatgi. Template Attacks. In Burton S. Kaliski Jr., Çetin Kaya Koç, and Christof Paar, editors, *Cryptographic Hardware and Embedded Systems – CHES 2002, 4th International Workshop, Redwood Shores, CA, USA, August 13-15, 2002, Revised Papers*, volume 2523 of *Lecture Notes in Computer Science*, pages 13–28. Springer, 2003.

[CT03]   Jean-Sébastien Coron and Alexei Tchulkine. A New Algorithm for Switching from Arithmetic to Boolean Masking. In Colin D. Walter, Çetin Kaya Koç, and Christof Paar, editors, *Cryptographic Hardware and Embedded Systems – CHES 2003, 5th International Workshop, Cologne, Germany, September 8-10, 2003, Proceedings*, volume 2779 of *Lecture Notes in Computer Science*, pages 89–97. Springer, 2003.

[CZ06]   Zhimin Chen and Yujie Zhou. Dual-Rail Random Switching Logic: A Countermeasure to Reduce Side Channel Leakage. In *Cryptographic Hardware* and Embedded Systems – CHES 2006, 8th International Workshop, Yokohama, Japan, October 10-13, 2006, Proceedings, Lecture Notes in Computer Science. Springer, 2006.

[EMV04]   EMVCo. EMV Integrated Circuit Card Specifications for Payment Systems – Book 2: Security and Key Management, June 2004. Available online at http://www.emvco.com/.

[ETS⁺05]   Reouven Elbaz, Lionel Torres, Gilles Sassatelli, Pierre Guillemin, C. Anguille, M. Bardouillet, Christian Buatois, and Jean-Baptiste Rigaud. Hardware Engines for Bus Encryption: A Survey of Existing Techniques. In *2005 Design, Automation and Test in Europe Conference and Exposition (DATE 2005), 7-11 March 2005, Munich, Germany*, pages 40–45. IEEE Computer Society, 2005.

[FG05]   Wieland Fischer and Berndt M. Gammel. Masking at Gate Level in the Presence of Glitches. In Josyula R. Rao and Berk Sunar, editors, *Cryptographic Hardware and Embedded Systems – CHES 2005, 7th International Workshop, Edinburgh, UK, August 29 - September 1, 2005, Proceedings*, volume 3659 of *Lecture Notes in Computer Science*, pages 187–200. Springer, 2005.

[FML⁺03]   Jacques J. A. Fournier, Simon Moore, Huiyun Li, Robert D. Mullins, and George S. Taylor. Security Evaluation of Asynchronous Circuits. In Colin D. Walter, Çetin Kaya Koç, and Christof Paar, editors, *Cryptographic Hardware and Embedded Systems – CHES 2003, 5th International Workshop, Cologne, Germany, September 8-10, 2003, Proceedings*, volume 2779 of *Lecture Notes in Computer Science*, pages 137–151. Springer, 2003.

[FMP03]   Pierre-Alain Fouque, Gwenaëlle Martinet, and Guillaume Poupard. Attacking Unbalanced RSA-CRT Using SPA. In Colin D. Walter, Çetin Kaya Koç, and Christof Paar, editors, *Cryptographic Hardware and Embedded Systems – CHES 2003, 5th International Workshop, Cologne, Germany, September 8-10, 2003, Proceedings*, volume 2779 of *Lecture Notes in Computer Science*, pages 254–268. Springer, 2003.

[FMPV04]   Pierre-Alain Fouque, Frédéric Muller, Guillaume Poupard, and Frédéric Valette. Defeating Countermeasures Based on Randomized BSD Representations. In Marc Joye and Jean-Jacques Quisquater, editors, *Cryptographic Hardware and Embedded Systems – CHES 2004, 6th International Workshop, Cambridge, MA, USA, August 11-13, 2004, Proceedings*, volume 3156 of *Lecture Notes in Computer Science*, pages 312–327. Springer, 2004.

[FP99]   Paul N. Fahn and Peter K. Pearson. IPA: A New Class of Power Attacks. In Çetin Kaya Koç and Christof Paar, editors, *Cryptographic Hardware and Embedded Systems – CHES'99, First International Workshop, Worcester, MA, USA, August 12-13, 1999, Proceedings*, volume 1717 of *Lecture Notes in Computer Science*, pages 173–186. Springer, 1999.

[FPP97]   David Freedman, Robert Pisani, and Roger Purves. *Statistics*. W. W. Norton & Company, 3rd edition, 1997. ISBN 0-393-97083-3.

[FS03]   Wieland Fischer and Jean-Pierre Seifert. Unfolded Modular Multiplication. In Toshihide Ibaraki, Naoki Katoh, and Hirotaka Ono, editors, *Algorithms and*

Computation, 14th International Symposium, ISAAC 2003, Kyoto, Japan, December 15-17, 2003, Proceedings*, volume 2906 of *Lecture Notes in Computer Science*, pages 726–735. Springer, 2003.

[FV03]   Pierre-Alain Fouque and Frédéric Valette. The Doubling Attack - *Why Upwards Is Better than Downwards*. In Colin D. Walter, Çetin Kaya Koç, and Christof Paar, editors, *Cryptographic Hardware and Embedded Systems – CHES 2003, 5th International Workshop, Cologne, Germany, September 8-10, 2003, Proceedings*, volume 2779 of *Lecture Notes in Computer Science*, pages 269–280. Springer, 2003.

[GHM⁺04]   Sylvain Guilley, Philippe Hoogvorst, Yves Mathieu, Renaud Pacalet, and Jean Provost. CMOS Structures Suitable for Secured Hardware. In *2004 Design, Automation and Test in Europe Conference and Exposition (DATE 2004), 16-20 February 2004, Paris, France*, volume 2, pages 1414–1415. IEEE Computer Society, 2004.

[GHMP05]   Sylvain Guilley, Philippe Hoogvorst, Yves Mathieu, and Renaud Pacalet. The "Backend Duplication" Method. In Josyula R. Rao and Berk Sunar, editors, *Cryptographic Hardware and Embedded Systems – CHES 2005, 7th International Workshop, Edinburgh, UK, August 29 - September 1, 2005, Proceedings*, volume 3659 of *Lecture Notes in Computer Science*, pages 383–397. Springer, 2005.

[GM04]   Jovan D. Golić and Renato Menicocci. Universal Masking on Logic Gate Level. *IEE Electronic Letters*, 40(9):526–527, April 2004. ISSN 0013-5194.

[GMO01]   Karine Gandolfi, Christophe Mourtel, and Francis Olivier. Electromagnetic Analysis: Concrete Results. In Çetin Kaya Koç, David Naccache, and Christof Paar, editors, *Cryptographic Hardware and Embedded Systems – CHES 2001, Third International Workshop, Paris, France, May 14-16, 2001, Proceedings*, volume 2162 of *Lecture Notes in Computer Science*, pages 251–261. Springer, 2001.

[GNS05]   Peter J. Green, Richard Noad, and Nigel P. Smart. Further Hidden Markov Model Cryptanalysis. In Josyula R. Rao and Berk Sunar, editors, *Cryptographic Hardware and Embedded Systems – CHES 2005, 7th International Workshop, Edinburgh, UK, August 29 - September 1, 2005, Proceedings*, volume 3659 of *Lecture Notes in Computer Science*, pages 61–74. Springer, 2005.

[GOK⁺05]   Frank K. Gürkaynak, Stephan Oetiker, Hubert Kaeslin, Norbert Felber, and Wolfgang Fichtner. Improving DPA Security by Using Globally-Asynchronous Locally-Synchronous Systems. In *31th European Solid-State Circuits Conference - ESSCIRC 2005, Grenoble, France, September 12-16, 2005, Proceedings*, pages 407–410. IEEE, September 2005.

[Gol03]   Jovan D. Golić. DeKaRT: A New Paradigm for Key-Dependent Reversible Circuits. In Colin D. Walter, Çetin Kaya Koç, and Christof Paar, editors, *Cryptographic Hardware and Embedded Systems – CHES 2003, 5th International Workshop, Cologne, Germany, September 8-10, 2003, Proceedings*, volume 2779 of *Lecture Notes in Computer Science*, pages 98–112. Springer, 2003.

[Gou01]   Louis Goubin. A Sound Method for Switching between Boolean and Arithmetic Masking. In Çetin Kaya Koç, David Naccache, and Christof Paar, editors, *Cryptographic Hardware and Embedded Systems – CHES 2001, Third International Workshop, Paris, France, May 14-16, 2001, Proceedings*, volume 2162 of *Lecture Notes in Computer Science*, pages 3–15. Springer, 2001.

[Gou03]   Louis Goubin. A Refined Power-Analysis Attack on Elliptic Curve Cryptosystems. In Yvo Desmedt, editor, *Public Key Cryptography - PKC 2003, 6th International Workshop on Theory and Practice in Public Key Cryptography, Miami, FL, USA, January 6-8, 2003, Proceedings*, volume 2567 of *Lecture Notes in Computer Science*, pages 199–210. Springer, 2003.

[GP99]   Louis Goubin and Jacques Patarin. DES and Differential Power Analysis – The Duplication Method. In Çetin Kaya Koç and Christof Paar, editors, *Cryptographic Hardware and Embedded Systems – CHES'99, First International Workshop, Worcester, MA, USA, August 12-13, 1999, Proceedings*, volume 1717 of *Lecture Notes in Computer Science*, pages 158–172. Springer, 1999.

[GT03]   Jovan D. Golić and Christophe Tymen. Multiplicative Masking and Power Analysis of AES. In Burton S. Kaliski Jr., Çetin Kaya Koç, and Christof Paar, editors, *Cryptographic Hardware and Embedded Systems – CHES 2002, 4th International Workshop, Redwood Shores, CA, USA, August 13-15, 2002, Revised Papers*, volume 2523 of *Lecture Notes in Computer Science*, pages 198–212. Springer, 2003.

[Has00]   M. Anwar Hasan. Power Analysis Attacks and Algorithmic Approaches to their Countermeasures for Koblitz Curve Cryptosystems. In Çetin Kaya Koç and Christof Paar, editors, *Cryptographic Hardware and Embedded Systems – CHES 2000, Second International Workshop, Worcester, MA, USA, August 17-18, 2000, Proceedings*, volume 1965 of *Lecture Notes in Computer Science*, pages 93–108. Springer, 2000.

[HGS01]   Nick Howgrave-Graham and Nigel P. Smart. Lattice Attacks on Digital Signature Schemes. *Designs, Codes and Cryptography*, 23(3):283–290, August 2001. ISSN 0925-1022.

[HKM$^+$05]   JaeCheol Ha, ChangKyun Kim, SangJae Moon, IlHwan Park, and HyungSo Yoo. Differential Power Analysis on Block Cipher ARIA. In Laurence T. Yang, Omer F. Rana, Beniamino Di Martino, and Jack Dongarra, editors, *High Performance Computing and Communcations, First International Conference, HPCC 2005, Sorrento, Italy, September 21-23, 2005, Proceedings*, volume 3726 of *Lecture Notes in Computer Science*, pages 541–548. Springer, 2005.

[HM02]   JaeCheol Ha and SangJae Moon. Randomized Signed-Scalar Multiplication of ECC to Resist Power Attacks. In Burton S. Kaliski Jr., Çetin Kaya Koç, and Christof Paar, editors, *Cryptographic Hardware and Embedded Systems – CHES 2002, 4th International Workshop, Redwood Shores, CA, USA, August 13-15, 2002, Revised Papers*, volume 2523 of *Lecture Notes in Computer Science*, pages 551–563. Springer, 2002.

[HOM06]   Christoph Herbst, Elisabeth Oswald, and Stefan Mangard. An AES Smart Card Implementation Resistant to Power Analysis Attacks. In Jianying Zhou, Moti

Yung, and Feng Bao, editors, *Applied Cryptography and Network Security, Second International Conference, ACNS 2006*, volume 3989 of *Lecture Notes in Computer Science*, pages 239–252. Springer, 2006.

[HP06]   Helena Handschuh and Bart Preneel. Blind Differential Cryptanalysis for Enhanced Power Attacks. In *Selected Areas in Cryptography, 13th International Workshop, SAC 2006, Montreal, Quebec, Canada, August 17-18, 2006*, Lecture Notes in Computer Science. Springer, 2006.

[IKV01]   Mary J. Irwin, Mahmut T. Kandemir, and Narayanan Vijaykrishnan. Simple-Power: A Cycle-Accurate Energy Simulator. IEEE TCCA Newsletter, January 2001.

[Int03]   International Electrotechnical Commission (IEC). IEC 61967: Integrated Circuits - Measurement of Electromagnetic Emissions, 150 kHz to 1 GHz, 2003. Available online at http://www.iec.ch.

[IPS02]   James Irwin, Daniel Page, and Nigel P. Smart. Instruction Stream Mutation for Non-Deterministic Processors. In *IEEE International Conference on Application-Specific Systems, Architectures and Processors, 2002, July 17-19, Proceedings*, pages 286–295. IEEE Computer Society, 2002.

[ISW03]   Yuval Ishai, Amit Sahai, and David Wagner. Private Circuits: Securing Hardware against Probing Attacks. In Dan Boneh, editor, *Advances in Cryptology - CRYPTO 2003, 23rd Annual International Cryptology Conference, Santa Barbara, California, USA, August 17-21, 2003, Proceedings*, volume 2729 of *Lecture Notes in Computer Science*, pages 463–481. Springer, 2003.

[ITT02]   Kouichi Itoh, Masahiko Takenaka, and Naoya Torii. DPA Countermeasure Based on the Masking Method. In Kwangjo Kim, editor, *Information Security and Cryptology - ICISC 2001, 4th International Conference Seoul, Korea, December 6-7, 2001, Proceedings*, volume 2288 of *Lecture Notes in Computer Science*, pages 440–456. Springer, 2002.

[Jaf06a]   Joshua Jaffe. Introduction to Differential Power Analysis, June 2006. Presented at ECRYPT Summerschool on Cryptographic Hardware, Side Channel and Fault Analysis.

[Jaf06b]   Joshua Jaffe. More Differential Power Analysis: Selected DPA Attacks, June 2006. Presented at ECRYPT Summerschool on Cryptographic Hardware, Side Channel and Fault Analysis.

[JPS05]   Marc Joye, Pascal Paillier, and Berry Schoenmakers. On Second-Order Differential Power Analysis. In Josyula R. Rao and Berk Sunar, editors, *Cryptographic Hardware and Embedded Systems – CHES 2005, 7th International Workshop, Edinburgh, UK, August 29 - September 1, 2005, Proceedings*, volume 3659 of *Lecture Notes in Computer Science*, pages 293–308. Springer, 2005.

[JQ01]   Marc Joye and Jean-Jacques Quisquater. Hessian Elliptic Curves and Side-Channel Attacks. In Çetin Kaya Koç, David Naccache, and Christof Paar, editors, *Cryptographic Hardware and Embedded Systems – CHES 2001, Third*

International Workshop, Paris, France, May 14-16, 2001, Proceedings, volume 2162 of *Lecture Notes in Computer Science*, pages 402–410. Springer, 2001.

[JT01]   Marc Joye and Christophe Tymen. Protections against Differential Analysis for Elliptic Curve Cryptography. In Çetin Kaya Koç, David Naccache, and Christof Paar, editors, *Cryptographic Hardware and Embedded Systems – CHES 2001, Third International Workshop, Paris, France, May 14-16, 2001, Proceedings*, volume 2162 of *Lecture Notes in Computer Science*, pages 377–390. Springer, 2001.

[Kay98]   Steven M. Kay. *Fundamentals of Statistical Signal Processing - Detection Theory*. Signal Processing Series. Prentice Hall, 1st edition, 1998. ISBN 0-13-504135-X.

[KJJ99]   Paul C. Kocher, Joshua Jaffe, and Benjamin Jun. Differential Power Analysis. In Michael Wiener, editor, *Advances in Cryptology - CRYPTO '99, 19th Annual International Cryptology Conference, Santa Barbara, California, USA, August 15-19, 1999, Proceedings*, volume 1666 of *Lecture Notes in Computer Science*, pages 388–397. Springer, 1999. This paper is included in Appendix A of this book.

[KJMV04] Sébastien Kunz-Jacques, Frédéric Muller, and Frédéric Valette. The Davies-Murphy Power Attack. In Pil Joong Lee, editor, *Advances in Cryptology - ASIACRYPT 2004, 10th International Conference on the Theory and Application of Cryptology and Information Security, Jeju Island, Korea, December 5-9, 2004, Proceedings*, pages 451–467. Springer, 2004.

[KK99]   Oliver Kömmerling and Markus G. Kuhn. Design Principles for Tamper-Resistant Smartcard Processors. In *USENIX Workshop on Smartcard Technology (Smartcard '99)*, pages 9–20, May 1999.

[KKT06]   Konrad J. Kulikowski, Mark G. Karpovsky, and Alexander Taubin. Power Attacks on Secure Hardware Based on Early Propagation of Data. In *12th IEEE International On-Line Testing Symposium (IOLTS 2006), July 10-12, 2006*, pages 131–138. IEEE Computer Society, July 2006.

[Koc96]   Paul C. Kocher. Timing Attacks on Implementations of Diffie-Hellman, RSA, DSS, and Other Systems. In Neal Koblitz, editor, *Advances in Cryptology - CRYPTO '96, 16th Annual International Cryptology Conference, Santa Barbara, California, USA, August 18-22, 1996, Proceedings*, number 1109 in Lecture Notes in Computer Science, pages 104–113. Springer, 1996.

[Koc05]   Paul C. Kocher. Design and Validation Strategies for Obtaining Assurance in Countermeasures to Power Analysis and Related Attacks. In *NIST Phyiscal Security Workshop, September 26-29, 2005*, 2005.

[KSS⁺05]   Konrad J. Kulikowski, Ming Su, Alexander B. Smirnov, Alexander Taubin, Mark G. Karpovsky, and Daniel MacDonald. Delay Insensitive Encoding and Power Analysis: A Balancing Act. In *11th International Symposium on Advanced Research in Asynchronous Circuits and Systems (ASYNC 2005), 14-16 March 2005, New York, NY, USA*, pages 116–125. IEEE Computer Society, March 2005.

[KST06] Konrad J. Kulikowski, Alexander B. Smirnov, and Alexander Taubin. Automated Design of Cryptographic Devices Resistant to Multiple Side-Channel Attacks. In *Cryptographic Hardware and Embedded Systems – CHES 2006, 8th International Workshop, Yokohama, Japan, October 10-13, 2006, Proceedings*, Lecture Notes in Computer Science. Springer, 2006.

[KW03] Chris Karlof and David Wagner. Hidden Markov Model Cryptoanalysis. In Colin D. Walter, Çetin Kaya Koç, and Christof Paar, editors, *Cryptographic Hardware and Embedded Systems – CHES 2003, 5th International Workshop, Cologne, Germany, September 8-10, 2003, Proceedings*, volume 2779 of *Lecture Notes in Computer Science*, pages 17–34. Springer, 2003.

[LD99] Julio López and Ricardo Dahab. Fast Multiplication on Elliptic Curves over $GF(2^m)$ without Precomputation. In Çetin Kaya Koç and Christof Paar, editors, *Cryptographic Hardware and Embedded Systems – CHES'99, First International Workshop, Worcester, MA, USA, August 12-13, 1999, Proceedings*, volume 1717 of *Lecture Notes in Computer Science*, pages 316–327. Springer, 1999.

[LMPV04] Joseph Lano, Nele Mentens, Bart Preneel, and Ingrid Verbauwhede. Power Analysis of Synchronous Stream Ciphers with Resynchronization Mechanisms. In *ECRYPT Workshop, SASC - The State of the Art of Stream Ciphers, 2004, October 14-15, Brugge, Belgium*, pages 327–333, October 2004.

[LMV04] Hervé Ledig, Frédéric Muller, and Frédéric Valette. Enhancing Collision Attacks. In Marc Joye and Jean-Jacques Quisquater, editors, *Cryptographic Hardware and Embedded Systems – CHES 2004, 6th International Workshop, Cambridge, MA, USA, August 11-13, 2004, Proceedings*, volume 3156 of *Lecture Notes in Computer Science*, pages 176–190. Springer, 2004.

[LS01] Pierre-Yvan Liardet and Nigel P. Smart. Preventing SPA/DPA in ECC Systems Using the Jacobi Form. In Çetin Kaya Koç, David Naccache, and Christof Paar, editors, *Cryptographic Hardware and Embedded Systems – CHES 2001, Third International Workshop, Paris, France, May 14-16, 2001, Proceedings*, volume 2162 of *Lecture Notes in Computer Science*, pages 391–401. Springer, 2001.

[LSP04] Kerstin Lemke, Kai Schramm, and Christof Paar. DPA on n-Bit Sized Boolean and Arithmetic Operations and Its Application to IDEA, RC6, and the HMAC-Construction. In Marc Joye and Jean-Jacques Quisquater, editors, *Cryptographic Hardware and Embedded Systems – CHES 2004, 6th International Workshop, Cambridge, MA, USA, August 11-13, 2004, Proceedings*, volume 3156 of *Lecture Notes in Computer Science*, pages 205–219. Springer, 2004.

[MA04] Sumio Morioka and Toru Akishita. A DPA-resistant Compact AES S-Box Circuit using Additive Mask. In *Computer Security Composium (CSS), October 16, 2004, Proceedings*, pages 679–684, September 2004. (In Japanese only).

[MAC+02] Simon Moore, Ross J. Anderson, Paul Cunningham, Robert D. Mullins, and George S. Taylor. Improving Smart Card Security using Self-timed Circuits. In *Eighth International Symposium on Asynchronous Circuits and Systems (ASYNC 2002), Proceedings*, pages 211–218. IEEE Computer Society, 2002.

[Man03a]   Stefan Mangard. A Simple Power-Analysis (SPA) Attack on Implementations of the AES Key Expansion. In Pil Joong Lee and Chae Hoon Lim, editors, *Information Security and Cryptology - ICISC 2002, 5th International Conference Seoul, Korea, November 28-29, 2002, Revised Papers*, volume 2587 of *Lecture Notes in Computer Science*, pages 343–358. Springer, 2003.

[Man03b]   Stefan Mangard. Exploiting Radiated Emissions – EM Attacks on Cryptographic ICs. In Timm Ostermann and Christoph Lackner, editors, *Austrochip 2003, Linz, Austria, October 1st, 2003, Proceedings*, pages 13–16, 2003.

[Man04]   Stefan Mangard. Hardware Countermeasures against DPA – A Statistical Analysis of Their Effectiveness. In Tatsuaki Okamoto, editor, *Topics in Cryptology - CT-RSA 2004, The Cryptographers' Track at the RSA Conference 2004, San Francisco, CA, USA, February 23-27, 2004, Proceedings*, volume 2964 of *Lecture Notes in Computer Science*, pages 222–235. Springer, 2004.

[May03]   Alexander May. *New RSA Vulnerabilities Using Lattice Reduction Methods*. PhD thesis, University of Paderborn, 2003.

[MDS99a]   Thomas S. Messerges, Ezzy A. Dabbish, and Robert H. Sloan. Investigations of Power Analysis Attacks on Smartcards. In *USENIX Workshop on Smartcard Technology (Smartcard '99)*, pages 151–162, May 1999.

[MDS99b]   Thomas S. Messerges, Ezzy A. Dabbish, and Robert H. Sloan. Power Analysis Attacks of Modular Exponentiation in Smartcards. In Çetin Kaya Koç and Christof Paar, editors, *Cryptographic Hardware and Embedded Systems – CHES'99, First International Workshop, Worcester, MA, USA, August 12-13, 1999, Proceedings*, volume 1717 of *Lecture Notes in Computer Science*, pages 144–157. Springer, 1999.

[Mes00a]   Thomas S. Messerges. Securing the AES Finalists Against Power Analysis Attacks. In Bruce Schneier, editor, *Fast Software Encryption, 7th International Workshop, FSE 2000, New York, NY, USA, April 10-12, 2000, Proceedings*, volume 1978 of *Lecture Notes in Computer Science*, pages 150–164. Springer, 2000.

[Mes00b]   Thomas S. Messerges. Using Second-Order Power Analysis to Attack DPA Resistant Software. In Çetin Kaya Koç and Christof Paar, editors, *Cryptographic Hardware and Embedded Systems – CHES 2000, Second International Workshop, Worcester, MA, USA, August 17-18, 2000, Proceedings*, volume 1965 of *Lecture Notes in Computer Science*, pages 238–251. Springer, 2000.

[Möl01]   Bodo Möller. Securing Elliptic Curve Point Multiplication against Side-Channel Attacks. In George I. Davida and Yair Frankel, editors, *Information Security Conference ISC'01, Malaga, Spain, October 1-3, 2001, Proceedings*, volume 2200 of *Lecture Notes in Computer Science*, pages 324–334. Springer, 2001.

[MMS01a]   David May, Henk L. Muller, and Nigel P. Smart. Non-deterministic Processors. In Vijay Varadharajan and Yi Mu, editors, *Information Security and Privacy, 6th Australasian Conference, ACISP 2001, Sydney, Australia, July 11-13, 2001, Proceedings*, volume 2119 of *Lecture Notes in Computer Science*, pages 115–129. Springer, 2001.

[MMS01b]   David May, Henk L. Muller, and Nigel P. Smart. Random Register Renaming to Foil DPA. In Çetin Kaya Koç, David Naccache, and Christof Paar, editors, *Cryptographic Hardware and Embedded Systems – CHES 2001, Third International Workshop, Paris, France, May 14-16, 2001, Proceedings*, volume 2162 of *Lecture Notes in Computer Science Lecture Notes in Computer Science*, pages 28–38. Springer, 2001.

[Mon87]   Peter L. Montgomery. Speeding the Pollard and Elliptic Curve Methods of Factorization. *Mathematics of Computation*, 48(177):243–264, January 1987. ISSN 0025-5718.

[MPG05]   Stefan Mangard, Thomas Popp, and Berndt M. Gammel. Side-Channel Leakage of Masked CMOS Gates. In Alfred Menezes, editor, *Topics in Cryptology - CT-RSA 2005, The Cryptographers' Track at the RSA Conference 2005, San Francisco, CA, USA, February 14-18, 2005, Proceedings*, volume 3376 of *Lecture Notes in Computer Science*, pages 351–365. Springer, 2005.

[MPO05]   Stefan Mangard, Norbert Pramstaller, and Elisabeth Oswald. Successfully Attacking Masked AES Hardware Implementations. In Josyula R. Rao and Berk Sunar, editors, *Cryptographic Hardware and Embedded Systems – CHES 2005, 7th International Workshop, Edinburgh, UK, August 29 - September 1, 2005, Proceedings*, volume 3659 of *Lecture Notes in Computer Science*, pages 157–171. Springer, 2005.

[MS00]   Rita Mayer-Sommer. Smartly Analyzing the Simplicity and the Power of Simple Power Analysis on Smartcards. In Çetin Kaya Koç and Christof Paar, editors, *Cryptographic Hardware and Embedded Systems – CHES 2000, Second International Workshop, Worcester, MA, USA, August 17-18, 2000, Proceedings*, volume 1965 of *Lecture Notes in Computer Science*, pages 78–92. Springer, 2000.

[MS06]   Stefan Mangard and Kai Schramm. Pinpointing the Side-Channel Leakage of Masked AES Hardware Implementations. In *Cryptographic Hardware and Embedded Systems – CHES 2006, 8th International Workshop, Yokohama, Japan, October 10-13, 2006, Proceedings*, Lecture Notes in Computer Science. Springer, 2006.

[MSH+04]   François Mace, François-Xavier Standaert, Ilham Hassoune, Jean-Jacques Quisquater, and Jean-Didier Legat. A Dynamic Current Mode Logic to Counteract Power Analysis Attacks. In *19th Conference on Design of Circuits and Integrated Systems (DCIS 2004), Bordeaux, France, November 2004, Proceedings*, pages 186–191, 2004.

[MTT+05]   Daniel Mesquita, Jean-Denis Techer, Lionel Torres, Gilles Sassatelli, Gaston Cambon, Michel Robert, and Fernando Moraes. Current Mask Generation: A Transistor Level Security Against DPA Attacks. In *Proceedings of the 18th Annual Symposium on Integrated Circuits and System Design SBCCI '05*, pages 115–120. ACM Press, 2005.

[MV06]   Frédéric Muller and Frédéric Valette. High-Order Attacks Against the Exponent Splitting Protection. In Moti Yung, Yevgeniy Dodis, Aggelos Kiayias, and Tal Malkin, editors, *Public Key Cryptography - PKC 2006, 9th International*

Conference on Theory and Practice in Public-Key Cryptography, New York, NY, USA, April 24-26, 2006, Proceedings, volume 3958 of *Lecture Notes in Computer Science*, pages 315–329. Springer, 2006.

[MvOV97]   Alfred J. Menezes, Paul C. van Oorschot, and Scott A. Vanstone. *Handbook of Applied Cryptography*. Series on Discrete Mathematics and its Applications. CRC Press, 1997. ISBN 0-8493-8523-7, Available online at http://www.cacr.math.uwaterloo.ca/hac/.

[MVZG05]   Radu Muresan, Haleh Vahedi, Yang Zhanrong, and Stefano Gregori. Power-Smart System-On-Chip Architecture for Embedded Cryptosystems. In *Proceedings of the 3rd IEEE/ACM/IFIP International Conference on Hardware/Software Codesign and System Synthesis*, pages 184–189. ACM Press, 2005.

[Nat01]   National Institute of Standards and Technology (NIST). FIPS-197: Advanced Encryption Standard, November 2001. Available online at http://www.itl.nist.gov/fipspubs/.

[Nov02]   Roman Novak. SPA-Based Adaptive Chosen-Ciphertext Attack on RSA Implementation. In David Naccache and Pascal Paillier, editors, *Public Key Cryptography, 5th International Workshop on Practice and Theory in Public Key Cryptosystems, PKC 2002, Paris, France, February 12-14, 2002, Proceedings*, volume 2274 of *Lecture Notes in Computer Science*, pages 252–262. Springer, 2002.

[NS02]   Phong Q. Nguyen and Igor E. Shparlinski. The Insecurity of the Digital Signature Algorithm with Partially Known Nonces. *Journal of Cryptology*, 15(3):151–176, June 2002. ISSN 0933-2790.

[NS03]   Phong Q. Nguyen and Igor E. Shparlinski. The Insecurity of the Elliptic Curve Digital Signature Algorithm with Partially Known Nonces. *Design, Codes and Cryptography*, 30(2):201–217, September 2003. ISSN 0925-1022.

[OA01]   Elisabeth Oswald and Manfred Aigner. Randomized Addition-Subtraction Chains as a Countermeasure against Power Attacks. In Çetin Kaya Koç, David Naccache, and Christof Paar, editors, *Cryptographic Hardware and Embedded Systems – CHES 2001, Third International Workshop, Paris, France, May 14-16, 2001, Proceedings*, volume 2162 of *Lecture Notes in Computer Science*, pages 39–50. Springer, 2001.

[OGOP04]   Siddika Berna Örs, Frank K. Gürkaynak, Elisabeth Oswald, and Bart Preneel. Power-Analysis Attack on an ASIC AES Implementation. In *International Conference on Information Technology: Coding and Computing (ITCC'04), April 5-7, 2004, Las Vegas, Nevada, USA, Proceedings*, volume 2, pages 546–552. IEEE Computer Society, April 2004.

[OM07]   Elisabeth Oswald and Stefan Mangard. Template Attacks on Masking—Resistance is Futile. In *Topics in Cryptology - CT-RSA 2007, The Cryptographers' Track at the RSA Conference 2007, San Francisco, CA, USA, February 5-9, 2007, Proceedings*, Lecture Notes in Computer Science. Springer, 2007.

[OMHT06]   Elisabeth Oswald, Stefan Mangard, Christoph Herbst, and Stefan Tillich. Practical Second-Order DPA Attacks for Masked Smart Card Implementations of Block Ciphers. In David Pointcheval, editor, *Topics in Cryptology - CT-RSA 2006, The Cryptographers' Track at the RSA Conference 2006, San Jose, CA, USA, February 13-17, 2006, Proceedings*, volume 3860 of *Lecture Notes in Computer Science*, pages 192–207. Springer, 2006.

[OMPR05]   Elisabeth Oswald, Stefan Mangard, Norbert Pramstaller, and Vincent Rijmen. A Side-Channel Analysis Resistant Description of the AES S-box. In Henri Gilbert and Helena Handschuh, editors, *Fast Software Encryption, 12th International Workshop, FSE 2005, Paris, France, February 21-23, 2005, Revised Selected Papers*, volume 3557 of *Lecture Notes in Computer Science*, pages 413–423. Springer, 2005.

[OOP03]   Siddika Berna Örs, Elisabeth Oswald, and Bart Preneel. Power-Analysis Attacks on FPGAs – First Experimental Results. In Colin D. Walter, Çetin Kaya Koç, and Christof Paar, editors, *Cryptographic Hardware and Embedded Systems – CHES 2003, 5th International Workshop, Cologne, Germany, September 8-10, 2003, Proceedings*, volume 2779 of *Lecture Notes in Computer Science*, pages 35–50. Springer, 2003.

[OS02]   Katsuyuki Okeya and Kouichi Sakurai. A Second-Order DPA Attack Breaks a Window-Method Based Countermeasure against Side Channel Attacks. In Agnes Hui Chan and Virgil D. Gligor, editors, *Information Security, 5th International Conference, ISC 2002 Sao Paulo, Brazil, September 30 - October 2, 2002, Proceedings*, volume 2433 of *Lecture Notes in Computer Science*, pages 389–401. Springer, 2002.

[OS06]   Elisabeth Oswald and Kai Schramm. An Efficient Masking Scheme for AES Software Implementations. In Jooseok Song, Taekyoung Kwon, and Moti Yung, editors, *Information Security Applications, 6th International Workshop, WISA 2005, Jeju Island, Korea, August 22-24, 2005, Revised Selected Papers*, volume 3786 of *Lecture Notes in Computer Science*, pages 292–305. Springer, 2006.

[OSB99]   Alan V. Oppenheim, Ronald W. Schafer, and John R. Buck. *Discrete-time Signal Processing*. Signal Processing Series. Prentice Hall, 2nd edition, 1999. ISBN 0-13-754920-2.

[Osw03]   Elisabeth Oswald. Enhancing Simple Power-Analysis Attacks on Elliptic Curve Cryptosystems. In Burton S. Kaliski Jr., Çetin Kaya Koç, and Christof Paar, editors, *Cryptographic Hardware and Embedded Systems – CHES 2002, 4th International Workshop, Redwood Shores, CA, USA, August 13-15, 2002, Revised Papers*, volume 2523 of *Lecture Notes in Computer Science*, pages 82–97. Springer, 2003.

[Osw05]   Elisabeth Oswald. *Advances In Elliptic Curve Cryptography*, volume 317 of *London Mathematical Society Lecture Note Series*, chapter IV, Side-Channel Analysis, pages 69–86. Cambridge University Press, 2005.

[Pes]   David Pescovitz. 1972: The release of SPICE, still the industry standard tool for integrated circuit design. http://www.coe.berkeley.edu/labnotes/0502/history.html.

[PGH+04]   Norbert Pramstaller, Frank K. Gürkaynak, Simon Haene, Hubert Kaeslin, Norbert Felber, and Wolfgang Fichtner. Towards an AES Crypto-chip Resistant to Differential Power Analysis. In *30th European Solid-State Circuits Conference - ESSCIRC 2004, Leuven, Belgium, September 21-23, 2004, Proceedings*, pages 307–310. IEEE, September 2004.

[PM05]   Thomas Popp and Stefan Mangard. Masked Dual-Rail Pre-Charge Logic: DPA-Resistance without Routing Constraints. In Josyula R. Rao and Berk Sunar, editors, *Cryptographic Hardware and Embedded Systems – CHES 2005, 7th International Workshop, Edinburgh, UK, August 29 - September 1, 2005, Proceedings*, volume 3659 of *Lecture Notes in Computer Science*, pages 172–186. Springer, 2005.

[PM06]   Thomas Popp and Stefan Mangard. Implementation Aspects of the DPA-Resistant Logic Style MDPL. In *International Symposium on Circuits and Systems (ISCAS 2006), Island of Kos, Greece, May 21-24, 2006, Proceedings*, pages 2913–2916. IEEE, May 2006.

[POM+04]   Norbert Pramstaller, Elisabeth Oswald, Stefan Mangard, Frank K. Gürkaynak, and Simon Haene. A Masked AES ASIC Implementation. In Erwin Ofner and Manfred Ley, editors, *Austrochip 2004, Villach, Austria, October 8th, 2004, Proceedings*, pages 77–82, 2004.

[Pro05]   Emmanuel Prouff. DPA Attacks and S-Boxes. In Henri Gilbert and Helena Handschuh, editors, *Fast Software Encryption, 12th International Workshop, FSE 2005, Paris, France, February 21-23, 2005, Revised Selected Papers*, volume 3557 of *Lecture Notes in Computer Science (LNCS)*, pages 424–441. Springer, 2005.

[PS04]   Daniel Page and Martijn Stam. On XTR and Side-Channel Analysis. In Helena Handschuh and M. Anwar Hasan, editors, *Selected Areas in Cryptography, 11th International Workshop, SAC 2004, Waterloo, Canada, August 9-10, 2004, Revised Selected Papers*, volume 3357 of *Lecture Notes in Computer Science*, pages 54–68. Springer, 2004.

[PSDQ05]   Eric Peeters, François-Xavier Standaert, Nicolas Donckers, and Jean-Jacques Quisquater. Improved Higher-Order Side-Channel Attacks with FPGA Experiments. In Josyula R. Rao and Berk Sunar, editors, *Cryptographic Hardware and Embedded Systems – CHES 2005, 7th International Workshop, Edinburgh, UK, August 29 - September 1, 2005, Proceedings*, volume 3659 of *Lecture Notes in Computer Science*, pages 309–323. Springer, 2005.

[PV04]   Daniel Page and Frederik Vercauteren. Fault and Side-Channel Attacks on Pairing Based Cryptography. Cryptology ePrint Archive (http://eprint.iacr.org/), Report 2004/283, 2004.

[QS01]   Jean-Jacques Quisquater and David Samyde. ElectroMagnetic Analysis (EMA): Measures and Counter-Measures for Smart Cards. In Isabelle Attali and Thomas P. Jensen, editors, *Smart Card Programming and Security, International Conference on Research in Smart Cards, E-smart 2001, Cannes, France, September 19-21, 2001, Proceedings*, volume 2140 of *Lecture Notes in Computer Science*, pages 200–210. Springer, 2001.

[Rab]   Jan M. Rabaey. The SPICE Home Page. http://bwrc.eecs.berkeley.edu/Classes/IcBook/SPICE/.

[RCCR01]   Patrick Rakers, Larry Connell, Tim Collins, and Dan Russell. Secure Contactless Smartcard ASIC with DPA Protection. *IEEE Journal of Solid-State Circuits*, 36(3):559–565, March 2001. ISSN 0018-9200.

[RCN03]   Jan M. Rabaey, Anantha Chandrakasan, and Borivoje Nikolić. *Digital Integrated Circuits – A Design Perspective*. Electronics and VLSI Series. Prentice Hall, 2nd edition, 2003. ISBN 0-13-090996-3.

[Ric94]   John A. Rice. *Mathematical Statistics and Data Analysis*. Statistics Series. Duxbury Press, 2nd edition, 1994. ISBN 0-534-20934-3.

[RO04]   Christian Rechberger and Elisabeth Oswald. Practical Template Attacks. In Chae Hoon Lim and Moti Yung, editors, *Information Security Applications, 5th International Workshop, WISA 2004, Jeju Island, Korea, August 23-25, 2004, Revised Selected Papers*, volume 3325 of *Lecture Notes in Computer Science*, pages 443–457. Springer, 2004.

[RS01]   Tanja Römer and Jean-Pierre Seifert. Information Leakage Attacks against Smart Card Implementations of the Elliptic Curve Digital Signature Algorithm. In Isabelle Attali and Thomas P. Jensen, editors, *Smart Card Programming and Security, International Conference on Research in Smart Cards, E-smart 2001, Cannes, France, September 19-21, 2001, Proceedings*, volume 2140 of *Lecture Notes in Computer Science*, pages 211–219. Springer, 2001.

[RSA78]   Ronald L. Rivest, Adi Shamir, and Leonard Adleman. A Method for Obtaining Digital Signatures and Public-Key Cryptosystems. *Communications of the ACM*, 21(2):120–126, February 1978. ISSN 0001-0782.

[RWB04]   Girish B. Ratanpal, Ronald D. Williams, and Travis N. Blalock. An On-Chip Signal Suppression Countermeasure to Power Analysis Attacks. *IEEE Transactions on Dependable and Secure Computing*, 1(3):179–189, July-September 2004. ISSN 1545-5971.

[SA03]   Sergei P. Skorobogatov and Ross J. Anderson. Optical Fault Induction Attacks. In Burton S. Kaliski Jr., Çetin Kaya Koç, and Christof Paar, editors, *Cryptographic Hardware and Embedded Systems – CHES 2002, 4th International Workshop, Redwood Shores, CA, USA, August 13-15, 2002, Revised Papers*, volume 2523 of *Lecture Notes in Computer Science*, pages 2–12. Springer, 2003.

[SA05]   Timmy Sundström and Atila Alvandpour. A comparative analysis of logic styles for secure IC's against DPA attacks. In *23rd NORCHIP Conference, November 21-22, 2005*, pages 297–300, November 2005.

[SC01]   Amit Sinha and Anantha Chandrakasan. JouleTrack – A Web Based Tool for Software Energy Profiling. In *38th Design Automation Conference, DAC 2001, Las Vegas, NV, USA, June 18-22, 2001, Proceedings*, pages 220–225. ACM Press, June 2001.

[Sha00]   Adi Shamir. Protecting Smart Cards from Passive Power Analysis with De-
               tached Power Supplies. In Çetin Kaya Koç and Christof Paar, editors, *Crypto-
               graphic Hardware and Embedded Systems – CHES 2000, Second International
               Workshop, Worcester, MA, USA, August 17-18, 2000, Proceedings*, volume
               1965 of *Lecture Notes in Computer Science*, pages 71–77. Springer, 2000.

[Sko05]   Sergei P. Skorobogatov. *Semi-invasive attacks - A new approach to hardware
               security analysis*. PhD thesis, University of Cambridge, 2005. Available online
               at http://www.cl.cam.ac.uk/TechReports/.

[SLFP04]   Kai Schramm, Gregor Leander, Patrick Felke, and Christof Paar. A Collision-
               Attack on AES: Combining Side Channel- and Differential-Attack. In Marc
               Joye and Jean-Jacques Quisquater, editors, *Cryptographic Hardware and Em-
               bedded Systems – CHES 2004, 6th International Workshop, Cambridge, MA,
               USA, August 11-13, 2004, Proceedings*, volume 3156 of *Lecture Notes in Com-
               puter Science*, pages 163–175. Springer, 2004.

[SLP05]   Werner Schindler, Kerstin Lemke, and Christof Paar. A Stochastic Model
               for Differential Side Channel Cryptanalysis. In Josyula R. Rao and Berk
               Sunar, editors, *Cryptographic Hardware and Embedded Systems – CHES 2005,
               7th International Workshop, Edinburgh, UK, August 29 - September 1, 2005,
               Proceedings*, volume 3659 of *Lecture Notes in Computer Science*, pages 30–46.
               Springer, 2005.

[SMBY04]   Danil Sokolov, Julian Murphy, Alex Bystrov, and Alex Yakovlev. Improving
               the Security of Dual-Rail Circuits. In Marc Joye and Jean-Jacques Quisquater,
               editors, *Cryptographic Hardware and Embedded Systems – CHES 2004, 6th
               International Workshop, Cambridge, MA, USA, August 11-13, 2004, Proceed-
               ings*, volume 3156 of *Lecture Notes in Computer Science*, pages 282–297.
               Springer, 2004.

[SMBY05]   Danil Sokolov, Julian Murphy, Alex Bystrov, and Alex Yakovlev. Design and
               Analysis of Dual-Rail Circuits for Security Applications. *IEEE Transactions
               on Computers*, 54(4):449–460, April 2005. ISSN 0018-9340.

[SOV+05]   Hendra Saputra, Ozcan Ozturk, Narayanan Vijaykrishnan, Mahmut T. Kan-
               demir, and Richard Brooks. A Data-Driven Approach for Embedded Security.
               In *IEEE Computer Society Annual Symposium on VLSI (ISVLSI 2005), New
               Frontiers in VLSI Design, 11-12 May 2005, Tampa, FL, USA*, pages 104–109.
               IEEE, 2005.

[SP06]   Kai Schramm and Christof Paar. Higher Order Masking of the AES. In David
               Pointcheval, editor, *Topics in Cryptology - CT-RSA 2006, The Cryptographers'
               Track at the RSA Conference 2006, San Jose, CA, USA, February 13-17, 2006,
               Proceedings*, volume 3860 of *Lecture Notes in Computer Science*, pages 208–
               225. Springer, 2006.

[SPAQ06]   François-Xavier Standaert, Eric Peeters, Cedric Archambeau, and Jean-
               Jacques Quisquater. Towards Security Limits in Side-Channel Attacks (With
               an Application to Block Ciphers). In *Cryptographic Hardware and Embedded
               Systems – CHES 2006, 8th International Workshop, Yokohama, Japan, Octo-
               ber 10-13, 2006, Proceedings*, Lecture Notes in Computer Science. Springer,
               2006.

[SSAQ02]   David Samyde, Sergei P. Skorobogatov, Ross J. Anderson, and Jean-Jacques Quisquater. On a New Way to Read Data from Memory. In *IEEE Security in Storage Workshop (SISW'02)*, pages 65–69. IEEE Computer Society, 2002.

[SSI04]   Daisuke Suzuki, Minoru Saeki, and Tetsuya Ichikawa. Random Switching Logic: A Countermeasure against DPA based on Transition Probability. Cryptology ePrint Archive (http://eprint.iacr.org/), Report 2004/346, 2004.

[SSI05]   Daisuke Suzuki, Minoru Saeki, and Tetsuya Ichikawa. DPA Leakage Models for CMOS Logic Circuits. In Josyula R. Rao and Berk Sunar, editors, *Cryptographic Hardware and Embedded Systems – CHES 2005, 7th International Workshop, Edinburgh, UK, August 29 - September 1, 2005, Proceedings*, volume 3659 of *Lecture Notes in Computer Science*, pages 366–382. Springer, 2005.

[SVK⁺03]   Hendra Saputra, Narayanan Vijaykrishnan, Mahmut T. Kandemir, Mary J. Irwin, and Richard Brooks. Masking the energy behaviour of encryption algorithms. *IEE Proceedings - Computers and Digital Techniques*, 150(5):274–284, September 2003. ISSN 1350-2387.

[SWP03]   Kai Schramm, Thomas J. Wollinger, and Christof Paar. A New Class of Collision Attacks and Its Application to DES. In Thomas Johansson, editor, *Fast Software Encryption, 10th International Workshop, FSE 2003, Lund, Sweden, February 24-26, 2003, Revised Papers*, volume 2887 of *Lecture Notes in Computer Science*, pages 206–222. Springer, 2003.

[Syn]   Synopsys. The Synopsys Website. http://www.synopsys.com/.

[TAV02]   Kris Tiri, Moonmoon Akmal, and Ingrid Verbauwhede. A Dynamic and Differential CMOS Logic with Signal Independent Power Consumption to Withstand Differential Power Analysis on Smart Cards. In *28th European Solid-State Circuits Conference - ESSCIRC 2002, Florence, Italy, September 24-26, 2002, Proceedings*, pages 403–406. IEEE, September 2002.

[Thé06]   Nicolas Thériault. SPA Resistant Left-to-Right Integer Recodings. In Bart Preneel and Stafford Tavares, editors, *Selected Areas in Cryptography, 12th International Workshop, SAC 2005, Kingston, Ontario, Canada, August 11-12, 2005, Revised Selected Papers*, volume 3897 of *Lecture Notes in Computer Science*, pages 345–358. Springer, 2006.

[THH⁺05]   Kris Tiri, David Hwang, Alireza Hodjat, Bo-Cheng Lai, Shenglin Yang, Patrick Schaumont, and Ingrid Verbauwhede. Prototype IC with WDDL and Differential Routing - DPA Resistance Assessment. In Josyula R. Rao and Berk Sunar, editors, *Cryptographic Hardware and Embedded Systems – CHES 2005, 7th International Workshop, Edinburgh, UK, August 29 - September 1, 2005, Proceedings*, volume 3659 of *Lecture Notes in Computer Science*, pages 354–365. Springer, 2005.

[TKL05]   Elena Trichina, Tymur Korkishko, and Kyung-Hee Lee. Small Size, Low Power, Side Channel-Immune AES Coprocessor: Design and Synthesis Results. In Hans Dobbertin, Vincent Rijmen, and Aleksandra Sowa, editors, *Advanced Encryption Standard - AES, 4th International Conference, AES 2004,*

Bonn, Germany, May 10-12, 2004, Revised Selected and Invited Papers, volume 3373 of *Lecture Notes in Computer Science*, pages 113–127. Springer, 2005.

[TL05]   Zeynep Toprak and Yusuf Leblebici. Low-Power Current Mode Logic for Improved DPA-Resistance in Embedded Systems. In *International Symposium on Circuits and Systems (ISCAS 2005), Kobe, Japan, May 23-26, 2005, Proceedings*, volume 2, pages 1059–1062. IEEE, 2005.

[TSG03]   Elena Trichina, Domenico De Seta, and Lucia Germani. Simplified Adaptive Multiplicative Masking for AES. In Burton S. Kaliski Jr., Çetin Kaya Koç, and Christof Paar, editors, *Cryptographic Hardware and Embedded Systems – CHES 2002, 4th International Workshop, Redwood Shores, CA, USA, August 13-15, 2002, Revised Papers*, volume 2523 of *Lecture Notes in Computer Science*, pages 187–197. Springer, 2003.

[TV03]   Kris Tiri and Ingrid Verbauwhede. Securing Encryption Algorithms against DPA at the Logic Level: Next Generation Smart Card Technology. In Colin D. Walter, Çetin Kaya Koç, and Christof Paar, editors, *Cryptographic Hardware and Embedded Systems – CHES 2003, 5th International Workshop, Cologne, Germany, September 8-10, 2003, Proceedings*, volume 2779 of *Lecture Notes in Computer Science*, pages 137–151. Springer, 2003.

[TV04a]   Kris Tiri and Ingrid Verbauwhede. A Logic Level Design Methodology for a Secure DPA Resistant ASIC or FPGA Implementation. In *2004 Design, Automation and Test in Europe Conference and Exposition (DATE 2004), 16-20 February 2004, Paris, France*, volume 1, pages 246–251. IEEE Computer Society, 2004.

[TV04b]   Kris Tiri and Ingrid Verbauwhede. Place and Route for Secure Standard Cell Design. In Jean-Jacques Quisquater, Pierre Paradinas, Yves Deswarte, and Anas Abou El Kadam, editors, *Sixth International Conference on Smart Card Research and Advanced Applications (CARDIS '04), 23-26 August 2004, Toulouse, France*, pages 143–158. Kluwer Academic Publishers, August 2004.

[TV04c]   Kris Tiri and Ingrid Verbauwhede. Secure Logic Synthesis. In Jürgen Becker, Marco Platzner, and Serge Vernalde, editors, *Field Programmable Logic and Application, 14th International Conference, FPL 2004, Leuven, Belgium, August 30-September 1, 2004, Proceedings*, volume 3203 of *Lecture Notes in Computer Science*, pages 1052–1056. Springer, August 2004.

[TV05a]   Kris Tiri and Ingrid Verbauwhede. A VLSI Design Flow for Secure Side-Channel Attack Resistant ICs. In *2005 Design, Automation and Test in Europe Conference and Exposition (DATE 2005), 7-11 March 2005, Munich, Germany*, pages 58–63. IEEE Computer Society, 2005.

[TV05b]   Kris Tiri and Ingrid Verbauwhede. Design Method for Constant Power Consumption of Differential Logic Circuits. In *2005 Design, Automation and Test in Europe Conference and Exposition (DATE 2005), 7-11 March 2005, Munich, Germany*, pages 628–633. IEEE Computer Society, 2005.

[TV05c]   Kris Tiri and Ingrid Verbauwhede. Simulation Models for Side-Channel Information Leaks. In William H. Joyner Jr., Grant Martin, and Andrew B. Kahng, editors, *42nd Design Automation Conference, DAC 2005, Anaheim, CA, USA, June 13-17, 2005, Proceedings*, pages 228–233. ACM Press, June 2005.

[TV06]   Kris Tiri and Ingrid Verbauwhede. A Digital Design Flow for Secure Integrated Circuits. *IEEE Transactions on Computer-Aided Design of Integrated Circuits and Systems*, 25(7):1197–1208, July 2006. ISSN 0278-0070.

[Uni]   University of California at Berkeley. The University of California at Berkeley Website. http://www.berkeley.edu/.

[Wal02a]   Colin D. Walter. MIST: An Efficient, Randomized Exponentiation Algorithm for Resisting Power Analysis. In Bart Preneel, editor, *Topics in Cryptology - CT-RSA 2002, The Cryptographers' Track at the RSA Conference 2002, San Jose, CA, USA, February 18-22, 2002, Proceedings*, volume 2271 of *Lecture Notes in Computer Science*, pages 53–66. Springer, 2002.

[Wal02b]   Colin D. Walter. Some Security Aspects of the MIST Randomized Exponentiation Algorithm. In Burton S. Kaliski Jr., Çetin Kaya Koç, and Christof Paar, editors, *Cryptographic Hardware and Embedded Systems – CHES 2002, 4th International Workshop, Redwood Shores, CA, USA, August 13-15, 2002, Revised Papers*, volume 2523 of *Lecture Notes in Computer Science*, pages 276–290. Springer, 2002.

[Wal03]   Colin D. Walter. Seeing through MIST Given a Small Fraction of an RSA Private Key. In Marc Joye, editor, *Topics in Cryptology - CT-RSA 2003, The Cryptographers' Track at the RSA Conference 2003, San Francisco, CA, USA, April 13-17, 2003, Proceedings*, volume 2612 of *Lecture Notes in Computer Science*, pages 391–402. Springer, 2003.

[Wal04]   Colin D. Walter. Simple Power Analysis of Unified Code for ECC Double and Add. In Marc Joye and Jean-Jacques Quisquater, editors, *Cryptographic Hardware and Embedded Systems – CHES 2004, 6th International Workshop, Cambridge, MA, USA, August 11-13, 2004, Proceedings*, volume 3156 of *Lecture Notes in Computer Science*, pages 191–204. Springer, 2004.

[Wie01]   Andreas Wiemers. Kollisionsattacken beim Comp128 auf Smartcards. ECC-Brainpool Workshop on Side-Channel-Attacks on Cryptographic Algorithms, Bonn, Germany, December 2001.

[WOL02]   Johannes Wolkerstorfer, Elisabeth Oswald, and Mario Lamberger. An ASIC implementation of the AES SBoxes. In Bart Preneel, editor, *Topics in Cryptology - CT-RSA 2002, The Cryptographers' Track at the RSA Conference 2002, San Jose, CA, USA, February 18-22, 2002, Proceedings*, volume 2271 of *Lecture Notes in Computer Science*, pages 67–78. Springer, 2002.

[WW04]   Jason Waddle and David Wagner. Towards Efficient Second-Order Power Analysis. In Marc Joye and Jean-Jacques Quisquater, editors, *Cryptographic Hardware and Embedded Systems – CHES 2004, 6th International Workshop, Cambridge, MA, USA, August 11-13, 2004, Proceedings*, volume 3156 of *Lecture Notes in Computer Science*, pages 1–15. Springer, 2004.

[YB04]   An Yu and David S. Brée. A Clock-less Implementation of the AES Resists to Power and Timing Attacks. In *International Conference on Information*

Technology: Coding and Computing (ITCC'04), April 5-7, 2004, Las Vegas, Nevada, USA, Proceedings*, volume 2, pages 525–532. IEEE Computer Society, April 2004.

[YFP03]   Zhong C. Yu, Stephen B. Furber, and Luis A. Plana. An Investigation into the Security of Self-Timed Circuits. In *9th International Symposium on Advanced Research in Asynchronous Circuits and Systems (ASYNC 2003), 12-16 May 2003, Vancouver, BC, Canada*, pages 206–215. IEEE Computer Society, 2003.

[YHM⁺06]   HyungSo Yoo, Christoph Herbst, Stefan Mangard, Elisabeth Oswald, and SangJae Moon. Investigations of Power Analysis Attacks and Countermeasures for ARIA. In *Information Security Applications, 7th International Workshop, WISA 2006, Jeju Island, Korea, August 28-30, 2006*, Lecture Notes in Computer Science. Springer, 2006.

[YKH⁺04]   HyungSo Yoo, ChangKyun Kim, JaeCheol Ha, SangJae Moon, and IlHwan Park. Side Channel Cryptanalysis on SEED. In Chae Hoon Lim and Moti Yung, editors, *Information Security Applications, 5th International Workshop, WISA 2004, Jeju Island, Korea, August 23-25, 2004, Revised Selected Papers*, volume 3325 of *Lecture Notes in Computer Science*, pages 411–424. Springer, 2004.

[YWV⁺05]   Shengqi Yang, Wayne Wolf, Narayanan Vijaykrishnan, Dimitrios N. Serpanos, and Yuan Xie. Power Attack Resistant Cryptosystem Design: A Dynamic Voltage and Frequency Switching Approach. In *2005 Design, Automation and Test in Europe Conference and Exposition (DATE 2005), 7-11 March 2005, Munich, Germany*, pages 64–69. IEEE Computer Society, 2005.

[YY92]   Masakazu Yamashina and Hachiro Yamada. An MOS Current Mode Logic (MCML) Circuit for Low-Power Sub-GHz Processors. *IEICE Transactions on Electronics*, E75-C(10):1181–1187, October 1992. ISSN 0916-8516.

Author Index

Topic Index